Clean Energy Production Technologies

Series Editors

Neha Srivastava, Department of Chemical Engineering and Technology
IIT (BHU) Varanasi
Varanasi, India

P. K. Mishra, Department of Chemical Engineering and Technology
IIT (BHU) Varanasi
Varanasi, India

The consumption of fossil fuels has been continuously increasing around the globe and simultaneously becoming the primary cause of global warming as well as environmental pollution. Due to limited life span of fossil fuels and limited alternate energy options, energy crises is important concern faced by the world. Amidst these complex environmental and economic scenarios, renewable energy alternates such as biodiesel, hydrogen, wind, solar and bioenergy sources, which can produce energy with zero carbon residue are emerging as excellent clean energy source. For maximizing the efficiency and productivity of clean fuels via green & renewable methods, it's crucial to understand the configuration, sustainability and techno-economic feasibility of these promising energy alternates. The book series presents a comprehensive coverage combining the domains of exploring clean sources of energy and ensuring its production in an economical as well as ecologically feasible fashion. Series involves renowned experts and academicians as volume-editors and authors, from all the regions of the world. Series brings forth latest research, approaches and perspectives on clean energy production from both developed and developing parts of world under one umbrella. It is curated and developed by authoritative institutions and experts to serves global readership on this theme.

Chaudhery Mustansar Hussain
Ghazanfar Hussain
Editors

Green Nanostructured Photocatalysts for Clean Energy

Springer

Editors
Chaudhery Mustansar Hussain
Department of Chemistry and Environmental Science
New Jersey Institute of Technology
Newark, NJ, USA

Ghazanfar Hussain
Division of Science and Technology
Department of Education Lahore
Lahore, Pakistan

ISSN 2662-6861 ISSN 2662-687X (electronic)
Clean Energy Production Technologies
ISBN 978-981-95-1617-9 ISBN 978-981-95-1618-6 (eBook)
https://doi.org/10.1007/978-981-95-1618-6

This Springer imprint is published by the registered company Springer Nature Singapore Pte Ltd.
The registered company address is: 152 Beach Road, #21-01/04 Gateway East, Singapore 189721, Singapore

I would like to dedicate this book to
My beloved GOD
"Meray Pyarey Allah (Subhana Wa Ta Alla)"

Preface

In the quest to address the dual challenges of energy scarcity and environmental preservation, green nanostructured photocatalysts emerge as a pivotal innovation, marrying the ingenuity of nanotechnology with the principles of sustainability. This book, titled "Green Nanostructured Photocatalysts for Clean Energy," going to be published by Springer, serves as a comprehensive resource for scientists, engineers, and researchers dedicated to exploring the frontiers of photocatalysis in the context of green energy applications.

The urgency to transition toward renewable energy sources has never been more acute. As the global community grapples with escalating environmental concerns and the looming threat of fossil fuel depletion, the development of sustainable technologies is imperative. Photocatalysis, particularly utilizing green nanostructured materials, offers a promising avenue not only for energy conversion but also for vital environmental remediation processes such as water and air purification.

This volume is curated by esteemed editors, Chaudhery Mustansar Hussain and Ghazanfar Hussain, who bring together a diverse array of perspectives and expert insights. The chapters within this book are meticulously structured to cover the entire spectrum of green nanostructured photocatalysts—from fundamental concepts and synthesis techniques to advanced applications and future perspectives.

The initial chapters provide a foundational understanding of photocatalysis and its significance in sustainable energy applications, setting the stage for more in-depth discussions on various types of green nanostructured materials, including those based on earth-abundant metals and innovative carbon-based materials. Subsequent sections delve into practical applications, such as water and air purification, showcasing the role of green photocatalysts in environmental cleanup.

Further, the book explores the synthesis and characterization of these materials, offering readers a detailed look at the methodologies and challenges involved in creating effective photocatalytic systems. The latter chapters focus on the safety, environmental impacts, and the future potential of these technologies, including emerging trends and prospective developments.

By providing a blend of theoretical knowledge and practical application, this book aims to inspire ongoing research and innovation in the field of green nanostructured photocatalysts. It is intended not only for those currently engaged in the field but also for newcomers and students who are eager to contribute to sustainable development through scientific research.

As the world continues to evolve, the insights and technologies discussed within these pages offer hope and direction for a cleaner, more sustainable future, harnessing the power of the sun and innovative materials to forge a path toward green energy solutions.

Newark, NJ, USA Chaudhery Mustansar Hussain
Lahore, Pakistan Ghazanfar Hussain

Acknowledgments

We would like to acknowledge this book to Our Lovely Sons & Daughters SARIM HUSSAIN CHAUDHERY, AZLAN HUSSAIN CHAUDHERY, MAIDA ALI, HAFIZA SAFA HUSSAIN CHAUDHERY, HAFIZA DUA HUSSAIN CHAUDHERY, AIRA HUSSAIN CHAUDHERY for their dedicated support during the compiling this book.
Editors: Chaudhery Mustansar Hussain & Ghazanfar Hussain

Contents

Editors and Contributors

About the Editors

Chaudhery Mustansar Hussain, PhD, is an adjunct professor and director of laboratories in the Department of Chemistry & Environmental Sciences at the New Jersey Institute of Technology (NJIT), Newark, NJ, USA. His research is focused on the applications of nanotechnology and advanced materials, environmental management, sustainability, analytical chemistry, and other various industries. He is the author of numerous papers in peer-reviewed journals as well as a prolific author and editor of around 200 books, including scientific monographs and handbooks in his research areas. He has published with Elsevier, American Chemical Society, Royal Society of Chemistry, John Wiley & Sons, CRC Press, and Springer.

Ghazanfar Hussain is a research scholar in science and technology, Department of Education, Punjab, Pakistan. His key areas of research are environment, nanotechnology, analytical chemistry, material science and engineering, sustainability, modern and smart technologies and industrial development. He is author of monographs on chemistry, sample pre-treatments, nanotechnology, and advanced technology. He is a professional and affiliated with several organizations.

Contributors

Janice Adamski Federal University of Rio Grande do Sul, Department of Chemical Engineering, Porto Alegre, RS, Brazil

Gözde Koşarsoy Ağçeli Department of Biology/Biotechnology, Hacettepe University, Ankara, Turkey

Junaid Aman School of Applied and Life Sciences, Uttaranchal University, Dehradun, Uttarakhand, India

Homam Anntar Indian Institute of Sustainability, Gujarat University, Ahmedabad, Gujarat, India

Delele Worku Ayele Bahir Dar Energy Center, Bahir Dar Institute of Technology, Bahir Dar University, Bahir Dar, Ethiopia

Department of Chemistry, College of Science, Bahir Dar University, Bahir Dar, Ethiopia

Suyanne A. L. Bachmann Federal Institute of Education, Science and Technology of Rio Grande do Sul, Feliz, RS, Brazil

Sasmita Chand Manipal School of Architecture and Planning, Manipal Academy of Higher Education, Manipal, India

Sony J. Chundattu Department of Sciences and Humanities, School of Engineering and Technology, CHRIST (Deemed to be University), Bangalore, India

Bao-Trong Dang Faculty of Chemical Engineering, Ho Chi Minh City University of Technology (HCMUT), Ho Chi Minh City, Vietnam

Vietnam National University Ho Chi Minh City, Linh Trung Ward, Thu Duc City, Ho Chi Minh City, Vietnam

Mthokozisi Dladla Human Molecular Biology Unit (School of Biomedical Sciences), Faculty of Health Sciences, University of the Free State, Bloemfontein, South Africa

Kanika Dulta School of Applied and Life Sciences, Uttaranchal University, Dehradun, Uttarakhand, India

Faculty of Materials Science and Technology, VSB-Technical University of Ostrava, Ostrava-Poruba, Czech Republic

Xolile Fuku Institute for Nanotechnology and Water Sustainability (iNanoWS), College of Science, Engineering and Technology, University of South Africa, Roodepoort, Florida, South Africa

Hamid Garmestani Georgia Institute of Technology, Department of Materials Science and Engineering, Atlanta, GA, USA

Soumya Ghosh Human Molecular Biology Unit (School of Biomedical Sciences), Faculty of Health Sciences, University of the Free State, Bloemfontein, South Africa

Natural and Medical Sciences Research Center, University of Nizwa, Nizwa, Oman

Sreeraj Gopi Molecules Biolabs, Koratty, Kerala, India

Sıla Hoşoğlu Department of Biology/Biotechnology, Hacettepe University, Ankara, Turkey

Muhammad Ibrahim State Key Laboratory of Geomicrobiology and Environmental Changes, Engineering Research Center of Nano-Geomaterials of the Ministry of Education, Faculty of Materials Science and Chemistry, China University of Geosciences, Wuhan, China

Sudhanshu Jangir Indian Institute of Sustainability, Gujarat University, Ahmedabad, Gujarat, India

Gururaj Kudur Jayaprakash Laboratory of Quantum Chemistry, School of Advanced Chemical Sciences, Shoolini University, Bajhol, Himachal Pradesh, India

Department of Chemistry, Nitte Meenakshi Institute of Technology, Bangalore, Karnataka, India

Kuppusamy Kanagaraj Center for Supramolecular Chemistry & Catalysis and Department of Chemistry, College of Science, Shanghai University, Shanghai, China

Kalimuthu Karuppanan Department of Biotechnology, School of Bioengineering, SRM Institute of Science and Technology, KTR Campus, Chennai, India

Radha Khirwar School of Applied and Life Sciences, Uttaranchal University, Dehradun, Uttarakhand, India

Hui Li State Key Laboratory of Geomicrobiology and Environmental Changes, Engineering Research Center of Nano-Geomaterials of the Ministry of Education, Faculty of Materials Science and Chemistry, China University of Geosciences, Wuhan, China

Shaoguang Li State Key Laboratory of Geomicrobiology and Environmental Changes, Engineering Research Center of Nano-Geomaterials of the Ministry of Education, Faculty of Materials Science and Chemistry, China University of Geosciences, Wuhan, China

Steven Y. Liang George W. Woodruff School of Mechanical Engineering, Georgia Institute of Technology, Atlanta, GA, USA

Mohd Mazhar Taqgene Training and Research Institute, Dehradun, Uttarakhand, India

Mothi Krishna Mohan Department of Sciences and Humanities, School of Engineering and Technology, CHRIST (Deemed to be University), Bangalore, India

M. Muthukumaran PG and Research Department of Botany, Ramakrishna Mission Vivekananda College (Autonomous), Chennai, India

M. Muthulakshmi Ethiraj College for Women (Autonomous), Chennai, India

Pham Thi Le Na Faculty of Chemical Engineering, Ho Chi Minh City University of Technology (HCMUT), Ho Chi Minh City, Vietnam

Vietnam National University Ho Chi Minh City, Linh Trung Ward, Thu Duc City, Ho Chi Minh City, Vietnam

Navid Nasajpour-Esfahani Georgia Institute of Technology, Department of Materials Science and Engineering, Atlanta, GA, USA

Minh-Kha Nguyen Faculty of Chemical Engineering, Ho Chi Minh City University of Technology (HCMUT), Ho Chi Minh City, Vietnam

Vietnam National University Ho Chi Minh City, Linh Trung Ward, Thu Duc City, Ho Chi Minh City, Vietnam

Prayag Bhavesh Patel Daniel Guggenheim School of Aerospace Engineering, Atlanta, GA, USA

Dephan Pinheiro Department of Chemistry, CHRIST (Deemed to be University), Bangalore, India

Kannan Raman Department of Pharmacology, St. John's College of Pharmaceutical Sciences & Research, Kattappana, Idukki, Kerala, India

Upasana Rana Uttaranchal School of Computing Sciences, Uttaranchal University, Dehradun, India

Morgana Rosset University of São Paulo, Polytechnic School, Department of Chemical Engineering, São Paulo, SP, Brazil

Kutloano Edward Sekhosana Institute for Nanotechnology and Water Sustainability (iNanoWS), College of Science, Engineering and Technology, University of South Africa, Roodepoort, Florida, South Africa

Vaishnavi Sharma Laboratory of Quantum Chemistry, School of Advanced Chemical Sciences, Shoolini University, Bajhol, Himachal Pradesh, India

Shweta Singh Indian Institute of Sustainability, Gujarat University, Ahmedabad, Gujarat, India

T. S. Subha PG and Research Department of Botany, Bharathi Women's College, Chennai, India

K. R. Sunaja Devi Department of Chemistry, CHRIST (Deemed to be University), Bangalore, India

Minbale Admas Teshager Department of Chemistry, College of Science, Bahir Dar University, Bahir Dar, Ethiopia

Duyen Phuc-Hanh Tran Department of Civil Engineering, Chung Yuan Christian University, Taoyuan, Taiwan, Republic of China

Tharmaraj Vairaperumal Institute of Clinical Medicine, College of Medicine, National Cheng Kung University, Tainan, Taiwan, Republic of China

Arun Varghese Department of Chemistry, CHRIST (Deemed to be University), Bangalore, India

Ababay Ketema Worku Bahir Dar Energy Center, Bahir Dar Institute of Technology, Bahir Dar University, Bahir Dar, Ethiopia

Fan Xia State Key Laboratory of Geomicrobiology and Environmental Changes, Engineering Research Center of Nano-Geomaterials of the Ministry of Education, Faculty of Materials Science and Chemistry, China University of Geosciences, Wuhan, China

Muhammad Yousuf State Key Laboratory of Geomicrobiology and Environmental Changes, Engineering Research Center of Nano-Geomaterials of the Ministry of Education, Faculty of Materials Science and Chemistry, China University of Geosciences, Wuhan, China

Martins Lateef Yusuf State Key Laboratory of Geomicrobiology and Environmental Changes, Engineering Research Center of Nano-Geomaterials of the Ministry of Education, Faculty of Materials Science and Chemistry, China University of Geosciences, Wuhan, China

Part I
Fundamentals of Green Nanostructured Photocatalysts

Chapter 1
Introduction to Photocatalysis and Its Significance in Green Energy Applications

Navid Nasajpour-Esfahani, Prayag Bhavesh Patel, Hamid Garmestani, and Steven Y. Liang

Abstract Photocatalysts play a vital role in addressing environmental challenges through various applications. This chapter offers an essential introduction to photocatalysts and their importance in advancing green energy solutions. It covers a range of applications, including carbon dioxide (CO_2) reduction, hydrogen evolution, pollutant degradation, and water splitting, all integral to the field of photocatalysis in green energy. Furthermore, this chapter explores the diverse materials used as photocatalysts, providing insights into their selection for each application. By offering a comprehensive understanding of material properties and their suitability for specific photocatalytic processes, readers gain essential knowledge for effectively implementing these solutions in green energy initiatives.

Keywords Photocatalysis · Green energy · CO_2 reduction · Water splitting · Hydrogen evolution · Pollutant degradation

N. Nasajpour-Esfahani (✉) · H. Garmestani
Georgia Institute of Technology, Department of Materials Science and Engineering, Atlanta, GA, USA

P. B. Patel
Daniel Guggenheim School of Aerospace Engineering, Atlanta, GA, USA

S. Y. Liang
George W. Woodruff School of Mechanical Engineering, Georgia Institute of Technology, Atlanta, GA, USA

C. M. Hussain, G. Hussain (eds.), *Green Nanostructured Photocatalysts for Clean Energy*, Clean Energy Production Technologies,
https://doi.org/10.1007/978-981-95-1618-6_1

1.1 Introduction

As technology advances in the age of industrialization and economic growth, it has brought about serious environmental challenges. Unchecked industrial and urban development has resulted in pollution of the air and water, shortages of natural resources, and concerns about energy availability. Researchers and engineers have turned to new methods like photocatalysis as a means of addressing environmental degradation and promoting sustainable development as a solution to these major problems. The ability of photocatalysis, a type of advanced oxidation process (AOP), to efficiently degrade persistent organic pollutants in water and air has garnered attention. Photocatalysis utilizes the unique electronic structures of specific metal oxide semiconductors, such as titanium dioxide (TiO_2), to catalyze the breakdown of organic compounds when exposed to light. These semiconductors generate extremely reactive species such as hydroxyl radicals when stimulated by light, and they have the ability to mineralize pollutants into benign compounds. The wide range of applications for photocatalysis highlights its versatility. In terms of water treatment, photocatalysis has proven to be incredibly effective at breaking down a wide range of contaminants, such as artificial coloring, biocides, endocrine disruptors, and medications. Furthermore, it has shown potential in the degradation of lignin and biological decontamination (Rizzo et al. 2009). Furthermore, complex mixtures of industrial pollutants and pharmaceuticals, as well as wastewater scenarios containing explosive substances, can be addressed using photocatalysis. Photocatalytic oxidation has become an effective method in air purification, rapidly eliminating odors, volatile organic compounds (VOCs), pathogens, and other airborne pollutants (Herrmann 1999). This technology is a versatile tool to improve public health and environmental quality as it can break down pollutants in both air and water. To maximize photocatalytic processes for air purification applications, researchers have experimented with various reactor designs, catalyst compositions, and operating parameters (Jeong et al. 2005). Additionally, photocatalytic water splitting is an efficient method for producing energy sustainably. By harnessing solar energy to drive the decomposition of water into hydrogen and oxygen, photocatalysis offers a renewable and environmentally friendly pathway for hydrogen generation. Through this process, solar energy can be stored as chemical bonds, providing a reliable and clean energy source for various uses, such as electricity production (Kudlek and Dudziak 2018). Although photocatalysis has great potential, there are a number of barriers that need to be addressed through additional research and innovation. These challenges include developing stable and effective photocatalysts, optimizing reactor designs for cost and scalability, and minimizing byproduct formation and catalyst deterioration (Lalitha et al. 2010). Furthermore, careful consideration of technical, financial, and regulatory aspects is necessary for the incorporation of photocatalytic systems into existing systems for water and air treatment. In this chapter, a brief review of current photocatalyst materials and their corresponding applications will be discussed.

1.2 CO_2 Reduction

1.2.1 Zinc Indium Sulfide ($ZnIn_2S_4$)/Titanium Dioxide (TiO_2) Photocatalysts

The rapid increase in carbon dioxide (CO_2) emissions into the atmosphere, stemming from industrial byproducts, has raised concerns about the future of the global climate. This issue has prompted many to seek ways to reduce the CO_2 concentration in the atmosphere, thereby sustaining the environment for future generations. From a photocatalytic perspective, a promising approach toward sustainable solar fuel involves extracting CO_2 from the atmosphere, combining it with water (H_2O), and reducing it into carbon monoxide (CO) and methane (CH_4). To achieve this, a heterojunction nanocomposite photocatalyst comprising $ZnIn_2S_4$ and TiO_2 was developed. The resulting photocatalyst exhibited high efficiency in reducing CO_2, yielding 1754 and 481 μmol/g of CO and CH_4, respectively. Throughout the study, the photocatalyst remained stable across five consecutive experiments while maintaining efficient photocatalytic activity. In this process, CO_2 serves as the input for producing CO for industrial purposes (Fig. 1.1). This device addresses multiple challenges faced by researchers, including the reduction of CO_2 in the atmosphere while concurrently producing CO, a crucial industrial material, as a byproduct. The process involves inputs of CO_2 and H_2O, which are then reduced by the photocatalyst, powered by light, yielding the outputs of CO and CH_4. This approach holds the potential to significantly diminish reliance on fossil fuels and mitigate carbon emissions. The catalyst represents cutting-edge technology and is unique due to its visible light sensitivity, chemical synthesis flexibility, synergistic charge separation, and efficient conversion of CO_2 into useful outputs. The study explores various combinations of composites and conditions, with the visible light active $ZnIn_2S_4$/TiO_2 nanocomposite photocatalyst emerging as the most efficient. This catalyst functions in a two-step process, utilizing in situ coprecipitation of TiO_2 with sol-gel hydrothermal treatment of $ZnIn_2S_4$. The efficiency of these materials stems from the

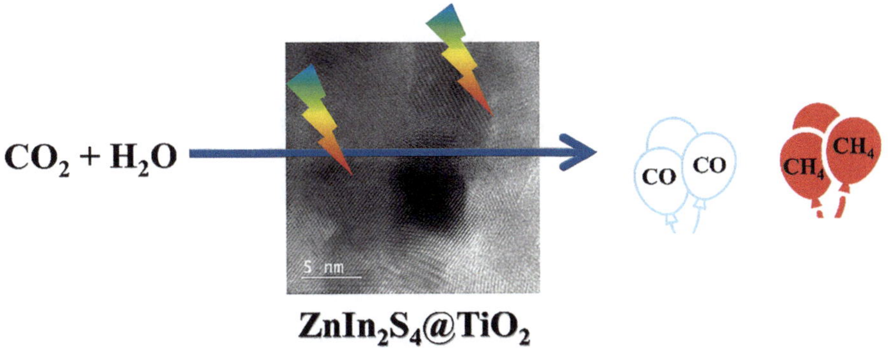

Fig. 1.1 Conversion of CO_2 into CO and CH_4 using $ZnIn_2S_4$/TiO_2 nanocomposite (Ge et al. 2023)

collaborative synergy between $ZnIn_2S_4$ and TiO_2 semiconductors within the photocatalyst, which prevents charge carrier recombination and enhances the utilization of photogenerated electrons at semiconductor interfaces, thereby intensifying the conversion of CO_2 into renewable solar fuel. This photocatalyst contributes to addressing the issue of increasing carbon emissions and generates valuable byproducts, enabling the powering of economies while safeguarding the Earth's environment and gradually mitigating the carbon cycle (Ge et al. 2023). Furthermore, under solar radiation, $ZnIn_2S_4$ shows promise in converting CO_2 into useful fuel compounds such as carbon monoxide and methane (Chen et al. 2014). To enhance its efficiency for CO_2 reduction, methods such as heterojunction construction and vacancy defect engineering have been employed (Guo et al. 2020). Zinc vacancies, for example, can be introduced into heterojunctions to enhance charge transfer and CO_2 conversion rates, as well as to augment the rate of charge separation and CO_2 adsorption (Wang et al. 2021).

1.2.2 Silver (Ag)-Based Photocatalysts

Silver, being a noble metal, possesses properties that render it an ideal candidate for photocatalytic applications, with these properties largely attributed to the localized surface plasmonic resonance (LSPR) effect. LSPR is a phenomenon observed in conductive nanomaterials, where surface plasmons are defined by the oscillation behavior of free charges on the material's surface. When subjected to incident light, the irradiation generates an electric field, inducing polarization of the free electrons within the nanoparticles. While positive charges remain stationary within the ionic core, the negative free electrons traverse the surface of the particle, resulting in an imbalance that triggers a linear restorative force opposing this motion. This process of motion and restoration iterates, leading to dipolar oscillation of the free electrons

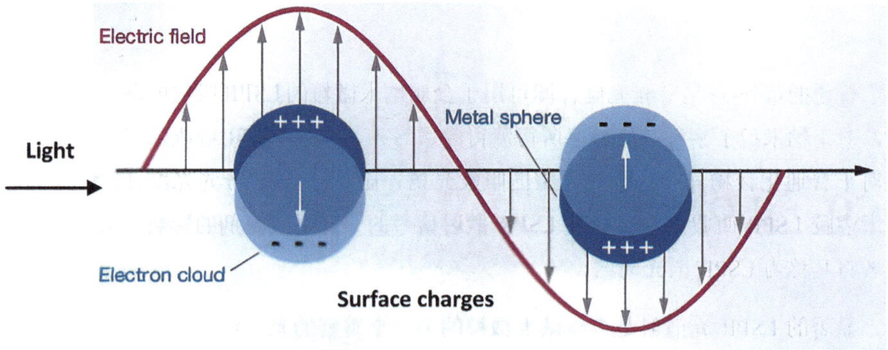

Fig. 1.2 The phenomenon known as surface-plasmon resonance (SPR) occurs in spherical nanometal particles. This involves the induction of a dipole which oscillates in synchronization with the electric field of the incident light (Liang et al. 2012)

on the nanoparticle's surface (Fig. 1.2). When the frequency of the oscillating excited electrons resonates with the frequency of the incident light, the LSPR effect ensues, manifesting as a plasmon-absorption band. The size of this band is contingent upon the nanoparticle's dimensions. The plasmon-absorption band exhibits exceedingly high enhancements of the electromagnetic field proximate to the particle's surface, along with a high molar extinction coefficient (Liang et al. 2012). Light with a lower frequency and light with a higher frequency behave differently. Light with a lower frequency than the plasma frequency is reflected due to the screening of electromagnetic radiation by the plasmons, while light with a higher frequency than the plasma frequency can be transmitted. This occurs because the screening effect takes longer to activate due to the slower electron response. However, when the light has the same frequency as the oscillating surface electrons, the surface plasmon resonance (SPR) effect becomes activated, resulting in electric fields that exceed the size of the original photons (Ingram et al. 2011). In mathematical terms, when both the frequency of the plasmon and the light coincide, the real part of the dielectric function approaches zero, inducing absorption resonance and thereby generating strong electric fields on the surfaces of the nanoparticles. Combining nanoparticles capable of producing this effect with a photocatalytically active material as a substrate is a promising method for applications in photocatalysis (Hou and Cronin 2013).

1.2.3 Tungsten Trioxide(WO_3)-Based Nanocomposite Photocatalysts

As the world becomes increasingly industrialized, the environment suffers from the release of pollutants resulting from industrial activities, posing a significant challenge. To address these issues, various methods, including photodecomposition using photocatalysts, have been employed. Among photodecomposition techniques, WO_3-based photocatalysts have exhibited considerable potential due to their superior oxidation potential, adsorption capacity, light-harvesting efficiency, and charge migration and separation capabilities. WO_3 octahedra and ultra-thin wet sheets have demonstrated high efficiency in photodegradation (Aslam et al. 2014). A subsequent study found that when silver was interfaced with an already efficient photocatalyst, both its efficiency and stability were improved (Yuan et al. 2017). One of the greatest threats to both humans and the environment is global warming. Scientists are actively seeking methods to convert CO_2 emissions into valuable solar fuels. WO_3 alone can only convert CO_2 into CH_4. Researchers have experimented with various shapes and combinations of WO_3 to identify the most efficient approach for converting CO_2 into fuel. Some successful methods include incorporating Molybdenum (Mo) into WO_3 nanorods and creating specialized structures such as Z-schemes and heterojunctions with other materials (Fig. 1.3) (Chen et al. 2012). To generate alternative fuels, scientists have incorporated additional composites into WO_3, including TiO_2, tungsten, and bronze, aiming to utilize sunlight more

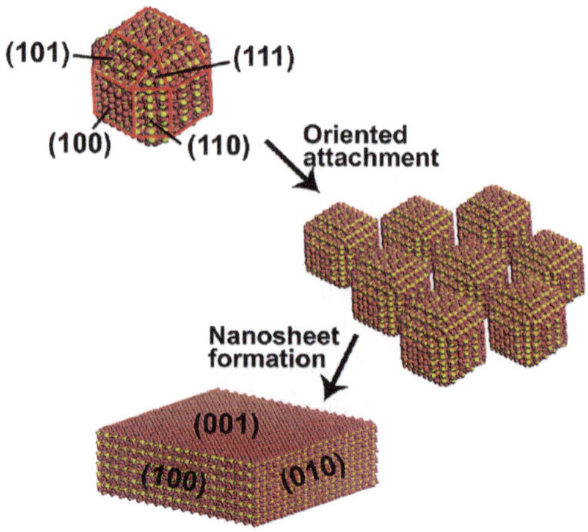

Fig. 1.3 Formation process of WO_3 nanosheets from small nanocrystals (Chen et al. 2012)

efficiently to produce various types of useful fuels (Wu et al. 2019). In summary, WO_3 nanocomposites have been utilized as photocatalysts to reduce CO_2 emissions for the next generation of green energy applications.

1.2.4 Bismuth-Based Photocatalysts

Bismuth-based photocatalysts can be employed to convert carbon dioxide into reusable fuels and chemicals utilized in industrial processes, including formic acid, methane, and carbon monoxide. The properties of bismuth oxyhalides ($Bi_4O_5I_2$) that render them ideal for this purpose are their high efficiency in absorbing visible light and their suitable band structures for CO_2 conversion. Research on these catalysts demonstrates their versatility and their potential to address some of the energy and pollution challenges we are confronting (Bai et al. 2018) (Fig. 1.4).

1.2.5 Conjugated Porous Polymer Photocatalysts

This section emphasizes the urgency of transitioning away from nonrenewable sources such as fossil fuels, which constitute the primary threat to the environment. This section also underscores the necessity of prioritizing solar energy and other renewable sources moving forward. Specifically, it highlights the potential of conjugated porous polymers (CPPs), encompassing covalent organic frameworks (COFs) and conjugated microporous polymers (CMPs), as effective materials for converting solar energy into cleaner and more sustainable sources of energy. This study delves into the intricate examination of the processes and structural concepts crucial in semiconductive COFs and CMPs. It elucidates the advantages and recent

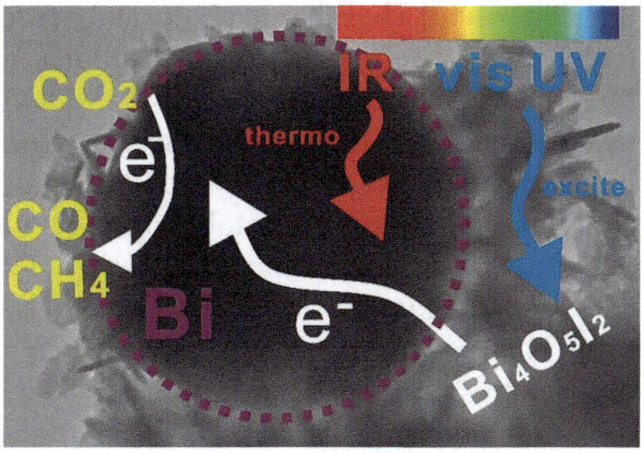

Fig. 1.4 Bismuth-decorated nanosheets composed of $Bi_4O_5I_2$ semimetal demonstrated remarkable photo-thermocatalytic performance in reducing CO_2 under UV-visible-infrared light (Bai et al. 2018)

breakthroughs in their applications across various photocatalytic processes, including hydrogen evolution, carbon dioxide reduction, pollution degradation, nitrogen fixation, and organic conversion. A pivotal aspect of any photocatalyst is the bandgap, ensuring it is suitable to facilitate the reactions. If the bandgap is suboptimal, the efficiency of the reaction diminishes significantly. However, researchers are exploring numerous strategies, including post-modification, donor-acceptor system construction, and statistical copolymerization, to address this issue (McDearmon et al. 2018). Water splitting entails a positive Gibbs free energy, indicating it as an uphill reaction. Consequently, it cannot proceed spontaneously and requires an external energy source. In this scenario, visible light irradiation serves as the energy form harnessed. While this reaction is efficient, researchers are exploring post-modifications using the aforementioned techniques to enhance efficiency further. Specific COFs and CMPs were highlighted, such as irreversible oxadiazole-linked (ODA)-COF and COFs based on pyrene, carbazole, and tetrastyryl (Yang et al. 2022). Carbon dioxide is one of the largest pollutants worldwide and poses a significant environmental challenge. Researchers are employing COFs and CMPs to break down CO_2 into fuels (such as formic acid (HCOOH), CO, or CH_4, among others) utilized in industrial processes. To enhance the efficiency of this process, the use of nitrogen-containing COFs has been implemented. Due to their capacity to effectively separate photogenerated charges, they prove efficient in reducing CO_2 (You et al. 2020).

1.2.6 Metal-Organic Framework Photocatalysts

Recent years have witnessed significant advancements in the field of photocatalysis, particularly with the introduction of metal-organic frameworks (MOFs) as viable options for various purposes, including energy conversion and environmental conservation. The degradation of pollutants through light-driven processes is among these applications that have garnered considerable attention, as it offers the potential for environmentally friendly and sustainable methods (Park and Oh 2017). The development of MOF-based photocatalysts for the degradation of organic pollutants under visible light irradiation constitutes an important and burgeoning area of research. Metal-linker MOFs, or ML-MOFs, possess unique structures and properties that render them effective in separating electrons and holes generated by photoinduced charge carriers, thereby aiding in the breakdown of organic pollutants. In a study, it was found that combining very stable porphyrin-MOFs (PCN-222(Zn)) with zirconium clusters linked by the Zn-TCPP linker exhibited efficient degradation of rhodamine B (RhB) dye. The addition of titanium dioxide nanoparticles (TiO_2 NPs) to PCN-222(Zn) to produce a hybrid photocatalyst (TP-222(Zn)) further enhanced its efficacy (Zhao et al. 2017). TiO_2 NPs were utilized as electron acceptors in this system, while PCN-222(Zn) facilitated better charge transfer and light absorption, leading to the generation of reactive oxygen species (ROS) responsible for dye degradation. Similarly, ML-MOFs have demonstrated potential in the photocatalytic reduction of carbon dioxide (CO_2) to produce useful chemicals through light absorption, owing to their structure and flexible properties. Visible light absorption enables the separation of electron-hole pairs (EHPs), followed by surface redox reactions that yield products such as carbon monoxide (CO), formate ions (HCOO-), formic acid (HCOOH), and methanol (CH_3OH) (Wang et al. 2019a). MOFs offer several advantages for photocatalytic CO_2 reduction, including the ability to position CO_2 molecules around active sites and efficiently utilize solar light due to their transition metal cations and organic linkers, which can be tailored to specific requirements (Li et al. 2020a).

1.2.7 Carbon-Based Photocatalysts

A potentially effective approach to combating both global warming and energy shortages is through the photocatalytic reduction of CO_2 into solar fuels, a method that has gained popularity in recent years. Carbon-based photocatalysts, possessing valuable electrochemical and physicochemical properties, are attracting significant interest. Graphene, carbon nanotubes (CNT), graphitic carbon nitride (GCN), and conducting polymers are among the materials serving as superior supports for photocatalysts, enhancing their capacity to absorb visible light and capture electrons. However, the photoconversion efficiency of pristine carbon materials is limited by the high rates of recombination of photogenerated charge carriers. Researchers have devised several enhancements to improve the photocatalytic performance of carbon-based photocatalysts, aiming to address this issue. These modifications include incorporating CO_2 capture materials to enhance the availability of reactant molecules at catalytic sites, as well as improving CO_2 adsorption efficiency and activation by introducing p-conjugated structures onto the carbon surface. For CO_2 adsorption and activation, nanocomposites such as Ag/RGO/CdS have demonstrated enhanced photocatalytic CO_2 reduction by leveraging the synergistic effects of reduced graphene oxide (RGO), silver (Ag), and cadmium sulfide (CdS) (Zhu et al. 2018). Additionally, the introduction of pores enhances CO_2 adsorption and increases effective photoreduction, as demonstrated by CdS nanorods deposited on nanoporous RGO. Moreover, ternary nanocomposites comprising graphene, noble metals, and principal photocatalysts are currently under investigation (Kandy and Gaikar 2019). These nanocomposites show multifunctional advancement effects that increase photocatalytic activities (Wang et al. 2019b). Graphene oxide (GO) surfaces that exhibit a uniform distribution of photocatalysts and enhanced CO_2 adsorption, leading to a high photocatalytic conversion of CO_2 to methanol, exemplify Bi_2O_3-graphene-ZnO ternary nanocomposites (Cho et al. 2017). Because of its enhanced adsorption capability, amine-functionalized graphitic carbon nitride (GCN) has demonstrated high photoreduction of CO_2. However, concerns regarding the cost-effectiveness of GCN as a photocatalyst arise due to its utilization of expensive chemicals and complex manufacturing processes (Wang et al. 2016). To tackle this issue, scientists have developed carbon/GCN hybrid composites and incorporated transparent ZIF-8 nanoclusters into GCN to enhance CO_2 adsorption capacity without compromising light absorption (Liu et al. 2017). The development of carbon-based nanomaterials holds immense potential for solar fuel conversion. These materials can significantly contribute to mitigating climate change and meeting global energy demands by enhancing CO_2 adsorption and activation, optimizing photocatalytic performance, and addressing cost concerns. To unlock the full potential of carbon-based photocatalysts for green energy, further research and development efforts in the field are necessary.

1.2.8 Bismuth Tungstate Photocatalysts

The advancement of clean energy production will necessitate the utilization of photocatalytic technology. The Aurivillius-phase perovskite Bismuth tungstate (Bi_2WO_6) is garnering considerable attention as a visible light-responsive photocatalyst due to its excellent optical and chemical characteristics, low cost, and flexible shape, in addition to being nontoxic. However, pure Bi_2WO_6 exhibits limited photocatalytic activity due to constraints such as a small specific surface area, low quantum yield, and rapid recombination of photogenerated carriers. Various approaches, including morphological control, metal deposition, semiconductor combination, dopant or defect introduction, and surface modification using carbon-based materials with conjugated π-structures, are under investigation to address these challenges. This section summarizes recent research progress on Bi_2WO_6-based photocatalysts to provide insights into the application of bismuth-based perovskite oxides in photocatalysis. Additionally, it examines the performance of Bi_2WO_6-based photocatalysts in three main applications: CO_2 reduction, pollutant removal, and water splitting. The review also outlines the possibilities, benefits, and challenges associated with Bi_2WO_6-based photocatalysts, aiming to offer ideas for the development of efficient photocatalysts for clean energy production and environmental cleanup. Current advancements in environmental technology present numerous opportunities, particularly in transforming environmental issues into energy sources. Nevertheless, significant challenges remain to be addressed, especially considering the rise in greenhouse gas emissions and the resulting environmental and human harm. One of the primary greenhouse gases targeted for reduction through photocatalysis is CO_2 (Sun et al. 2018). By harnessing the sun as an endless solar energy source, these technologies have the potential to convert atmospheric carbon dioxide into clean energy fuels. It has been demonstrated that structural manipulation of Bi_2WO_6 can enhance CO_2 reduction efficiency. Photocatalytic reduction of CO_2 to methanol has been achieved using three-dimensional Bi_2WO_6 quantum dots with hollow structures (Jiang et al. 2017). This was possible due to the dots' high specific surface area, optimized positions of the valence band maximum (VBM) and conduction band minimum (CBM), and pore size. The quantum size effect of hollow-structured Bi_2WO_6 increased charge transfer kinetics and accelerated the separation rate of photogenerated carriers, resulting in increased activity (Wang et al. 2015). To sum up, the development of photocatalysts based on Bi_2WO_6 offers a means to advance clean energy production and address environmental issues. Further advancements and research can help make Bi_2WO_6-based photocatalysts a more appealing option for clean energy generation and environmental cleanup.

1.3 Hydrogen Evolution

1.3.1 Cadmium Sulfide Photocatalysts

In recent years, there has been a significant movement among scientists, politicians, and even the general population to transition to greener practices, whether by better utilizing existing resources or creating new sources of renewable green energy. In a study, surface-functionalized cadmium sulfide (CdS) nanoparticles were used to separate minerals, methylene violet (MV), and rhodamine B (RhB), which are required for hydrogen evolution. The study investigated the change in photocatalytic hydrogen production based on whether sulfur was vacant. The sulfur vacancy on the surface of CdS quantum dots (QDs) showed a promising new method to enhance charge separation and improve photocatalytic performance. Many variables impact the properties of vacancy formation on the QDs, including average strain, photoluminescence, electrostatic repulsion, bandgap, crystallite size, dislocation density, and texture coefficient. The kinetics of growth inside the CdS were analyzed using UV-vis, Fourier transform infrared, and photoluminescence spectra for multiple capping agents. These results, combined with theoretical calculations, were used to examine the behavior of the CdS-ligand and its opto-electrical/structural properties. Thioglycolic acid functionalization essentially renders the surface passive and unreactive, providing the proper optical and textural properties for the purposes of the study. Density functional theory and time-dependent density functional theory were used to examine optimized nanocluster geometry, stability, and the bandgap. Subsequently, the water mineralization of MV and RhB was done using solar light, and then the water was used to create hydrogen using the CdS quantum dots for a carbon-neutral method. This study on methylene dye was conducted using a thin film photocatalyst. First, 5 mg of MV dye was added to 50 ml of deionized water, and then the film was submerged. One xenon bulb, simulating the sun, was used as the light source, and the amount of dye absorbed was observed and recorded at different intervals. The photocatalytic evolution of hydrogen through water splitting used a quartz reactor volume of 150 ml, where the degraded water was used to generate hydrogen gas (H_2). The xenon light was placed 10 cm away from the reactor. The film was placed into a solution of 100 ml containing 10% NaCl/water. The sample was illuminated and stirred for 25 hours, and the gas produced was measured. It was then compared to an external control group to measure the rate of H_2 evolution. This study demonstrates a new way to reuse wastewater to generate solar power. It explores how sulfur vacancies affect the process of hydrogen and MV dye degradation. The study also illustrates the physical qualities of a photocatalyst that improve its effectiveness, along with how different capping agents affect hydrogen and dye. Using both theoretical calculations and measured data, it is shown how different variables impact this process. The goal is to further improve this method and increase its efficiency so that wastewater can be used to create cleaner energy (Roshan et al. 2023).

1.3.2 Bismuth-Based Photocatalysts

In addition to CO_2 reduction, bismuth-based catalysts are being utilized to produce hydrogen using solar energy through a process known as water splitting. In this process, the catalyst absorbs light and generates electron-hole pairs, which are then involved in a photocatalytic process resulting in the production of hydrogen and oxygen. New materials such as $Bi_4O_5Br_2$/TL-Ti_3C_2 nanosheets have demonstrated enhanced hydrogen evolution activity due to their responsiveness to visible light and efficient electron transfer (Xi et al. 2020).

1.3.3 Zinc Indium Sulfide Photocatalysts

The use of semiconductor photocatalysis, especially zinc indium sulfide ($ZnIn_2S_4$), has garnered interest as a potential solution for harmful emissions and worldwide energy shortages. $ZnIn_2S_4$ exhibits good durability and absorbs visible light well, but its photocatalytic activity is limited by fast electron-hole pair recombination and a narrow visible light absorption range (Hou and Zhang 2020). Several methods for modification, such as metal deposition, element doping, and defect introduction, have been investigated to improve its performance. One important application of $ZnIn_2S_4$ photocatalysts is the production of hydrogen from water splitting, which is vital for generating sustainable clean energy. Platinum (Pt) and other noble-metal co-catalysts have been employed to increase the hydrogen evolution rate of $ZnIn_2S_4$, but their use is limited by the high cost of Pt. To overcome this limitation, scientists have explored combining $ZnIn_2S_4$ with carbon-based materials like graphene and carbon nanotubes, which enhance the efficiency of electron transfer and reduce recombination (Shen et al. 2012).

1.4 Pollutant Degradation

1.4.1 $ZnIn_2S_4$ Photocatalysts

Recently developed Z-scheme heterojunctions of the compound are being used in pollutant degradation, water splitting, and toxic metal remediation. The techniques used for these processes are highlighted in this chapter, offering readers insight into applications for environmental and other energy-related purposes based on recent research. Studies on pollutant degradation using $ZnIn_2S_4$ have been highly effective. Methylene blue (MB), Congo red (CR), and metronidazole (MTZ) were all efficiently and rapidly broken down by $ZnIn_2S_4$-g-C_3N_4/$BiVO_4$ nanocomposites. A major environmental issue is pharmaceutical wastewater, which contains large amounts of antibiotics and other chemicals harmful to human and animal life (Schwarzenbach et al. 2006). These compounds are toxic to life, and chromium Cr(IV), in particular, is known to be mutagenic and highly toxic. Utilizing different photocatalysts to break down these compounds and reduce them into less dangerous materials is an important area of research. Recently, an $AgVO_3$/$ZnIn_2S_4$ catalyst improved photocatalytic activity and prevented the photo-corrosion of $ZnIn_2S_4$ and $AgVO_3$. It was found that over time, the solution became much less toxic as the Cr(IV) was converted into the less toxic Cr(III) (Li et al. 2020b). Other studies used nanosheets and other modifications to solve the draw backs of $ZnIn_2S_4$ and increase the efficiency of photodegradation (Zhang et al. 2020a). $ZnIn_2S_4$ photocatalysts have been utilized in energy-related applications as well as in the removal of environmental pollutants, with a focus on degrading organic compounds, heavy metals, and dyes (Yu et al. 2019). $ZnIn_2S_4$-based materials have demonstrated excellent efficiency in degrading pollutants under visible light irradiation due to their photocatalytic properties. To enhance the rate at which pollutants degrade, methods such as coupling with carbon dots or graphene and creating heterojunctions with other materials have proven effective (Tian et al. 2017). Overall, $ZnIn_2S_4$-based photocatalysts offer promising solutions for addressing the world's energy and environmental problems due to their versatility and efficiency. The field of semiconductor photocatalysis will continue to progress with further research into innovative modification techniques and practical applications.

1.4.2 Graphitic Carbon Nitride Photocatalysts

Creating sustainable green energy is one of the largest questions that has yet to be answered in the modern era. Many different applications of photocatalysis have been viewed as possible solutions to this issue. This study looks at graphitic carbon nitride ($g\text{-}C_3N_4$), a nonmetallic semiconductor. This compound, in particular, is promising due to its high stability, nontoxicity, and efficient photocatalysis. However, this compound is not perfect and has major drawbacks including poor photoexcited charge separation, narrow absorption of visible light, and a low specific surface area. By combining $g\text{-}C_3N_4$ with other composites or modifying it to improve some of its shortcomings, it becomes viable for photocatalytic use. This article highlights the various uses for $g\text{-}C_3N_4$-based photocatalysts (GPBs) and the challenges currently being faced. Photocatalytic reduction of heavy metal ions (HMIs) is another method that has been increasing in popularity. Chromium is a major pollutant, and its isotope Cr(IV) is a very toxic metal, so traditionally, reducing it into Cr(III) before removing it is necessary. GPBs have been shown to be a very efficient and cost-effective way to do this. In the process of reducing heavy metals, pH is a very important factor. Lower pH levels benefit the reduction process, while higher pH levels reduce its efficiency. Higher pH levels increase the likelihood that the Cr's active sites would be shielded by other parts of the pollutants, making the reaction much less efficient (Zhang et al. 2017).

1.4.3 Nanofiber Photocatalysts

Improving membrane filtration and creating efficient artificial photosynthesis systems are major projects aimed at solving the world's energy and environmental problems. While electrospun nanofiber (NF) photocatalysts have shown promise, several obstacles must be overcome before they can be used by the public. This study clarifies major concepts in photocatalysis and discusses several tactics to improve NF photocatalyst performance. Increasing the lifetimes of the electrons and holes produced by light absorption in NFs is one of the primary obstacles of this method. To overcome this, scientists have investigated new approaches such as the use of heterojunctions and defective formulations, which can increase the lifetime of charge carriers and improve photocatalytic activity. Additionally, membrane filtration technologies are needed for water treatment procedures since they enable efficient pollutant removal. However, there are still issues with eliminating pathogens and organic contaminants. By using nanoparticles and making modifications to the production process, membrane properties can be adjusted to overcome these restrictions (Moustakas et al. 2014). These changes enable more efficient water treatment techniques and enhance the longevity and efficiency of the membrane. Sustainable water treatment can be facilitated by hybrid membrane materials, which combine the advantages of photocatalysis with the structural stability of nanomaterials. Membranes containing photocatalytic materials, such as TiO_2, can significantly improve the efficiency of contaminant removal while reducing fouling (Nyamutswa et al. 2018). Additionally, new techniques such as integrating filtration-photocatalysis systems show promise in preventing membrane fouling and removing organic matter present in the water, thereby improving its quality. Adsorption technologies are also essential for water treatment because they can effectively eliminate various contaminants. Adsorption processes successfully absorb pollutants using materials such as activated carbon and metal-organic frameworks (Huang et al. 2018). Additionally, incorporating nanoparticles into adsorbent materials enhances their ability to bond with pollutants, thereby improving removal efficiency (Georgieva et al. 2015). In conclusion, the advancement of membrane filtration, adsorption, and photocatalysis technologies is essential for sustainable water purification. It is important to continue the research in these areas to address the numerous drawbacks of water pollution and ensure that everyone has access to clean, safe drinking water.

1.4.4 Zeolitic Imidazolate Framework (ZIF) Photocatalysts

To address global warming and energy crisis, researchers have looked into photocatalysts as the solution to the major problem. New age materials are being looked at; in particular, zeolitic imidazolate frameworks (ZIFs), belonging to a sub-class of metal-organic frameworks (MOFs) have gained a lot of interest. Due to their huge specific surface area, tuneable porosity, diverse structure, ease of designing, and controllable properties ZIFs are being highly sought after as a method of pollutant degradation and clean energy generation. Like any material, ZIFs have their own unique advantages and shortcomings. Researchers have been looking at different methods such as thermal processing to improve light responsiveness and electron transfer. By using these processes to improve the shortcomings of ZIFs, many functional materials and better photocatalysts can be prepared. By thermal treatment of ZIFs, formation of composites, heterostructures with other semiconductors or electron mediators which may have similar structural characteristics but better light capture abilities, charge transfer capabilities and strong redox properties can be fabricated. This section looks at these processes as well as the usages of ZIFs and how they can be used to solve the energy and pollution problems that face the world. Additionally, the properties of ZIFs, bottlenecks, and future research are also discussed. Imidazole is a heterocyclic compound that is a five-part ring containing three carbon atoms and two nitrogen atoms and can come in two different forms (Shalini et al. 2010). It is a white, colorless solid with a melting point of 88.9 °C and a boiling point of 267.8 °C. Discovered in the 1840s, imidazole has become important in pharmaceuticals, being used to create various drugs, including anticancer, antibacterial, anti-inflammatory, and diagnostic agents. Imidazole is so versatile due to its hydrogen bonding ability. It can form H-bonds with drugs and proteins, allowing it to do many different things when combined with other compounds (Anderson and Long 2010). Zinc-based ZIFs are a popular compound used in photocatalysis. ZIF-8, in particular, is a water-stable compound with an area and pore size that make it ideal for the separation of gaseous mixtures (Ortiz et al. 2014). Another compound that is used with ZIFs is cadmium. Because of its shorter bandgap, it is suitable for some photocatalytic applications when interfaced with ZIFs to create Cd-ZIF-8 (Xie et al. 2010). Like zinc-based ZIFs, CD-ZIF-8 is an ideal candidate to degrade dyes using visible light. This is the case because of its high thermal stability and unique cubic phase structure. Combining ZIFs with other low bandgap materials allows for better photocatalytic performance. Due to its low cost, ease of access, and stability, copper is an ideal candidate. This lower bandgap helps to move electrons to ZIF's broad bandgap and slows down the charge recombination rate (Zoolfakar et al. 2014). Similarly, other metals such as cobalt and nickel can be interfaced with ZIFs to be used in different ways, depending on the properties of the metal that the ZIF is interfaced with. ZIFS are also being used for water splitting and hydrogen evolution reactions. Different ZIF-derived materials, such as ZnS nanosheets and ZIF-67 @CoAl LDH, are explored for their photocatalytic efficiency in water splitting (Jin et al. 2022). In summary, the unique properties of ZIFs such as its high porosity, ease of bonding with other compounds, and ability to act as charge carriers for catalytic materials make them an ideal solution to the energy crisis that we are facing.

1.4.5 TiO_2-Based Photocatalysts

In recent decades, there have been many different types of nanomaterial photocatalysts that have been researched for their energy and environmental applications. One of the most researched is TiO_2. The main issue with this material is that it has a large bandgap which limits the amount of light absorption, and the relatively high charge-carrier recombination, which restricts the photocatalytic performance. Recently, studies have tried to modify the engineering of TiO_2 to broaden the photo response and increase charge separation efficiency. To improve the photocatalytic performance even more, it is important to identify the progress of defective TiO_2 photocatalysts. In this article, various synthesis strategies for defective TiO_2, especially black TiO_2 nanomaterials, are looked into as well as the way that they are used. The types of defects and characteristics as well as the photocatalytic mechanism are discussed. In the textile industry, dyes are a very common source of waste and are sometimes put into the water (Guettaï and Ait Amar 2005). TiO_2 is commonly used to detoxify water and that makes it ideal to use in this case. Various methods, including high-pressure hydrogen treatment and plasma synthesis, have been used to create black TiO_2 materials with excellent stability and enhanced degradation rates for organic pollutants (Wang et al. 2013a). Defective TiO_2 nanomaterials produced through oxidation and other types of reduction are used mostly for UV and light-driven photocatalytic reduction reactions of water contaminants. In particular, they present promising and efficient methods to use light to treat the wastewater of the textile industry, which is filled with toxic organic dyes. Defective black TiO_2 without a platinum co-catalyst has been shown to be very efficient at hydrogen production. Additionally, TiO_2 nanotubes and nanoparticles are being used in solar fuel production and creating artificial photosynthesis for renewable energy. They are also used to create nanocrystals and films that have shown high efficiency at converting CO_2 into CH_4 and CO. In summary, TiO_2 materials, especially defective and black TiO_2, are being explored as a way to reduce the pollution in the environment whether that be water cleaning, CO_2 reduction, and hydrogen production (Liu et al. 2014).

1.5 Water Splitting

1.5.1 Gadolinium-Based Photocatalysts

In recent years, there has been more interest in gadolinium (Gd)-based photocatalysts for water treatment to address environmental concerns. This section covers various ways in which Gd-based nanomaterials are utilized and enhanced for more efficient application in photocatalysis. Furthermore, the section describes synthesis methods, optimization, and modification techniques for Gd photocatalysts. As with any material, there are some drawbacks, but this section also discusses how researchers are overcoming these challenges. Gadolinium is a silver-white, malleable, and ductile earth metal. It has a hexagonal close-packed (alpha) structure and then transitions into a body-centered cubic form (beta) at 1235 °C (Pyykkö 2015). Certain isotopes of Gd have many unpaired electrons, giving them a high total spin and a large magnetic moment, which makes them susceptible to magnets and gives them the potential to absorb and emit light. Gd is also paramagnetic and acts as a Heisenberg ferromagnet (Dan'kov et al. 1998). All the properties mentioned before vary based on the type of Gd and its purity. A common compound used for photocatalysis is Gd_2O_3, and its main drawback is that its crystalline phase varies greatly, depending on the environment, ranging from tetragonal, monoclinic, trigonal, and more. Hydrogen production by splitting water is one of the most promising methods for creating sustainable energy (Xia et al. 2021). In most cases, Gd was combined with titanium or other mixed-base semiconductor systems. Photocatalytic reduction of CO_2 is a significant method to combat the current crisis caused by global warming. A study was conducted on Pd-doped $Gd(OH)_3$ NPs for the photocatalytic conversion of CO_2 to methanol. It was found that the Pd forms Schottky barriers on the $Gd(OH)_3$, facilitating charge mobility and hindering recombination. The NP was also shown to have excellent efficiency and visible light harvesting and was reusable for up to five cycles (Al-Mhyawi and Abdel Salam 2018). Another major problem caused by industrialization is the pollutants that are put into the water supply no matter what the regulations against it. Specifically, heavy metals are a major problem; a way to fix this is its removal by using photocatalyst. It was found that Cr(VI) can be removed by using $Gd_{0.2}La_{0.2}Ce_{0.2}Hf_{0.2}Zr_{0.2}O_2$ (GLCHZ) and $Gd_{0.2}La_{0.2}Y_{0.2}Hf_{0.2}Zr_{0.2}O_2$ (GLYHZ) for reduction of Cr(VI) under sunlight (Anandkumar et al. 2021). Both photocatalysts showed narrow bandgaps, optimum numbers of oxygen vacancies, and suppressed recombination and led to efficient reduction of Cr(VI). Further studies showed that other photocatalysts (GO-Gd_2O_3) were also successful at removing other heavy metals such as Pb(II) and As(V) (Raizada et al. 2019).

1.5.2 Strontium Titanate-Based Photocatalysts

This section focuses on converting solar energy to chemical energy to solve the impending energy crisis facing us. Scientists believe that a possible way to do this is through photocatalytic water splitting. The basic principle of solar splitting is when valence band electrons of semiconductors travel to the conduction band. Under light irradiation, electron-hole pairs are created. These pairs then participate in hydrogen evolution and oxygen evolution reactions (Kudo and Miseki 2009). This reaction has a positive Gibbs free energy change, meaning that the reactions are non-spontaneous and require energy input. This emphasizes the importance of developing more efficient catalysts for the future of this area of research, especially in terms of large-scale use. One approach attempting to solve this issue is the use of semiconductor powder photocatalysts, which are dispersed in water under sunlight to produce gas. This method could potentially represent the future of large-scale fuel production. However, its realization is hindered by the unavailability of proper photoelectrode materials possessing suitable band structures, chemical stability, and the ability to be mass-produced (Faraji et al. 2019). The article then delves further into $SrTiO_3$, a compound utilized for water splitting due to its low cost and stability. However, there are some drawbacks to using this material: its bandgap is not optimal for efficiently splitting water under visible light. Only about 5% of the light can be utilized for photocatalytic water splitting due to $SrTiO_3$'s wide bandgap. Additionally, although the structure of $SrTiO_3$ is suitable for the redox reaction of H_2O splitting, it cannot perform the reaction alone due to the rapid recombination of the generated charge carriers (Iwashina and Kudo 2011). Recently, aluminum was combined with $SrTiO_3$. When tested, this compound was found to be up to 96% efficient for water splitting under UV light irradiation. Other compounds are being tested in a similar manner to find the most efficient solution to the energy crisis by identifying a clean and effective method to make the process of water splitting more commercially available (Takata et al. 2020).

1.5.3 WO_3-Based Nanocomposite Photocatalysts

Another way that researchers have been exploring to shift away from fossil fuels is through water splitting. To accelerate the oxidation and reduction of water, the bandgap of the semiconductor has to meet certain redox potential conditions. A study found that sulfur, when added to WO_3, boosted the efficiency of photocatalytic activity, but too much sulfur ended up acting as recombination sites, thereby reducing the overall efficiency (Li et al. 2012). A similar result was found when W^{5+} was used to fill the oxygen vacancies. A proper amount would increase efficiency, but beyond a certain point, the presence of W^{5+} would begin to have a negative impact on photocatalysis (Liu et al. 2018). An issue with the oxygen-doped WO_3 is that they were less stable than desired. To address this, researchers created heterostructures by combining WO_3 with other semiconductors. These heterostructures, specifically CdS-WO_3, greatly increase the stability of the photocatalyst (Lou et al. 2018). Other attempts to increase the efficiency of WO_3 have been made using Z-scheme heterojunctions and photocatalysts, both of which have resulted in improved photocatalytic reactions and hydrogen evolution (Ding et al. 2019).

1.5.4 $ZnIn_2S_4$ Photocatalysts

Due to their strong redox abilities, efficient charge transport, and enhanced charge separation, Z-scheme heterostructures of $ZnIn_2S_4$-based photocatalysts are being used to generate renewable energy. Recently, $ZnIn_2S_4$-based Z-scheme photocatalysis has been employed to effectively degrade hazardous organic compounds. The $ZnIn_2S_4$ photocatalyst exhibited an efficiency of 64.2% in degrading Methylene blue (MB) within 300 minutes. Adding Bi_2S_3 onto $ZnIn_2S_4$ improved the efficiency to 95.4% simultaneously. This improvement is attributed to optimizing the weight ratios of both chemicals to ensure an equal space charge region and allowing for equal light penetration, which is necessary for controlling charge recombination (Ben Assaker et al. 2015). Water splitting is another promising method for creating renewable and clean fuel. The challenge lies in the fact that water splitting is an endothermic reaction, making it highly challenging. One approach is to use semiconductors as photoelectrodes to split water into its gas components. In this process, conduction band electrons are utilized to reduce the water, while valence band holes oxidize the water for the O_2 evolution reaction. However, the selection of semiconductors for this application is limited due to the requirement of a bandgap of at least 1.23 eV. $ZnIn_2S_4$ possesses an ideal bandgap for this purpose, but it lacks efficient charge separation and the proper structure. To address this issue, nanocarbon was added to the $ZnIn_2S_4$ to serve as a solid electron mediator, enabling more efficient charge separation and thus more efficient photocatalysis. Despite progress being made, there remains the question of optimizing the oxygen evolution reaction and overcoming inherent energy constraints (Abe 2010).

1.5.5 Bismuth-Based Photocatalysts

In recent years, there have been constant improvements and advancements in bismuth oxyhalides ($Bi_xO_yX_z$). These bismuth compounds are ideal for photocatalysis due to their bandgap, stability, and electrical conductivity. These properties make them effective in pollutant degradation, clean energy creation, and other applications. Researchers are exploring ways to synthesize these compounds and tailor them to further increase efficiency. They are primarily focusing on structural modifications and heterostructure formations. Bismuth-based photocatalysts are particularly effective at breaking down pollutants. They achieve this by creating reactive oxygen species when exposed to visible light, which oxidize and break down compounds such as 4-tert-butylphenol, rhodamine B, and others (Wang et al. 2013b). To further improve this breakdown, researchers have tried to use microwaves in addition to visible light to aid this process and increase efficiency in both air and water (Xiao et al. 2012). In water, bismuth-based photocatalysts are limited in their efficiency. To fix this, researchers have created heterojunctions, such as $Bi_5O_7Br/BiOBr$, to combat these weaknesses and improve photocatalytic efficiency in water (Zhang et al. 2020b).

1.6 Conclusion

In conclusion, photocatalysis is a technology that holds enormous potential for addressing environmental issues and promoting environmentally friendly initiatives. It has the potential to significantly reduce pollution, ensure access to clean water and air, and accelerate the transition to renewable energy sources through interdisciplinary cooperation and ongoing research. By utilizing light and semiconductor materials, photocatalysis offers a feasible path toward a cleaner and more sustainable future. This chapter explored the introduction of photocatalysts with their applications related to the green energy concept. In addition, various photocatalyst materials were discussed in each application to show the capability of different materials for this technology. The readers can be familiar with the introduction of photocatalysts by reading this chapter and understand the basic mechanism of photocatalysts in green energy applications.

References

Abe R (2010) Recent progress on photocatalytic and photoelectrochemical water splitting under visible light irradiation. J Photochem Photobiol C: Photochem Rev 11(4):179–209. https://doi.org/10.1016/j.jphotochemrev.2011.02.003

Al-Mhyawi SR, Abdel Salam M (2018) Enhancement of photocatalytic activity of Gd(OH)3 nanoparticles by Pd deposition for reduction of CO2 to methanol. J Photochem Photobiol A Chem 367:89–93. https://doi.org/10.1016/j.jphotochem.2018.08.024

Anandkumar M, Lathe A, Palve AM, Deshpande AS (2021) Single-phase Gd0.2La0.2Ce0.2Hf0.2Zr0.2O2 and Gd0.2La0.2Y0.2Hf0.2Zr0.2O2 nanoparticles as efficient photocatalysts for the reduction of Cr(VI) and degradation of methylene blue dye. J Alloys Compd 850:156716. https://doi.org/10.1016/j.jallcom.2020.156716

Anderson EB, Long TE (2010) Imidazole- and imidazolium-containing polymers for biology and material science applications. Polymer 51(12):2447–2454. https://doi.org/10.1016/j.polymer.2010.02.006

Aslam I et al (2014) Synthesis of three-dimensional WO3 octahedra: characterization, optical and efficient photocatalytic properties. RSC Adv 4(71):37914–37920. https://doi.org/10.1039/C4RA05724D

Bai Y, Yang P, Wang P, Xie H, Dang H, Ye L (2018) Semimetal bismuth mediated UV–vis-IR driven photo-thermocatalysis of Bi4O5I2 for carbon dioxide to chemical energy. J CO_2 Util 23:51–60. https://doi.org/10.1016/j.jcou.2017.11.007

Ben Assaker I et al (2015) Electrodeposited ZnIn2S4 onto TiO2 thin films for semiconductor-sensitized photocatalytic and photoelectrochemical applications. Appl Surf Sci 351:927–934. https://doi.org/10.1016/j.apsusc.2015.06.038

Chen X, Zhou Y, Liu Q, Li Z, Liu J, Zou Z (2012) Ultrathin, single-crystal WO3 nanosheets by two-dimensional oriented attachment toward enhanced photocatalystic reduction of CO2 into hydrocarbon fuels under visible light. ACS Appl Mater Interfaces 4(7):3372–3377. https://doi.org/10.1021/am300661s

Chen Y et al (2014) Hierarchical core–shell carbon nanofiber@ZnIn2S4 composites for enhanced hydrogen evolution performance. ACS Appl Mater Interfaces 6(16):13841–13849. https://doi.org/10.1021/am5032083

Cho KM et al (2017) Amine-functionalized graphene/CdS composite for photocatalytic reduction of CO2. ACS Catal 7(10):7064–7069. https://doi.org/10.1021/acscatal.7b01908

Dan'kov SY, Tishin AM, Pecharsky VK, Gschneidner KA (1998) Magnetic phase transitions and the magnetothermal properties of gadolinium. Phys Rev B 57(6):3478–3490. https://doi.org/10.1103/PhysRevB.57.3478

Ding Y, Wei D, He R, Yuan R, Xie T, Li Z (2019) Rational design of Z-scheme PtS-ZnIn2S4/WO3-MnO2 for overall photo-catalytic water splitting under visible light. Appl Catal B 258:117948. https://doi.org/10.1016/j.apcatb.2019.117948

Faraji M et al (2019) Two-dimensional materials in semiconductor photoelectrocatalytic systems for water splitting. Energy Environ Sci 12(1):59–95. https://doi.org/10.1039/C8EE00886H

Ge Y, Zhang C, Yu Z, Wang D (2023) ZnIn2S4/TiO2 photocatalyst for CO2 photoreduction: advancing sustainable energy conversion to renewable solar fuels. J Ind Eng Chem. https://doi.org/10.1016/j.jiec.2023.11.027

Georgieva VG, Tavlieva MP, Genieva SD, Vlaev LT (2015) Adsorption kinetics of Cr(VI) ions from aqueous solutions onto black rice husk ash. J Mol Liq 208:219–226. https://doi.org/10.1016/j.molliq.2015.04.047

Guettaï N, Ait Amar H (2005) Photocatalytic oxidation of methyl orange in presence of titanium dioxide in aqueous suspension. Part I: parametric study. Desalination 185(1–3):427–437. https://doi.org/10.1016/j.desal.2005.04.048

Guo S et al (2020) Oxygen vacancy mediated bismuth stannate ultra-small nanoparticle towards photocatalytic CO2-to-CO conversion. Appl Catal B 276:119156. https://doi.org/10.1016/j.apcatb.2020.119156

Herrmann J-M (1999) Heterogeneous photocatalysis: fundamentals and applications to the removal of various types of aqueous pollutants. Catal Today 53(1):115–129. https://doi.org/10.1016/S0920-5861(99)00107-8

Hou W, Cronin SB (2013) A review of surface plasmon resonance-enhanced photocatalysis. Adv Funct Mater 23(13):1612–1619. https://doi.org/10.1002/adfm.201202148

Hou H, Zhang X (2020) Rational design of 1D/2D heterostructured photocatalyst for energy and environmental applications. Chem Eng J 395:125030. https://doi.org/10.1016/j.cej.2020.125030

Huang Y et al (2018) Bilayered antimicrobial nanofiber membranes for wound dressings via in situ cross-linking polymerization and electrospinning. Ind Eng Chem Res 57(50):17048–17057. https://doi.org/10.1021/acs.iecr.8b03122

Ingram DB, Christopher P, Bauer JL, Linic S (2011) Predictive model for the design of plasmonic metal/semiconductor composite photocatalysts. ACS Catal 1(10):1441–1447. https://doi.org/10.1021/cs200320h

Iwashina K, Kudo A (2011) Rh-doped SrTiO 3 photocatalyst electrode showing cathodic photocurrent for water splitting under visible-light irradiation. J Am Chem Soc 133(34):13272–13275. https://doi.org/10.1021/ja2050315

Jeong J, Sekiguchi K, Lee W, Sakamoto K (2005) Photodegradation of gaseous volatile organic compounds (VOCs) using TiO2 photoirradiated by an ozone-producing UV lamp: decomposition characteristics, identification of by-products and water-soluble organic intermediates. J Photochem Photobiol A Chem 169(3):279–287. https://doi.org/10.1016/j.jphotochem.2004.07.014

Jiang Z et al (2017) Photocatalytic reduction of CO2 to methanol by three-dimensional hollow structures of Bi2WO6 quantum dots. Appl Catal B 219:209–215. https://doi.org/10.1016/j.apcatb.2017.07.023

Jin Z, Yang M, Hao X, Zhu P (2022) ZIF-67 dodecahedron coupled with CoAl-layered double hydroxide as S-scheme heterojunction for efficient visible-light-driven hydrogen evolution. Appl Surf Sci 592:153300. https://doi.org/10.1016/j.apsusc.2022.153300

Kandy MM, Gaikar VG (2019) Continuous photocatalytic reduction of CO2 using nanoporous reduced graphene oxide (RGO)/cadmium sulfide (CdS) as catalyst on porous anodic alumina (PAA)/aluminum support. J Nanosci Nanotechnol 19(8):5323–5331. https://doi.org/10.1166/jnn.2019.16817

Kudlek E, Dudziak M (2018) Degradation pathways of pentachlorophenol and benzo(a)pyrene during heterogeneous photocatalysis. Water Sci Technol 77(10):2407–2414. https://doi.org/10.2166/wst.2018.192

Kudo A, Miseki Y (2009) Heterogeneous photocatalyst materials for water splitting. Chem Soc Rev 38(1):253–278. https://doi.org/10.1039/B800489G

Lalitha K, Reddy JK, Phanikrishna Sharma MV, Kumari VD, Subrahmanyam M (2010) Continuous hydrogen production activity over finely dispersed Ag2O/TiO2 catalysts from methanol:water mixtures under solar irradiation: a structure–activity correlation. Int J Hydrog Energy 35(9):3991–4001. https://doi.org/10.1016/j.ijhydene.2010.01.106

Li W, Li J, Wang X, Chen Q (2012) Preparation and water-splitting photocatalytic behavior of S-doped WO3. Appl Surf Sci 263:157–162. https://doi.org/10.1016/j.apsusc.2012.09.021

Li G et al (2020a) In-situ growth UiO-66-NH2 on the Bi2WO6 to fabrication Z-scheme heterojunction with enhanced visible-light driven photocatalytic degradation performance. Colloids Surf A Physicochem Eng Asp 603:125256. https://doi.org/10.1016/j.colsurfa.2020.125256

Li C, Che H, Yan Y, Liu C, Dong H (2020b) Z-scheme AgVO3/ZnIn2S4 photocatalysts: 'One Stone and Two Birds' strategy to solve photocorrosion and improve the photocatalytic activity and stability. Chem Eng J 398:125523. https://doi.org/10.1016/j.cej.2020.125523

Liang A, Liu Q, Wen G, Jiang Z (2012) The surface-plasmon-resonance effect of nanogold/silver and its analytical applications. TrAC Trends Anal Chem 37:32–47. https://doi.org/10.1016/j.trac.2012.03.015

Liu N et al (2014) Black TiO_2 nanotubes: cocatalyst-free open-circuit hydrogen generation. Nano Lett 14(6):3309–3313. https://doi.org/10.1021/nl500710j

Liu S, Chen F, Li S, Peng X, Xiong Y (2017) Enhanced photocatalytic conversion of greenhouse gas CO2 into solar fuels over g-C3N4 nanotubes with decorated transparent ZIF-8 nanoclusters. Appl Catal B 211:1–10. https://doi.org/10.1016/j.apcatb.2017.04.009

Liu Q et al (2018) Surface oxygen vacancy and defect engineering of WO 3 for improved visible light photocatalytic performance. Cat Sci Technol 8(17):4399–4406. https://doi.org/10.1039/C8CY00994E

Lou Z et al (2018) Continual injection of photoinduced electrons stabilizing surface plasmon resonance of non-elemental-metal plasmonic photocatalyst CdS/WO3−x for efficient hydrogen generation. Appl Catal B 226:10–15. https://doi.org/10.1016/j.apcatb.2017.12.023

McDearmon B, Page ZA, Chabinyc ML, Hawker CJ (2018) Organic electronics by design: the power of minor atomic and structural changes. J Mater Chem C Mater 6(14):3564–3572. https://doi.org/10.1039/C7TC05052F

Moustakas NG, Katsaros FK, Kontos AG, Romanos GE, Dionysiou DD, Falaras P (2014) Visible light active TiO2 photocatalytic filtration membranes with improved permeability and low energy consumption. Catal Today 224:56–69. https://doi.org/10.1016/j.cattod.2013.10.063

Nyamutswa L, Zhu B, Navaratna D, Collins S, Duke M (2018) Proof of concept for light conducting membrane substrate for UV-activated photocatalysis as an alternative to chemical cleaning. Membranes (Basel) 8(4):122. https://doi.org/10.3390/membranes8040122

Ortiz AU, Freitas AP, Boutin A, Fuchs AH, Coudert F-X (2014) What makes zeolitic imidazolate frameworks hydrophobic or hydrophilic? The impact of geometry and functionalization on water adsorption. Phys Chem Chem Phys 16(21):9940–9949. https://doi.org/10.1039/C3CP54292K

Park J, Oh M (2017) Construction of flexible metal–organic framework (MOF) papers through MOF growth on filter paper and their selective dye capture. Nanoscale 9(35):12850–12854. https://doi.org/10.1039/C7NR04113F

Pyykkö P (2015) Magically magnetic gadolinium. Nat Chem 7(8):680–680. https://doi.org/10.1038/nchem.2287

Raizada P, Sudhaik A, Singh P, Hosseini-Bandegharaei A, Thakur P (2019) Converting type II AgBr/VO into ternary Z scheme photocatalyst via coupling with phosphorus doped g-C3N4 for enhanced photocatalytic activity. Sep Purif Technol 227:115692. https://doi.org/10.1016/j.seppur.2019.115692

Rizzo L, Meric S, Kassinos D, Guida M, Russo F, Belgiorno V (2009) Degradation of diclofenac by TiO2 photocatalysis: UV absorbance kinetics and process evaluation through a set of toxicity bioassays. Water Res 43(4):979–988. https://doi.org/10.1016/j.watres.2008.11.040

Roshan R et al (2023) Photocatalytic waste-to-renewable energy nexus using solar light induced quantum dots. Energy Convers Manag 283:116917. https://doi.org/10.1016/j.enconman.2023.116917

Schwarzenbach RP et al (2006) The challenge of micropollutants in aquatic systems. Science (1979) 313(5790):1072–1077. https://doi.org/10.1126/science.1127291

Shalini K, Sharma P, Kumar N (2010) Imidazole and its biological activities: a review. Chem Sin 1(3):36–47. [Online]. Available: http://pelagiaresearchlibrary.com/der-chemica-sinica/vol1-iss3/DCS-2010-1-3-36-47.pdf

Shen J, Zai J, Yuan Y, Qian X (2012) 3D hierarchical ZnIn2S4: the preparation and photocatalytic properties on water splitting. Int J Hydrog Energy 37(22):16986–16993. https://doi.org/10.1016/j.ijhydene.2012.08.038

Sun Z et al (2018) Catalysis of carbon dioxide photoreduction on nanosheets: fundamentals and challenges. Angew Chem Int Ed 57(26):7610–7627. https://doi.org/10.1002/anie.201710509

Takata T et al (2020) Photocatalytic water splitting with a quantum efficiency of almost unity. Nature 581(7809):411–414. https://doi.org/10.1038/s41586-020-2278-9

Tian Q et al (2017) Dimensional heterostructures of 1D CdS/2D ZnIn2S4 composited with 2D graphene: designed synthesis and superior photocatalytic performance. Dalton Trans 46(9):2770–2777. https://doi.org/10.1039/C7DT00018A

Wang Z et al (2013a) H-doped black titania with very high solar absorption and excellent photocatalysis enhanced by localized surface plasmon resonance. Adv Funct Mater 23(43):5444–5450. https://doi.org/10.1002/adfm.201300486

Wang J, Yu Y, Zhang L (2013b) Highly efficient photocatalytic removal of sodium pentachlorophenate with Bi3O4Br under visible light. Appl Catal B 136–137:112–121. https://doi.org/10.1016/j.apcatb.2013.02.009

Wang G et al (2015) A bismuth-based metal–organic framework as an efficient visible-light-driven photocatalyst. Chem Eur J 21(6):2364–2367. https://doi.org/10.1002/chem.201405047

Wang Y et al (2016) Facile one-step synthesis of hybrid graphitic carbon nitride and carbon composites as high-performance catalysts for CO2 photocatalytic conversion. ACS Appl Mater Interfaces 8(27):17212–17219. https://doi.org/10.1021/acsami.6b03472

Wang L, Jin P, Duan S, She H, Huang J, Wang Q (2019a) In-situ incorporation of Copper(II) porphyrin functionalized zirconium MOF and TiO2 for efficient photocatalytic CO2 reduction. Sci Bull (Beijing) 64(13):926–933. https://doi.org/10.1016/j.scib.2019.05.012

Wang S et al (2019b) Porous hypercrosslinked polymer-TiO2-graphene composite photocatalysts for visible-light-driven CO2 conversion. Nat Commun 10(1):676. https://doi.org/10.1038/s41467-019-08651-x

Wang X et al (2021) 3D macropore carbon-vacancy g-C3N4 constructed using polymethylmethacrylate spheres for enhanced photocatalytic H2 evolution and CO2 reduction. J Energy Chem 53:139–146. https://doi.org/10.1016/j.jechem.2020.05.001

Wu C-M, Naseem S, Chou M-H, Wang J-H, Jian Y-Q (2019) Recent advances in tungsten-oxide-based materials and their applications. Front Mater 6. https://doi.org/10.3389/fmats.2019.00049

Xi Q et al (2020) Facile synthesis of 2D Bi4O5Br2/2D thin layer-Ti3C2 for improved visible-light photocatalytic hydrogen evolution. J Solid State Chem 289:121470. https://doi.org/10.1016/j.jssc.2020.121470

Xia C et al (2021) The emerging covalent organic frameworks (COFs) for solar-driven fuels production. Coord Chem Rev 446:214117. https://doi.org/10.1016/j.ccr.2021.214117

Xiao X et al (2012) Oxygen-rich bismuth oxyhalides: generalized one-pot synthesis, band structures and visible-light photocatalytic properties. J Mater Chem 22(43):22840. https://doi.org/10.1039/c2jm33556e

Xie Y, Ali G, Yoo SH, Cho SO (2010) Sonication-assisted synthesis of CdS quantum-dot-sensitized TiO_2 nanotube arrays with enhanced photoelectrochemical and photocatalytic activity. ACS Appl Mater Interfaces 2(10):2910–2914. https://doi.org/10.1021/am100605a

Yang S, Lv H, Zhong H, Yuan D, Wang X, Wang R (2022) Transformation of covalent organic frameworks from N-acylhydrazone to oxadiazole linkages for smooth electron transfer in photocatalysis. Angew Chem 134(10). https://doi.org/10.1002/ange.202115655

You S-Q et al (2020) A hydrazone-based covalent organic framework/iridium (III) complex for photochemical CO2 reduction with enhanced efficiency and durability. J Catal 392:49–55. https://doi.org/10.1016/j.jcat.2020.09.029

Yu H et al (2019) Modulation of Bi2MoO6 -based materials for photocatalytic water splitting and environmental application: a critical review. Small 15(23). https://doi.org/10.1002/smll.201901008

Yuan X et al (2017) Highly efficient visible-light-induced photoactivity of Z-scheme Ag2CO3/Ag/WO3 photocatalysts for organic pollutant degradation. Environ Sci Nano 4(11):2175–2185. https://doi.org/10.1039/C7EN00713B

Zhang F, Zhang Y, Zhou C, Yang Z, Xue H, Dionysiou DD (2017) A new high efficiency visible-light photocatalyst made of SnS2 and conjugated derivative of polyvinyl alcohol and its application to Cr(VI) reduction. Chem Eng J 324:140–153. https://doi.org/10.1016/j.cej.2017.05.009

Zhang G et al (2020a) Hierarchical Z -scheme g-C3N4/Au/ZnIn2S4 photocatalyst for highly enhanced visible-light photocatalytic nitric oxide removal and carbon dioxide conversion. Environ Sci Nano 7(2):676–687. https://doi.org/10.1039/C9EN01325C

Zhang L et al (2020b) Facile synthesis of Bi5O7Br/BiOBr 2D/3D heterojunction as efficient visible-light-driven photocatalyst for pharmaceutical organic degradation. Sep Purif Technol 231:115917. https://doi.org/10.1016/j.seppur.2019.115917

Zhao Y et al (2017) Coordinative integration of a metal-porphyrinic framework and TiO_2 nanoparticles for the formation of composite photocatalysts with enhanced visible-light-driven photocatalytic activities. J Mater Chem A Mater 5(29):15380–15389. https://doi.org/10.1039/C7TA03840B

Zhu Z, Han Y, Chen C, Ding Z, Long J, Hou Y (2018) Reduced graphene oxide-cadmium sulfide nanorods decorated with silver nanoparticles for efficient photocatalytic reduction carbon dioxide under visible light. ChemCatChem 10(7):1627–1634. https://doi.org/10.1002/cctc.201701573

Zoolfakar AS, Rani RA, Morfa AJ, O'Mullane AP, Kalantar-zadeh K (2014) Nanostructured copper oxide semiconductors: a perspective on materials, synthesis methods and applications. J Mater Chem C 2(27):5247–5270. https://doi.org/10.1039/C4TC00345D

Chapter 2
Synthesis and Characterization of Green Nanostructured Photocatalysts: A General Approach

Janice Adamski, Morgana Rosset, and Suyanne A. L. Bachmann

Abstract Addressing the challenges in the synthesis and characterization of green photocatalysts is essential, and it is important to emphasize the need to combine various analytical techniques to obtain comprehensive and complete data. In this context, this chapter provides a comprehensive overview of methodologies and approaches for the synthesis and characterization of nanostructured photocatalysts, with a focus on efficiency and sustainability. Initially, a variety of approaches are presented for the synthesis and preparation of photocatalytic nanomaterials. Synthesis techniques, such as the sol-gel method, solvothermal, hydrothermal, polymer precursor method, chemical deposition techniques, and electrochemical methods, are explored, emphasizing the importance of selecting environmentally friendly solvents, reagents, and processes. Subsequently, it delves into the characterization approaches for the photocatalysts. Conventional and advanced techniques are analyzed, such as scanning electron microscopy (SEM), transmission electron microscopy (TEM), X-ray photoelectron spectroscopy (XPS), Rutherford backscattering spectrometry (RBS), X-ray diffraction (XRD), Brunauer-Emmett-Teller (BET) surface area analysis, and ultraviolet-visible (UV-vis) spectroscopy. These techniques provide detailed insights into the photocatalysts' physicochemical and morphological characteristics. This knowledge is crucial in advancing photocatalysis and charting a course for renewed efforts in environmental preservation and efficient energy usage.

J. Adamski
Federal University of Rio Grande do Sul, Department of Chemical Engineering, Porto Alegre, RS, Brazil

M. Rosset
University of São Paulo, Polytechnic School, Department of Chemical Engineering, São Paulo, SP, Brazil

S. A. L. Bachmann (✉)
Federal Institute of Education, Science and Technology of Rio Grande do Sul, Feliz, RS, Brazil
e-mail: suyanne.bachmann@feliz.ifrs.edu.br

C. M. Hussain, G. Hussain (eds.), *Green Nanostructured Photocatalysts for Clean Energy*, Clean Energy Production Technologies,
https://doi.org/10.1007/978-981-95-1618-6_2

Keywords Sol-gel method · Solvothermal · Hydrothermal · Polymer precursor method · Chemical deposition techniques · Electrochemical methods · Synthesis · Solvents · Chemical method · Nanoparticles · Oxides · Particles properties · Scanning electron microscopy (SEM) · Transmission electron microscopy (TEM) · X-ray photoelectron spectroscopy (XPS) · Rutherford backscattering spectrometry (RBS) · X-ray diffraction (XRD) · Brunauer-Emmett-Teller surface area analysis (BET) · Ultraviolet-visible spectroscopy (UV-Vis)

2.1 Synthesis of Green Nanostructured Photocatalysts

The search for methods to synthesize green nanostructured photocatalysts is driven by the need to develop more sustainable and environmentally friendly technologies for photocatalysis applications. In this section, we will explore the main strategies and approaches used in the synthesis of photocatalysts that meet these environmental criteria.

2.1.1 Principles of Green Synthesis of Photocatalysts

The synthesis of green nanocatalysts is a burgeoning field within contemporary nanotechnology, owing to its myriad benefits, such as heightened efficiency, sustainability, cost-effectiveness, biocompatibility, and eco-friendliness (Abduljabbar et al. 2024). The green synthesis of nanostructured photocatalysts follows the fundamental principles of green chemistry, aiming to reduce environmental impact and promote sustainability. These principles are essential to ensure that the manufacturing process of photocatalysts is environmentally friendly and safe for humans.

- Minimization of the use of toxic chemicals: The careful choice of precursors and solvents with low environmental impact is crucial to reducing the toxicity of the synthesis process. Opting for natural reagents such as herbs, bacteria, fungi, and agricultural waste instead of harmful chemicals helps mitigate risks to human health and the environment.
- Efficient use of resources: Synthesis methods should be designed to minimize energy and material consumption. This can be achieved through process optimization techniques such as using renewable energy, employing low-temperature synthesis methods, and maximizing the efficiency of reagents.
- Waste minimization: Synthesis processes should be developed to generate the least amount of hazardous waste possible. This includes adopting strategies for reagent recycling and implementing efficient purification methods to minimize the generation of unwanted by-products.

By adhering to these principles, the green synthesis of nanostructured photocatalysts not only contributes to environmental preservation but also offers significant advantages in terms of cost, efficiency, and safety, making it a promising approach for producing materials with unique properties and diverse applications.

With the rising popularity of biological approaches, various sources such as bacteria, fungi, algae, and plants have been employed for the production of nanoparticles (NPs). Plant extracts serve as valuable resources in NP synthesis due to their cost-effectiveness and environmentally safe nature. Various works on the use of plant extract to synthesize NPs have been reported (Rasli et al. 2020).

2.1.2 *Sustainable Synthesis Methods*

Nanomaterials can be made from a variety of materials, including metals, metal oxides, polymers, and ceramics, and can exist in different forms like nanoparticles, nanotubes, and nanowires. There are several methods of nanomaterial synthesis including chemical, physical, and, recently, biological approaches that include green-based materials. These methods include two possible approaches: top-down, starting from larger particles to nanoparticles, or bottom-up, starting from atoms to more complex structures (Joseph and Vijayanandan 2021; Rani et al. 2023; Tejashwini et al. 2024). Figure 2.1 shows a schematic summary of the main nanomaterial synthesis methods.

It is important to highlight that chemical and physical methods can also present a green approach through replacement with green solvents, waste minimization, or the use of more environmentally friendly precursors. The choice of a synthesis approach relies on the desired properties of the final nanoparticle product, as each approach presents its own set of advantages and limitations.

The search for greener methods of synthesizing nanophotocatalysts has intensified in recent decades, driven by the need to reduce the environmental impact of energy conversion technologies (Khatami and Iravani 2021; Balakrishnan et al. 2024; Yergaziyeva et al. 2024). In this topic, we will explore the main methods for synthesizing green nanostructured photocatalysts, highlighting their environmental advantages and potential applications.

2.1.2.1 Chemical Methods

Green chemistry is an approach to the synthesis, processing, and use of chemicals that is more environmentally benign. The green synthesis of nanoparticles commonly entails the utilization of environmentally friendly solvents, which may encompass water, organic solvents derived from natural sources, or even no solvents at all. The selection of solvent in green nanoparticle synthesis is contingent upon the particular synthesis method and the characteristics of the nanoparticles being produced (Gupta et al. 2023). Below are the main chemical methods for the synthesis of green nanophotocatalysts.

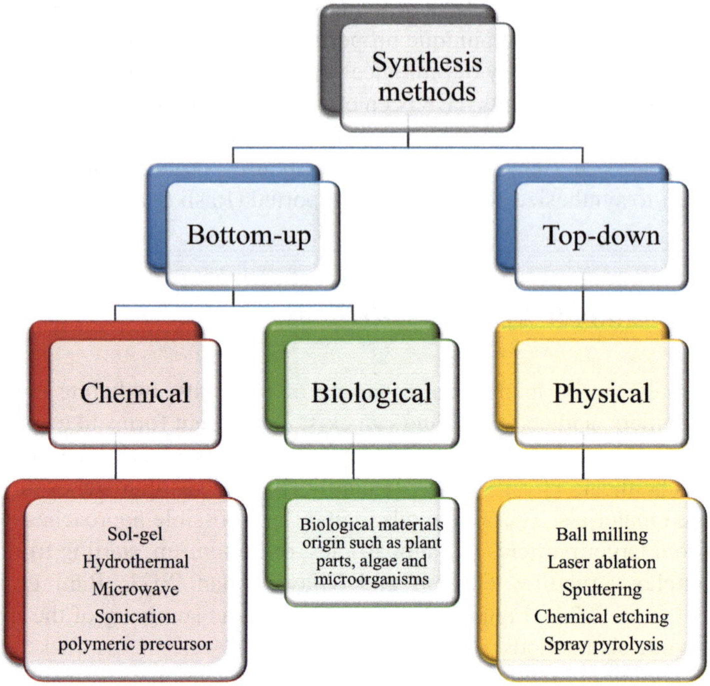

Fig. 2.1 Schematic summary of the main nanomaterial synthesis methods. (Adapted from Tejashwini et al. (2024))

- *Sol-Gel Method: A Precise and Controlled Approach*

The sol-gel method derives from various synthesis methods, where a reaction occurs in a homogeneous medium (sol) that destabilizes in a controlled manner to form particles by agglutination of the components (gel) (Sakka 2022). It usually involves a hydrolytic process, where an anhydrous component, such as a colloidal suspension of particles or reactive polymeric chains in an aqueous medium, bonds together forming networks. The gelation process, or sol-gel transition, allows for good homogeneity and produces particles of metastable or amorphous phases, with interstices that encapsulate the solvent of the medium. With the heating and drying of the gel, the solvent's evaporation leads to an increase in capillary pressure in the interstices, which can collapse and result in pore closure (Sheibani 2019; Ahmad et al. 2020).

Among the categories of molecules used in this process are alkoxides, compounds containing alkoxy groups (RO^-) that interact with metal cations or cations bound to metal centers, such as silicon (Si) and titanium (Ti). The hydrolytic instability, that is, the ease of having bonds broken by reacting with water, of alkoxysilanes is an essential property for their use in the sol-gel process (Sakka 2022).

The sol-gel method is among those capable of producing materials with ease in controlling the morphology and physical properties of the material solely by controlling simple components of the method, such as temperature, precursor concentration, stirring speed, pH, as well as stabilizers, and the nature and amount of the solvent (Liang et al. 2020; Pradhan et al. 2022).

However, since the process produces metastable and/or amorphous forms, post-synthesis treatment, either thermal or hydrothermal, is necessary to acquire desirable characteristics. Thermal treatment and calcination can increase crystallinity, but it can also result in a decrease in surface area and pore volume and an increase in crystallite size and alter the type of crystalline phase. The degree of crystallinity can influence the electron/hole pair recombination process and thus the photocatalytic activity. Furthermore, preparation methods can influence the physical properties of oxides, such as morphology, surface area, and crystallinity (Tang et al. 2022).

This method offers precise control over the morphology and structure of photocatalysts, allowing for the obtainment of materials with optimal properties for photocatalytic applications. Additionally, the use of more environmentally friendly sol-gel precursors can contribute to reducing the environmental impact of nanophotocatalyst synthesis.

- *Polymeric Precursor Method: Controlled Synthesis of Nanomaterials*

The polymeric precursor method, also known as the Pechini method, is an effective approach for the preparation of nanomaterials, thin films, and inorganic oxides, based on the formation of a polymeric network in which metal ions are uniformly distributed (Wiranwetchayan et al. 2020; Cerón-Urbano et al. 2023), which emerged as an alternative to the sol-gel method. This method typically involves three distinct steps: the formation of metal chelates, polyesterification, and ester formation.

In the first step, a chelating agent, commonly a hydroxycarboxylic acid such as citric acid, is employed to form chelates with the dissolved metal ions. Next, a crosslinking agent of the polyhydric alcohol type, such as glycerol or polyethylene glycol, is added to initiate polyesterification, forming the polymeric resin. Citric acid and ethylene glycol are the most frequently used reagents in this process due to their effectiveness in resin formation. The second phase involves thermal treatment at high temperatures (from 400 °C onward), which causes the decomposition of organic matter and the formation of properly ordered crystallites. This result is especially relevant when seeking to obtain highly crystalline materials with a controlled distribution of constituents in the crystal structure. However, a significant drawback of this method is that the heating required for the complete decomposition of the organic matter can lead to particle growth (Samriti et al. 2023).

For the synthesis of perovskite-type oxides, the polymeric precursor method has proven particularly successful in obtaining monophasic powders at relatively low temperatures (Al-Hartomy et al. 2012). In this case, the formation of chelates between the dissolved cations and the α-hydroxycarboxylic acid, such as citric acid, is followed by polyesterification in the presence of a polyhydroxy alcohol (Ren et al. 2020). Nitrates are generally the preferred salts due to their easy elimination during the calcination step.

The polymeric precursor method offers a versatile and controlled route for the synthesis of nanomaterials and inorganic oxides, allowing for the obtainment of products with adjustable properties suitable for a variety of applications, from catalysis to electronics. Its ongoing success in producing materials with high purity and homogeneity contributes to significant advancements in the field of functional and nanostructured materials (Lima et al. 2021).

- *Hydrothermal Method: Controlled Synthesis in Aqueous Medium*

The hydrothermal method is a widely employed synthesis technique in the production of nanomaterials, including nanophotocatalysts. In this process, synthesis occurs in an aqueous medium at elevated temperatures and controlled pressure, allowing for the formation of controlled nanostructures with specific properties (Guan et al. 2013).

One of the distinct advantages of the hydrothermal method is its ability to offer precise control over the morphology, size, and distribution of the produced particles. This is due to the variety of parameters that can be adjusted during the process, such as reagent concentration, pH, temperature, and reaction time (Duy Trinh et al. 2019; Li et al. 2023). This flexibility enables the obtainment of nanomaterials with tailored characteristics for various applications, including photocatalysis.

Furthermore, the hydrothermal method is recognized for its ecological efficiency, especially when compared to conventional methods that utilize direct heating. By employing relatively lower temperatures and avoiding the use of organic solvents, the hydrothermal method minimizes energy consumption and reduces the environmental impact of nanomaterial synthesis (Shanmugaratnam et al. 2024).

Precise adjustments to process parameters enable the obtainment of different crystalline structures and phases of the synthesized materials. This allows for the production of nanophotocatalysts with optimized optical, electronic, and structural properties for the desired application, such as organic pollutant degradation in wastewater or solar energy conversion (Qi et al. 2021).

- *Solvothermal Method: Controlled Synthesis in Organic Medium*

The solvothermal method is a versatile and effective strategy for the synthesis of nanomaterials, including nanostructured photocatalysts, through the reaction between precursors dissolved in organic solvents. In this process, the reaction takes place in a sealed reactor or closed test tube, subjected to controlled heating and pressure, providing favorable conditions for the formation of microcrystals (Ling and Dai 2020; Senasu et al. 2022).

A distinctive feature of the solvothermal method is its ability to control the nucleation and growth of crystals, resulting in final products with controlled size, morphology, and particle distribution. This is possible due to the synergy between slow nucleation and rapid crystal growth under specific temperature and pressure conditions (Iqbal et al. 2017; Cappelluti et al. 2020).

Several parameters, such as pH, solvents and their proportions, temperature, additives, and reaction time, significantly influence the characteristics of materials synthesized by the solvothermal method. Therefore, it is essential to optimize these parameters to obtain nanophotocatalysts with the desired properties for a wide range of applications, from pollutant degradation to solar energy conversion (Samriti et al. 2023).

Although the solvothermal method offers significant benefits, such as a high reaction rate and good yield, it may also present some challenges (Rafiq et al. 2020; Vinodhkumar et al. 2020). The high temperature and pressure used can lead to the formation of oxides and impurities, affecting the quality of the synthesized material. To overcome these issues, alternative methods that employ thermal energy in a non-conventional way, such as microwaves (Chang et al. 2021; Wang et al. 2021) and sonochemistry (Huang et al. 2020), are being increasingly utilized.

2.1.2.2 Biological Methods

Although chemical synthesis processes have advantages, there is the use of toxic solvents and surfactants (Pantidos and Horsfall 2014), the generation of toxic and carcinogenic by-products (Hoag et al. 2009), and significant energy requirements (Horwat et al. 2011). Thus, physical and chemical methods have been replaced by green synthesis approaches that employ natural substitutes such as parts of plants (Singh et al. 2019; Yashni et al. 2021; Gaur et al. 2023), fungi (Li et al. 2019; Gupta and Chundawat 2019), bacteria (Srivastava and Mukhopadhyay 2014), and algae (Singaravelu et al. 2007; Shankar et al. 2016), which have been used in photocatalyst synthesis. Therefore, photocatalysts synthesized by biological routes are of great interest, as they contribute to clean technology and reduce the need for excessive chemicals (Joseph and Vijayanandan 2021).

Natural substitutes have biomolecules such as polysaccharides, pigments, phenolic compounds, and enzymes that have chemical groups that can actively participate in the synthesis of nanophotocatalysts (Goutam et al. 2018; Paradva et al. 2023). Among natural substitutes, plants have stood out for the synthesis of photocatalysts, since extracts from microorganisms require special conditions to obtain, including the following (Narayanan and Sakthivel 2010; Joseph and Vijayanandan 2021):

- Need for nutrition and, consequently, control of temperature and environmental conditions, such as pH, for example
- Limited cell viability depending on cell reproduction time and previous conditions
- Greater risk of contamination, requiring aseptic conditions and long incubation periods.

Furthermore, there is little literature supporting the use of algae in the synthesis of nanoparticles (Vijayaraghavan and Ashokkumar 2017). In this way, we will explore the use of phytoextracts in the process of synthesis of nanophotocatalysts.

In chemical approaches, nanoparticles are synthesized through the interaction of three fundamental components, in this case, precursor, reducing agent, and stabilizing agent (Vijayaraghavan and Ashokkumar 2017). Normally, the precursor is made up of a metallic salt; the reducing agent is a chemical substance such as solvent or organic salt, and the stabilizing agent is a surfactant. The reducing agent's function is to supply electrons to the precursor metal ion to form atoms, which later aggregate to form particles, while the function of the surfactant is to stabilize and limit the size of the particles (Paradva et al. 2023).

The main difference in the synthesis route mediated by plant extract is the replacement of the chemical reducing agent, such as sodium borohydride, polyvinyl alcohol, potassium bitartrate, and elemental hydrogen, with plant extract biomolecules (Vijayaraghavan and Ashokkumar 2017; Rani et al. 2023). Several chemical constituents of the plant have the potential to act as a reducing agent, notably phytochemicals such as alkaloids, tannins, steroids, terpenes, phenols, alcohol, and saponins (Vijayaraghavan and Ashokkumar 2017; Achudhan et al. 2020; Paradva et al. 2023).

The synthesis process of green nanophotocatalysts begins with the addition of metallic salt to the plant extract. The plant extract is generally obtained through a simple extraction, which involves boiling the plant part in distilled water. Normally, there is a need for preconditioning by drying the part of the plant before the extraction stage (Rani et al. 2023).

The synthesis mechanism of nanophotocatalysts occurs in three stages: activation phase, growth phase, and termination phase. Figure 2.2 shows the synthesis mechanism.

In the activation phase, the reduction of metal ions occurs, followed by the nucleation of the reduced ions. Biomolecules transfer electrons, reducing metal ions to a less oxidized state, mainly in the zero valence form (Singh et al. 2018). Then, the reduced metal atoms undergo nucleation, moving on to the next step. In the growth phase, spontaneous and nonselective aggregation of particles formed in the activation phase occurs, resulting in distinct morphologies (Iravani 2011). As the growth phase extends, the particles can acquire different and irregular shapes (Ghodake et al. 2010). The termination phase, and last step, define the shape of the nanophotocatalyst. Some chemical species present in biomolecules, such as hydroxyls, provide stability to the particles (Nasrollahzadeh and Sajadi 2015; Xu et al. 2022 *apud* Langa and Hintsho-Mbita 2023). Therefore, the biomolecules are capable of acting as a reducing agent and also contributing to the stabilization of the photocatalyst nanoparticles.

It is important to highlight that biomolecules are made up of phytochemicals, whose stability and reducing action depend on the synthesis conditions. According to Joseph and Vijayanandan (2021), different factors influence the synthesis, stability, and properties of nanophotocatalysts, including the nature and concentration of the phytochemical used, metal salt concentration, pH, temperature, electrochemical potential, and reaction time. For example, the pH of the medium can affect the charges, and the degree of ionization of these substances, as well as high temperatures, could cause their degradation. Therefore, it is very important to know the factors and how they influence the synthesis of nanophotocatalysts and, consequently, how they affect their properties.

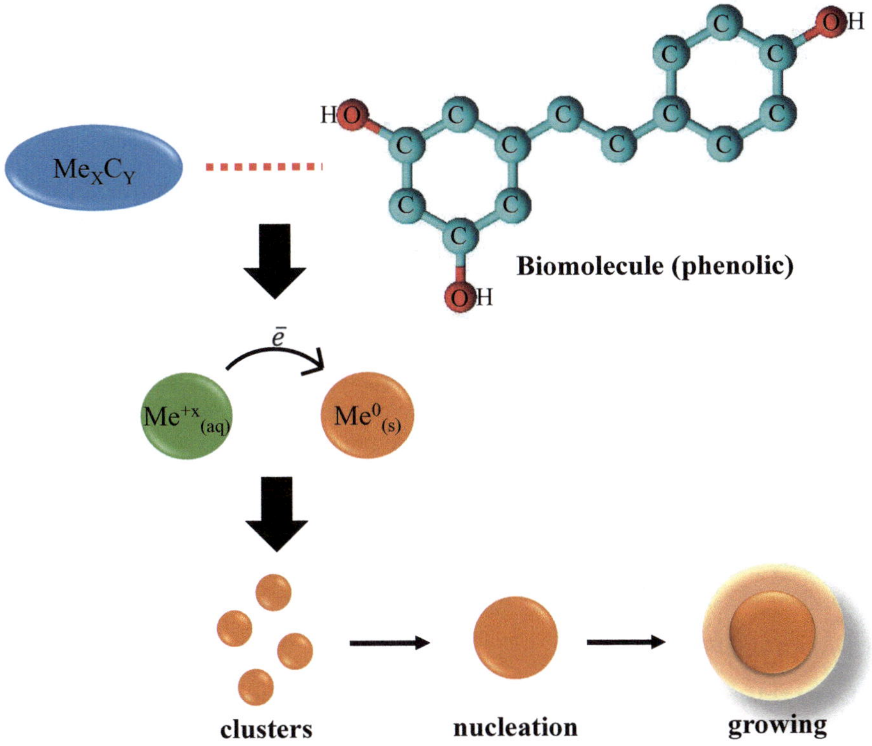

Fig. 2.2 Synthesis mechanism of nanomaterial photocatalysts. (Adapted from Joseph and Vijayanandan (2021))

2.1.2.3 Physical Methods

Several physical techniques have been successful in the production of nanophotocatalysts, each presenting its advantages and characteristics. Among these techniques are ball milling, laser ablation, sputtering, and spray pyrolysis. Each of these approaches provides a unique and efficient way to manufacture nanophotocatalysts with customized properties for a variety of applications.

- *Ball Milling*

Ball milling is a widely used mechanical technique for reducing powders to fine particles and mixing materials. This environmentally friendly and cost-effective approach has found applications in various fields, including metallurgy, mineral processing, construction, and organic compound synthesis. Currently, it is experiencing a resurgence due to its successful synthesis of various organic, inorganic, and hybrid nanomaterials. This technique offers advantages such as short reaction time, large-scale preparation, and minimal or no solvent usage (Tao and Wang 2021).

Liu et al. (2024) employed ball milling to introduce oxygen vacancies in WO_{3-x} photocatalysts. The common method for introducing such vacancies in metal oxides involves thermal treatment in reducing atmospheres such as hydrogen, nitrogen, argon, or vacuum. Due to its direct operation, gentle conditions, economic viability, environmental friendliness, and large-scale processing capability, ball milling is highly suitable for industrial and research applications.

The preparation of photocatalytic materials composed of ball milling has been extensively studied. Zhang et al. (2023) prepared the La^{3+}/TiO_2 photocatalyst using this method, showing an eightfold increase in degradation rate compared to the unmilled material. Additionally, Kato and Shirai (2022) proposed a novel and simple bandgap engineering method to fabricate homo/heterojunctions in WO_3 at room temperature using ball milling, resulting in a significant improvement in photocatalytic performance in degrading aqueous pollutants of azoic dyes under visible light.

Ball milling provides mechanical energy to powders, inducing a bonding state on the particle surface and changes in physicochemical properties, such as crystalline structure and particle size (specific surface area). The photocatalytic reaction on a semiconductor particle is triggered by photogenerated electrons and holes, which migrate from the bulk to the surface and react with adsorbed substrates. Therefore, it is widely recognized that a large specific surface area leads to high activity, with the amount of reactants influenced by this specific area (Shinoda et al. 2022).

- *Laser Ablation*

Laser ablation is an innovative and unconventional technique that provides uniform nanostructures in significantly less time than conventional methods. This technique has been applied in various technologies, including the preparation of microstructures and nanostructures directly from bulk materials. In this process, the nanostructures are removed from the surface of the target material and placed in a liquid medium (Motakef-Kazemi et al. 2023).

Mostafa et al. (2021) prepared the ZnO/MWCNTsO nanocomposite for methylene blue (MB) degradation using the pulsed laser ablation in liquids (PLAL) method, highlighting that this method is the most effective physical route for nanofabrication due to its simplicity and reagent economy. Jasbi et al. (2021) utilized the pulsed laser ablation in liquid method to prepare the structure of graphene/TiO_2 nanocomposites by decorating graphene with TiO_2 nanoparticles.

Pulsed laser ablation in liquid can be employed as an efficient way to produce nanostructures and has attracted significant interest due to its various advantages, such as the simplicity of the experimental setup and the ability to control the size of the produced material by altering different parameters, such as laser pulse duration, laser fluence, laser wavelength, solution pH, and liquid environment temperature, among others. In this method, it is possible to create pure colloidal solutions without the formation of additional by-products. Additionally, various other properties of nanoparticles, such as surface charges, size, and accumulation, can be controlled (Jasbi et al. 2021; Alamro et al. 2022).

- *Sputtering*

Sputtering is a widely used plasma-based physical vapor deposition (PVD) technique for depositing thin films. The vapor of the thin film material is created through physical means from a solid deposition source. The most commonly used tool for sputtering deposition is magnetron sputtering (MS), essentially a diode sputtering configuration where a magnetic pack is placed behind the target (cathode) to better confine the plasma near the sputtering region (Lundin and Sarakinos 2012).

Sputtering allows for fine control of film composition by adjusting the sputtering target and atmosphere. The film thickness can be easily controlled by altering the sputtering time (Pan et al. 2024).

Yu et al. (2020) used sputtering to form the Ag_2O/TiO_2 heterojunction. Sputtering is widely used for material deposition, ion doping, and the formation of heterostructures, among others. For instance, it is used for doping precious metals to enhance the photocatalytic activity in semiconductors, reducing the recombination of photogenerated electrons and holes. Silver (Ag^0/Ag^+) is one of the most investigated noble metal dopants for TiO_2 photocatalysts (Yu et al. 2020).

Alfaro Cruz et al. (2023) produced SnO_2-ZnO photocatalytic films for hydrogen production. The authors used the sputtering technique to deposit Zn metal onto the SnO_2 film. Sputtering deposition serves to modify the physicochemical properties of the film, generating defects and impurities that can create intermediate states between the valence and conduction bands. In these intermediate states, photogenerated charges capture electrons and enhance the photocatalytic activity (Alfaro Cruz et al. 2023).

Lelis et al. (2023) employed reactive magnetron sputtering to produce a fine powder of magnetic photocatalytic (MP) materials consisting of bilayered particles, with carbon-doped TiO_2 or zinc-rich ZnO as the photocatalyst on one side and a nickel ferromagnetic metal layer on the other side. The activity of the synthesized MP powder was tested for the photocatalytic inactivation of *Salmonella typhimurium* bacteria under UV and visible light.

Sputtering is a low-temperature technique compared to chemical methods such as sol-gel, hydrolysis-precipitation, and solvothermal techniques and thus can be used for the synthesis of nanostructures at temperatures close to ambient atmosphere, compatible with temperature-sensitive substrates such as polystyrene, polyethylene, and other polymers (Varnagiris et al. 2021).

- *Spray Pyrolysis*

The spray pyrolysis method, or spray pyrolysis, is a deposition technique that involves spraying a precursor solution through a neutral gas, such as nitrogen, resulting in extremely fine droplets that are deposited onto a substrate. These droplets contain the necessary constituents to react and form a desired chemical compound on the substrate. Chemical reagents are carefully selected so that any undesired products besides the desired compound are volatile at the deposition temperature (Patil 1999).

The properties of the film produced by this method can be controlled by a variety of parameters, including spray rate, substrate temperature, ambient atmosphere, carrier gas, droplet size, and cooling rate after deposition. These parameters directly influence the structure, morphology, and composition of the final film.

Recent research has highlighted the spray pyrolysis method as an effective and scalable approach for synthesizing thin films and nanoparticles. For example, Boningari et al. (2018) used the method to synthesize S-doped TiO_2, emphasizing its speed and scalability compared to conventional wet methods. Additionally, it is noted that this method is carried out in a single step, simplifying the fabrication process.

Another study, conducted by Rahaman and Singha (2022), employed ultrasonic spray pyrolysis as a low-cost alternative to deposit films on glass substrates in a single step, demonstrating the versatility and applicability of this method in various research and development contexts.

Despite the advantages in terms of cost and efficiency, it is important to note that the spray pyrolysis method generally requires higher deposition temperatures than other techniques, such as sputtering deposition. However, the unique characteristics of this method make it a viable choice for various applications, especially those requiring large-scale production of thin films and nanoparticles.

2.1.3 Green Nanostructured Photocatalysts

Most of the nanocatalysts produced are made up of metals and metallic oxides (CuO, ZnO, Fe_3O_4, TiO_2, etc.), although noble metals such as Ag, Au, Pt, and Pd have received attention due to the versatility of the application of their nanoparticles in different fields of science (Vijayaraghavan and Ashokkumar 2017), such as medicine (Jain and Mehata 2017; Yeshchenko et al. 2024; Seçme et al. 2024) and effluent treatment (Adamski et al. 2024; Mostafazadeh et al. 2024; Khan et al. 2024).

ZnO and TiO_2 are the nanoparticles that have stood out the most. ZnO is an attractive semiconductor material for photonic and nanoelectronic applications, while TiO_2 has wide application in the paint and cosmetics industries and mainly in industrial photocatalytic processes and effluent treatment, due to its chemical, physical, and biological properties (Vijayaraghavan and Ashokkumar 2017). Table 2.1 summarizes some synthesis conditions of plant-mediated nanophotocatalysts and their characteristics.

Different metals have been used in the synthesis of green nanophotocatalysts (Table 2.1); however, Ti^+ ions are preferable over Ni^+ and Ag^+ because biomolecules reduce metal ions with higher ionization energy more effectively (Haverkamp and Marshall 2009). Therefore electrochemical potential is also an important factor in the synthesis of nanophotocatalysts as well as pH, contact time, and temperature.

The pH interferes with charge transformations and stabilization of biomolecules, which affects their ability to chelate metal ions and, consequently, results in morphological changes in nanophotocatalysts (Shamaila et al. 2016; Jain and Mehata

Table 2.1 Synthesis conditions and characteristics of various metal/metal oxide nanoparticles obtained from different plant species

Green-based material	Nanoparticles	Synthesis conditions			Functional groups	Shape	Size (nm)
		pH	Contact time	Temperature			
Lawsonia inermis leaf	Cu	11	*	*	Hydroxyl and carboxylic groups	Spherical	83
Gum karaya	CuO	*	1 h	75 °C	Hydroxyl and carboxylate groups	Spherical	2–10
Dodonaea viscosa	Fe	*	*	*	Carbonyl, polyols, hydroxyl, and amide I group	Spherical	50–60
Plantain peel extract	Fe_3O_4	*	2 h	70 °C	Polyphenolic, carbohydrates, carboxylic acids, and amine groups	Spherical	>50
Jatropha curcas L. latex	TiO_2	*	12 h	50 °C	Amide II, amide III, carboxylic acid, and amine groups	Spherical	25–50
Solanum trilobatum L. leaf	TiO_2	*	6 h	Room temperature	Hydroxyl, alkanes, monosubstituted alkynes, vinyl ethers, aldehydes, beta lactones, and aliphatic amine groups	*	
Aloe barbadensis Miller leaf extract	ZnO	*	5–6 h	150 °C	Phenolic compounds, terpenoids, proteins, free amino, and carboxylic groups	Spherical and hexagonal	25–55

Adapted from Vijayaraghavan and Ashokkumar (2017)
*Unreported values

2017). According to the literature, at a pH above 7.0, the synthesis of smaller particles is favored, compared to acidic environments (Vadlapudi et al. 2013). Therefore, basic media favor the increase in the specific surface area (m^2/m^3), which is favorable for the catalytic activity of the nanophotocatalyst (Joseph and Vijayanandan 2021; Mahlaule-Glory et al. 2022). Besides, the pH of the solution medium also influences the size, shape, and rate of the synthesized nanoparticles (Armendariz et al. 2004, as cited in Vijayaraghavan and Ashokkumar 2017).

The contact time is defined as the contact time necessary between the precursor salt and the biomolecule for the formation of photocatalyst nanoparticles to occur. It depends on several factors such as pH, temperature, light intensity, reduction potential, and concentration of phytochemicals (Balavandy et al. 2014; Lade and Shanware 2020, as cited in Joseph and Vijayanandan 2021). This explains why the contact time is so variable, ranging from 1 to 12 h (Table 2.1).

Temperature also interferes with the formation of nanophotocatalysts. According to Cruz et al. (2010), increasing temperature favors nucleation and also improves the crystallinity of nanophotocatalysts. However, this factor must also be optimized, as temperatures beyond ideal promote the degradation of biomolecules, such as active amino acids and other phytochemicals, inactivating the reducing potential (Liu et al. 2020). A positive aspect of the biological synthesis of nanophotocatalysts is that they require less temperature when compared to physical and chemical methods and can be synthesized at room temperature in the vast majority of cases (Li et al. 2011). This improves the energy efficiency of the synthesis process, as there is no need for high temperature or pressure conditions (Langa and Hintsho-Mbita 2023; Paradva et al. 2023).

Table 2.1 shows that the catalyst nanoparticles obtained have a spherical shape and variable sizes between 2 and 80 nm, although other studies indicate sizes greater than 300 nm (Goutam et al. 2018; Masekela et al. 2022) and oval (Masekela et al. 2022), tetragonal (Goutam et al. 2018), and hexagonal (Sangeetha et al. 2011) morphologies. It is important to highlight that both the morphological properties of nanophotocatalysts and the surface chemistry are influenced by the synthesis conditions.

The morphology and size of the nanophotocatalysts are influenced by the concentration of the plant extract (Zheng et al. 2013). Different parts of plants have been used for the plant-mediated synthesis of nanophotocatalysts, such as flowers, seeds, leaves, stems, and roots, which normally result in spherical-shaped particles (Vijayaraghavan and Ashokkumar 2017; Langa and Hintsho-Mbita 2023). A study on the synthesis of TiO_2 nanophotocatalysts from (Bopape et al. 2023) showed that as plant mass increases, the specific surface area also increases, and the nanoparticles also become less rough and more spherical.

Also, particle size is related to catalytic activity. Studies report that the smaller the particle size of the photocatalyst, the greater the photodegradation (Winthachai et al. 2015, as cited in Joseph and Vijayanandan 2021). This is because as the size decreases, the surface-to-volume ratio increases, which, in turn, results in better photocatalytic activity. Therefore, one must opt for a plant extract capable of producing smaller nanophotocatalysts.

Metal oxides such as SnO_2, TiO_2, and ZnO are the main semiconductors used as heterogeneous photocatalysts; however, TiO_2 has the highest activity among all other semiconductors, which is why it is preferred for photocatalytic processes (Langa and Hintsho-Mbita 2023). According to the literature, the antioxidant and catalytic properties of the green nanophotocatalyst are related to the amount of mass of the flat part used. This is because the biomolecules present in the plant depend on the mass; therefore, the greater the mass, the greater the availability of phytochemicals and, consequently, the greater the antioxidant and catalytic potential of the nanophotocatalyst (Joseph and Vijayanandan 2021).

However, the mass required for the synthesis process must be optimized from the point of view of the best cost-benefit, since it may be economically unfeasible to indefinitely increase the concentration of the plant extract. The influence of excess concentration of the plant extract still needs to be further investigated, as does its relationship with the concentration of the metallic salt used. The optimal concentration of metal salt also needs to be determined, as it affects the shape and morphology of the nanophotocatalysts. Furthermore, if the salt concentration is excessive, the reduction of metal ions will be incomplete (Gopinath et al. 2013).

The interference of factors that influence the synthesis and consequently the physical and chemical properties of the photocatalysts can be evaluated by determining some of their properties and characteristics. For example, the concentration of phytoextracts affects the shape and morphology of nanophotocatalysts. These characteristics can be monitored through the use of some analytical techniques such as X-ray diffraction and scanning electron microscopy. Therefore, the following topic explores the main analytical techniques used to characterize and determine the structural and surface properties of nanophotocatalysts.

2.2 Characterization

Characterization is important to determine the physical and chemical properties, and morphological aspects, optimize the synthesis process, and estimate the catalytic activity of photocatalysts. Knowledge of the physical and chemical properties of a photocatalyst allows us to understand the phenomena involved in the catalytic process and is decisive for the interaction of the chemical species and the active sites of the photocatalyst, being able to define its application. This section will cover the main conventional and advanced characterization techniques for determining the properties of green nanophotocatalysts, highlighting the importance of evaluating and understanding these materials.

2.2.1 Conventional Characterization Methods

Conventional characterization techniques play a key role in the analysis and understanding of photocatalysts, providing crucial information about their structure, composition, and properties. These techniques include:

- *Zeta Potential*

Zeta potential refers to the electrical potential difference between the dispersion medium and the stationary layer of fluid adhered to the dispersed particle, providing insights into the surface charge characteristics of the particles (Delgado et al. 2005). This factor holds significant importance in assessing the stability of colloidal systems and can be leveraged to adjust the formulation of colloidal systems like medications and paints. It is generally considered that the zeta potential values greater than +30 mV or smaller than −30 mV result in stable suspensions (Meléndrez et al. 2010). A high zeta potential indicates that the particles have a strong mutual repulsion, which leads to the stability of the suspension, preventing the particles from agglomerating. On the other hand, a low zeta potential may indicate a tendency for the particles to agglomerate, resulting in the instability of the colloidal system (Noruzi 2015).

The zeta potential of the CaO-NPs, produced using the biosynthesis technique employing golden linseed (*Linum usitatissimum* L.) extract, displayed a positive surface charge of +12.7 ± 0.2 mV, signifying compatibility with anionic organic contaminants, such as the tartrazine yellow dye. This facilitates enhanced electrostatic interaction between the TY dye molecules and the surface of CaO-NPs (Bôlla de Menezes et al. 2024). The synthesized ZnO-NPs have zeta potential values of 0.909 mV (Abduljabbar et al. 2024); zeta potential analyses were run for the prepared metal nanoparticles using *Euphorbia retusa* plant extract to investigate the surface charge in suspension. All AgNPs (*Pterodon emarginatus* extract) obtained by Toledo et al. (2020) showed negative ZP; this can be attributed to the presence of polyphenols, flavonoids, and proteins from the plant extract. The values obtained varied between −27 and −39 mV, which indicates a good degree of stability. The zeta potential value of −15.3 mV for the nanoparticles synthesized indicates that the particles are moderately stable (Selvarajan and Mohanasrinivasan 2013). The zeta potential of the synthesized zinc oxide using *Ixora coccinea* leaf extract nanoparticles was found to be −49.19 mV. The high value confirms the repulsion among the particles and thereby increases the stability of the formulation (Yedurkar et al. 2016).

- *Zero Point Charge (ZPC)*

Zero point charge (ZPC) is the pH of a solution at which the net charge on the surface of the sample becomes neutral (pHzpc). The final pH of each sample is measured, and a graph is plotted between initial and final pH. The intersection of this curve where it crosses the initial pH = final pH line is known as pHzpc. When the pH of the reaction exceeds the nanoparticle's pHzpc, it undergoes protonation, whereas when the pH of the reaction falls below the pHzpc, it undergoes deprotonation (Jigyasa et al. 2021). The ZPC of the CaO-NPs, using the biosynthesis with *Linum usitatissimum* L. extract was of pHzpc = 7.74 (Bôlla de Menezes et al. 2024). pHzpc of magnetic iron oxide nanoparticles (MIONPs) synthesized using the fruit extract of *Cynometra ramiflora* is 8.1 (Bishnoi et al. 2018).

- *Microscopy Analysis*

The microscopy approach is a vital instrument for imaging and characterizing nanomaterials and serves as an ideal complement to spectroscopic methods. Techniques such as transmission electron microscopy (TEM), scanning electron microscopy (SEM), atomic force microscopy (AFM), and fluorescence microscopy are commonly employed in nanotechnology (Saad Algarni et al. 2022).

Electron microscopy provides an accurate assessment of the size, shape, spatial resolution, and differences in composition and structure of the NPs but has also a certain disadvantage—such techniques often require specific complicated sample preparation steps which can create NP agglomeration during the drying process (Król et al. 2017).

The SEM analysis results of $ZnFe_2O_4$ indicate that all samples exhibit broadly spherical morphologies with numerous agglomerations of flakes. Treating the $ZnFe_2O_4$ nanoparticles with various plant extracts resulted in products with distinct morphologies. Specifically, the particles treated with *Ziziphus mauritiana* extracts appeared coarser, whereas those treated with *Salvadora persica* extracts were finer and exhibited a more spherical shape, as shown in Fig. 2.3 (Bayahia 2023). Jamdar et al. (2024) studied samples of $SmMnO_3/Sm_2O_3$ nanocomposites prepared with citric acid and oxalic acid at temperatures of 800 and 900 °C using SEM images. They identified different morphologies and sizes of the nanocomposites. At a temperature of 900 °C using citric acid, nanocomposites with sizes between 50 and 60 nm and a similar shape were produced. Therefore, both temperature and fuel are important for forming ultrafine nanocomposites.

To see the effect of temperature on the texture of ZnO NPs synthesized using an aqueous extract of *Rosmarinus officinalis* leaves, characterizations by SEM and TEM were carried out by (Saad Algarni et al. 2022). In the ZnO-80 (synthesized at 80 °C) sample, individual particles with irregular morphology and an agglomerated spherical particle are observed in the micrograph. Conversely, the SEM imaging of the ZnO-180 (synthesized at 180 °C) sample reveals flaky agglomerated spherical structures. Notably, both ZnO-80 and ZnO-180 samples exhibit pores in the agglomerated particles, likely resulting from the release of volatile substances or gases during calcination. The results of the analysis show that the particles ZnO-80 are mainly spherical with sizes varying from 18 to 40 nm. On the other hand, NPs (ZnO-180) mainly have an elongated structure with sizes varying from 20 to 123 nm.

The TEM image (Fig. 2.4a) of the synthesized CuSNPs with *Chlorella ellipsoidea* alga extract depicted the formation of a dense aggregation of nanoparticles, predominantly spherical, with sizes ranging from 9 to 23 nm. Additionally, a small number of elongated and rod-shaped nanoparticles were observed. The

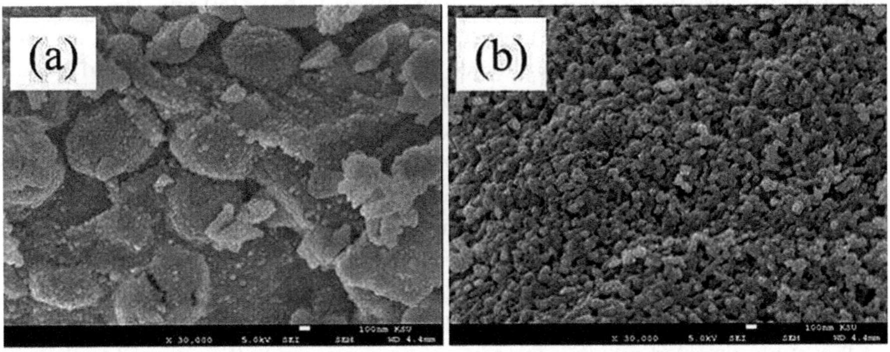

Fig. 2.3 SEM images of $ZnFe_2O_4$ modified with (**a**) *Ziziphus mauritiana*. (**b**) *Salvadora persica*. (Reprinted under permission (license number–5755510597630) from reference Bayahia (2023))

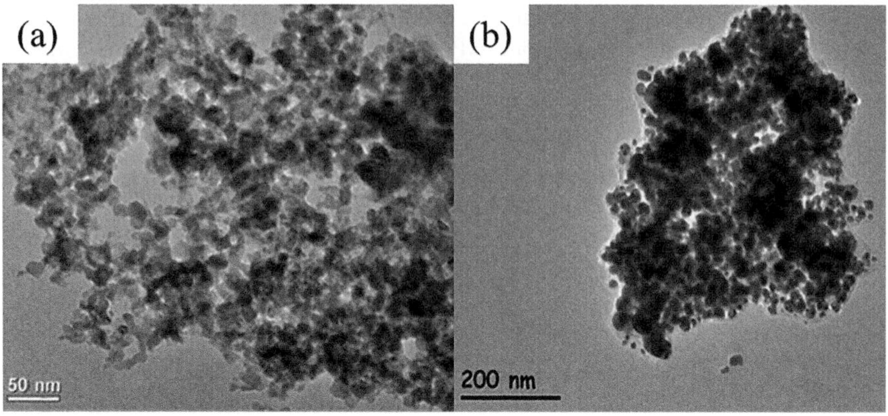

Fig. 2.4 (**a**) TEM image at a scale of 50 nm from synthesized CuSNPs with *Chlorella ellipsoidea*. (Reprinted under permission (license number–5755510895926) from reference Borah et al. (2022)). (**b**) TEM AgNPs, obtained by Reishi mushroom mediated. (Adapted from Aygün et al. (2020))

phytochemicals from algae were found to play a crucial role in governing the size, shape, and stability of the nanoparticles produced (Borah et al. 2022). TEM images (Fig. 2.4b) revealed that AgNPs synthesized from Reishi mushroom extract were dispersed, although some agglomeration between particles was also observed (Aygün et al. 2020). This phenomenon may arise from the tendency of small nanoparticles to aggregate due to Van der Waals forces of attraction but can be mitigated by controlling the concentration of the capping agent if necessary. The particles exhibited a spherical shape, with a mean particle size ranging between 9 and 21 nm.

Fluorescence microscopy enables not only the measurement of the physical properties of nanoparticles but also the comprehension of interactions between engineered nanomaterials and living matter, such as cells or tissues (Król et al. 2017).

The photoluminescence spectra of broc-ZnO and nb-ZnO (prepared using broccoli extract) were recorded at an excitation wavelength of 250 nm. The spectra of broc-ZnO displayed three emission peaks centered at approximately 268, 298, and 417 nm. The first two peaks are situated in the UV region, while the third extends into the visible region, which is typical of ZnO nanoparticles. It has been observed that the intensity of the emission peak decreases with an increase in the particle size of nanoparticles. The higher peak intensity of broc-ZnO is consistent with the smaller particle size (Osuntokun et al. 2019). To establish the correlation between the size of ZnO nanoparticles (NPs) and emission intensity, Saad Algarni et al. (2022) performed a quantitative investigation on both ZnO-80 and ZnO-180 samples. The fluorescence spectrum of the aqueous suspension of ZnO NPs was measured at room temperature using an excitation wavelength of 380 nm. The findings of this study revealed a higher intensity of peaks in ZnO-80 compared to those in ZnO-180, indicating that the crystallites of ZnO-80 are smaller than those of ZnO-180.

- *UV-Vis Spectroscopy*

UV-vis spectroscopy is typically the initial method employed for characterizing metallic nanoparticles. UV-visible analysis is used to confirm the formation of metallic nanoparticles, which exhibit characteristic peaks at different absorptions derived from their respective metallic salts. Typically, wavelengths ranging from 200 to 800 nm are employed to characterize nanoparticles (Hussain et al. 2016; Almatroudi 2020).

Noble metallic nanoparticles like gold (Au) and silver (Ag) exhibit pronounced absorption in the visible spectrum, peaking within the ranges of 500–600 nm for Au and 400–450 nm for Ag. This absorption is attributed to the surface plasmon resonance (SPR) phenomenon inherent in metallic nanoparticles (Noruzi 2015). According to Yakoup et al. (2024), the color change indicates the reduction of Ag^+ to Ag^0 in the $AgNO_3$ solution and the formation of biofilm-AgNPs (*P. aeruginosa* biofilm). The UV-vis spectrum of the solution reveals a single peak of maximum absorption rate at 364 nm, which indicates the possibility of biofilm-AgNPs synthesized in the solution. UV-vis spectra of ZnO particles synthesized using *Aloe vera* extract displayed strong UV absorption spectra with the absorption peak ranging from 358 to 375 nm due to its surface plasmon resonance (Sangeetha et al. 2011). The UV-vis spectrum of zinc oxide nanoparticles using *Ixora coccinea* leaf extract showed an absorbance peak at 340 nm, corresponding to the characteristic band of zinc oxide nanoparticles (Yedurkar et al. 2016). By employing a UV-visible spectrophotometer, the findings indicated that the peak absorbance values of ZnO-NPs and *Euphorbia retusa* plant extract were detected at 245 nm, suggesting the presence of the corresponding nanoparticles in the solution (Abduljabbar et al. 2024).

The *bandgap* energy (Eg) is determined from the absorption spectra and using the Tauc method, in which an extrapolation of the linear region of the $(\alpha h\nu)^2 \times h\nu$ graph provides the value of the *gap* energy (Khan et al. 2024). Generally, in the *bandgap* between 0.5 and 4.0 eV, materials are called semiconductors. This occurs when there is a separation between the energy bands, leading to the use of thermal or electrical energy to obtain some conduction (Loddo et al. 2018).

The optical properties of CuSNPs (using *Chlorella ellipsoidea* extract) in aqueous dispersion were assessed by using UV-vis-NIR spectroscopy. The obtained Eg value by absorption peak 650 nm for CuSNPs is 2.39 eV, which is higher than that for bulk CuS (1.85 eV) results from a quantum size effect (Borah et al. 2022). The bandgap was found to be ~3.33 eV for ZnO nanoparticles prepared with *Cassia fistula* leaf extract which is smaller than that of the bulk ZnO (3.45 eV). This bandgap enhancement occurs due to the size effect of the nanoparticles in the thin film (Suresh et al. 2015).

- *X-Ray Diffraction (XRD)*

The X-ray diffraction (XRD) technique allows for obtaining structural information about crystalline metallic nanoparticles (NPs) and purity of the photocatalysts and also enables the verification of the formation of zero-valent nanoparticles. If the nanoparticles are in an amorphous structure, no diffraction peak will be observed, rendering the use of XRD unfeasible for sample identification (Noruzi 2015). The broadening of peaks in X-ray diffraction confirms the formation of nano-sized particles, with smaller nanoparticles resulting in broader reflections in XRD (Vijayaraghavan and Ashokkumar 2017). The size of nanoparticles can be calculated using the Debye-Scherrer equation.

$$D = 0.9\lambda / (\beta \cos\theta)$$

where D is the crystalline size, λ is the wavelength (1.54, β is the width of the XRD peak posted at half maximum height (FWHM)) and θ is the Bragg diffraction angle.

XRD analysis showed that all ZnO products have the wurtzite structure. The XRD patterns of the products exhibit good crystallinity for all samples with and without aloe extract. It was observed that the orientation of ZnO nanoparticles shifted slightly to the left when aloe extract was used, compared to ZnO prepared without aloe extract, due to inter-particle distance (Sangeetha et al. 2011). In the study by (Bayahia 2023), pure $ZnFe_2O_4$ exhibits a crystalline structure and sharp X-ray diffraction peaks, whereas the modification of $ZnFe_2O_4$ with plant extracts (*Salvadora persica* and *Ziziphus mauritiana*) reduced the crystallinity of $ZnFe_2O_4$, resulting in an amorphous structure.

The current X-ray diffraction (XRD) analysis showcases the characterization of $SmMnO_3/Sm_2O_3$ nanocomposites. Two chelates, oxalic acid and citric acid, produce small and desirable nanoparticles. The XRD patterns indicate the absence of additional phases or impurities, underscoring the purity of the samples (Jamdar et al. 2024).

The synthesis route determines the properties of the NPs produced in terms of their crystalline growth, morphology, size, size distribution, stability, and aggregation (Król et al. 2017). The different sources of natural resources for biosynthesis, the so-called extracts, strongly influence the size of the crystallites.

According to Rasli et al.'s review (2020), crystallite diameters vary between 2 and 70 nm, depending on the extract used in the synthesis. The crystal sizes of the $SmMnO_3/Sm_2O_3$ nanocomposites were determined, yielding a range of 18–25 nm (Jamdar et al. 2024). The calculated crystallite size of the powder nanoparticles of ZnO, obtained from *A. indica* leaf extract, is about 18 nm (Elumalai and Velmurugan 2015). It was reported by Qu et al. (2011) that ZnO nanoparticles, synthesized using *Physalis alkekengi*, showed a polydisperse behavior and had a crystalline mean size of 72.5 nm. The XRD patterns enabled the determination of the average crystallite size of ZnO nanoparticles, measuring 43.3 nm for Zn-AL and 44.9 nm for Zn-ST (Primo et al. 2020). These findings indicate that the crystalline size is influenced by the polysaccharide employed in the synthesis, despite using the same calcination temperature.

- *Specific Surface Area*

It is well-known that the photocatalytic activity of nanoparticles strongly depends on their specific surface area and the volume and diameter of their pores. The specific surface area of the is measured by Brunauer-Emmett-Teller (BET) method through isotherms adsorption/desorption of nitrogen.

Synthesis of magnetic iron oxide nanoparticles (MIONPs) utilizing the fruit extract of *Cynometra ramiflora*, an inedible fruit waste, was applied as a photocatalytic (Bishnoi et al. 2018). Based on the BET analysis, the specific surface area and pore volumes were determined as 107.97 $m^2 g^{-1}$. The large surface area and mesoporous structure of the Fe_2O_3 nanoparticles obtained in the study can provide many active sites and thus can be used as a catalyst in many applications. On the other hand, the specific surface area of the Fe_2O_3 nanoparticles synthesized by chemical methods is 18.8 $m^2 g^{-1}$; in addition, the surface area of the commercial Fe_2O_3 is just 7.7 $m^2 g^{-1}$ (Liu et al. 2015). The different extracts resulted in $ZnFe_2O_4$ adopting different textural properties.

Using the *Salvadora persica* extract gave the nanoparticles a significantly larger surface area than using the *Ziziphus mauritiana* extract 3.583 and 0.965 $m^2 g^{-1}$, respectively (Bayahia 2023). The photocatalysts presented as mesoporous structures. The specific surface area for the Ni/Fe_3O_4 NPs using *Euphorbia maculata* extract was calculated to be almost 46.86 $m^2 g^{-1}$. In addition, the BET surface was measured as 3067 $m^2 g^{-1}$ for the CuO NPs (Pakzad et al. 2019).

2.2.2 *Advanced Characterization Methods*

In addition to conventional techniques, we will address advanced methods that provide more detailed insights into green nanostructured photocatalysts, including:

- *Dynamic Light Scattering (DLS)*

Dynamic light scattering (DLS), also known as photon correlation spectroscopy (PCS), is a widely used technique for evaluating the dimensions of particles in suspension. In this method, when a particle is exposed to a light source, such as a laser, it scatters light in multiple directions. Analysis of the fluctuations in the intensity of scattered light at a specific angle provides information about the particle's movement (Seto 2023).

For monodisperse and diluted samples, the particles do not interact with each other. They are in random motion (Brownian motion), resulting from interaction with the molecules that make up the dispersing medium. In the Stokes-Einstein theory of Brownian motion, the movement of the particle at very low concentration can be determined by the viscosity and temperature of the dispersing medium and the size of the particle (Feng et al. 2023).

This technique is widely used in the characterization of nano- and micrometer dispersions, emulsions, and colloidal solutions. The size of the dispersed particles in a liquid medium plays a fundamental role in various properties, such as reactivity, dissolution kinetics, stability of colloidal dispersions, and efficacy of delivery in carrier systems (Sdobnov et al. 2024).

Particle size analysis by DLS is particularly useful for investigating nanostructured photocatalysts, allowing for a detailed understanding of their dimensions and behavior in suspension (Bairamis and Konstantinou 2021; Tănase et al. 2021).

- *Photoluminescence Spectroscopy (PL)*

Photoluminescence spectroscopy (PL) is a technique employed in the characterization of nanophotocatalysts, allowing for the study of photogenerated charges in the particles and their optical and electronic properties (Rao et al. 2024).

The phenomenon of photoluminescence refers to the emission of light by a sample after being excited by an external light source. Typically, monochromatic light sources, such as lasers and lamps followed by monochromators, are used to allow sample excitation with known energy. When photons strike the sample, they are absorbed, exciting electrons to higher energy states. Subsequently, the electronic system relaxes to the ground state, emitting photons of lower or equal energy to the received photon (Kaminska et al. 2020).

In semiconductor solids, such as nanophotocatalysts, photoluminescence occurs through the absorption of a photon from the excitation light, creating an electron-hole pair that will recombine, generating another photon. The absorbed photon has energy greater than the bandgap of the semiconductor material, and to reach energy equilibrium, electrons must attain the minimum energy of the conduction band, which occurs through phonon emission. After thermalization, the electron-hole pair can recombine, emitting photons corresponding to the energy difference between the two occupied states (Rao et al. 2024).

Temperature is a factor that influences photoluminescence spectra. At low temperatures, charge carriers exhibit little mobility, making excitonic emissions more efficient. At room temperature, carriers can acquire thermal energy and escape to non-radiative recombination centers, resulting in emission peaks of lower intensity and greater full width at half maximum (Shi et al. 2021).

Therefore, photoluminescence spectroscopy provides a detailed understanding of the optical and electronic properties of nanophotocatalysts, being an essential technique in the characterization of these promising materials (Sang et al. 2020; Qi et al. 2021; Swain and Basu 2021; Kumar 2023).

- *X-Ray Photoelectron Spectroscopy (XPS)*

X-ray photoelectron spectroscopy (XPS) is an essential technique in the analysis of sample surfaces, based on the photoelectric effect. Its significant development occurred from the 1960s onward, with the advancement of ultrahigh vacuum (UHV) technology, allowing for the production and maintenance of clean surfaces at atomic levels for analysis (Isaacs et al. 2021; Krishna and Philip 2022).

When a surface is irradiated with X-rays of energy $h\nu$, photoelectrons are emitted with kinetic energy determined by the equation $E_k = h\nu - |E_B| - \varphi$, where E_k is the kinetic energy of the photoelectron, $h\nu$ is the photon energy, E_B is the binding energy of a given electronic level, and φ is the work function of the material (Isaacs et al. 2021).

Analysis of the kinetic energy of the photoelectrons allows for the determination of the sample's chemical composition, as electronic levels are specific to each element. However, due to the strong interaction of electrons with matter and their short mean free path, the information obtained by XPS pertains only to the first atomic layers of the surface (Sadhukhan et al. 2023).

XPS spectra are employed to identify the presence of elements and quantify the chemical composition of the surface. The energy range considered is scanned multiple times, and the software averages the values to obtain the final spectrum, reducing statistical variations (Lima et al. 2021; Pradhan et al. 2022; Kumar 2023).

- *Electrochemical Impedance Spectroscopy (EIS)*

Photocatalysis, an electrochemical process, involves the transport of charges (electrons and holes) participating in redox chemical reactions. Therefore, it is important to also study the behavior of these charges in photocatalytic materials. Electrochemical impedance spectroscopy (EIS) emerges as an essential tool for such purpose, allowing for the determination of the kinetics of chemical processes and the evaluation of the electrical response of chemical systems (Bredar et al. 2020).

The electrochemical impedance measurement is performed by scanning at different frequencies of alternating current, resulting in contributions of electrical resistance (Z': abscissa axis, real) and reactance (Z'': ordinate axis, imaginary). This technique provides a detailed analysis of the electrical properties of the material under study (Srinivasan and Fasmin 2021).

The material's microstructure is directly related to its conductivity, and it is possible to correlate EIS results with intragranular and grain boundary conductivity. This relationship between electrical response and material structure offers valuable insights into its efficiency and performance in photocatalytic processes (He 2019).

Thus, EIS can be employed in the characterization of photocatalysts, providing an analysis of their electrical properties and their correlation with microstructure, significantly contributing to the development and optimization of photocatalytic materials for a variety of industrial and environmental applications.

2.3 Advances in the Synthesis and Characterization of Green Photocatalysts

The field of synthesis and characterization of green photocatalysts has been revolutionized by nanotechnology, providing a vast array of tools and techniques for designing materials with enhanced properties. Since their inception, photocatalysts have been crucial in various applications, ranging from water purification to renewable energy production. However, the pursuit of more efficient and sustainable materials has driven the integration of nanotechnology into this landscape, enabling significant advancements in the efficiency and selectivity of photocatalysts.

In recent years, there has been a growing interest in the sustainable synthesis of photocatalysts, driven by the need to develop more environmentally friendly technologies. A promising approach has been the utilization of renewable sources as raw materials in photocatalyst synthesis (Zuliani and Cova 2021; Song et al. 2023; Yue et al. 2023). Materials derived from natural sources such as lignin, cellulose, and biomass are being explored as viable alternatives, thereby reducing dependence on nonrenewable resources and minimizing the environmental impact of photocatalyst production.

Furthermore, sustainable synthesis methods are being developed to minimize energy consumption and the use of toxic solvents. Techniques such as sol-gel synthesis, hydrothermal, and microwave and ultrasound-assisted synthesis are being adapted for the production of green photocatalysts, aiming to reduce the use of harmful chemicals and promote a more eco-efficient approach (Yousefi et al. 2021; Ahmed and Mohamed 2024). Simultaneously, there is an ongoing effort to create biodegradable and biocompatible photocatalysts, utilizing natural organic materials such as biodegradable polymers and proteins as substrates for photocatalytic material synthesis (Zhao et al. 2021; Song et al. 2023). This approach aims not only to minimize waste and environmental pollution but also to expand the applications of photocatalysts in areas such as biomedicine and environmental catalysis.

The biological synthesis of green nanophotocatalysts can be mediated by plants, algae, or microorganisms, such as fungi and bacteria. Recently, studies have used residues in the synthesis of green nanophotocatalysts. Khazaei et al. (2017) used eggshells to obtain a palladium nanocomposite. The synthesis process was mediated by barberry extract. Other examples of research also used waste to obtain nanocomposites of metal oxides, notably, chicken bone (Mohammad Sajadi et al. 2019), recycled glass (Salam et al. 2020), rice husk silica gel and peach leaves (AbuKhadra et al. 2020), corncob (Xie et al. 2021), pomelo peel (Huang et al. 2022), and shells of shrimps (Usman et al. 2024). Research has also used residues as reducing agents, such as *Chrysanthemum* spp. floral waste to obtain $ZnFe_2O_4$@ZnO (Nguyen et al. 2023) and *Phoenix dactylifera* waste as a bioreductant to obtain ZnO nanoparticles (Rambabu et al. 2021). Therefore, waste can perform different functions, either as a support medium or as a raw material for the production of composite nanophotocatalysts.

Nanotechnology plays a fundamental role in the synthesis and characterization of photocatalysts with improved properties. It makes it possible to manipulate and control the structural and morphological characteristics of nanoparticles, which directly influences their performance as photocatalysts (Almatroudi 2020; Abduljabbar et al. 2024; Ashraf et al. 2024). Here are some ways in which nanotechnology has contributed. Nanotechnology allows the synthesis of nanoparticles with specific sizes and shapes, optimized for optimal photocatalytic activity. For example, nanoparticles of different shapes (such as nanowires, nanotubes, and spherical nanoparticles) can have different photocatalytic properties. Nanostructures have a significantly larger surface area compared to their macroscale counterparts, which increases reaction efficiency. This is because more reaction sites are available to interact with the reactants, thus increasing the reaction rate. Nanotechnology allows the design of specific energy band structures, which can increase the quantum efficiency of photocatalysis. By adjusting the energy bands, it is possible to reduce the recombination of electrons and holes, thus improving the overall efficiency of the photocatalysis process.

2.4 Challenges and Future Perspectives

Given the increasingly scarce availability of natural resources, a good nanophotocatalyst must present some desirable properties, such as stability, reusability, and low cost. The main difficulty in controlling the stability of nanophotocatalysts is due to the presence of biomolecules on their surface. The functional groups present in biomolecules, such as hydroxyls, carboxyls, and amines (Vijayaraghavan and Ashokkumar 2017), result in surface charges, attracting other charged particles, or even adjacent biomolecules, resulting in agglomeration (Haseena et al. 2019). Furthermore, if the metal ion that gives rise to the photocatalyst nanoparticle has magnetic properties, agglomeration can also occur due to magnetic dipole interaction (Somanathan et al. 2019). Nanophotocatalysts with magnetic properties can be interesting from the point of view of recovering the nanophotocatalyst, which could be done with the help of a magnet. Therefore, the relationship between the technical feasibility of recovery versus stability must be carefully evaluated.

A negative aspect of the synthesis of nanophotocatalysts is related to the size of their particles (1–100 nm) (Joseph and Vijayanandan 2021). This makes it difficult to separate the nanophotocatalyst after use, especially in the photodegradation of organic substances, in addition to facilitating interaction with other biomolecular cellular components (Palencia et al. 2015 *apud* Paradva et al. 2023), which can bioaccumulate in living organisms (Ma et al. 2010). Therefore, there is a need to carry out toxicological assessments to enable its scale-up application (Paradva et al. 2023).

Besides, physical methods usually require a lot of energy for the nanophotocatalyst synthesis process, as is the case with spray pyrolysis, or even require expensive vacuum equipment, as is the case with the sputtering technique (Vijayaraghavan and Ashokkumar 2017; Langa and Hintsho-Mbita 2023). Therefore, these physical methods for synthesizing nanophotocatalysts make application on an industrial scale difficult.

The characterization of green photocatalysts also presents several challenges due to the complexity of the materials involved and the need to understand their structural, optical, and surface properties to optimize their efficiency and applicability. Characterizing green photocatalysts faces specific challenges related to their composition and ecological nature. Green photocatalysts are often composed of organic materials or derivatives from renewable sources, which can make analysis more complex due to the presence of multiple components and interactions among them (Vijayaraghavan and Ashokkumar 2017; Rasli et al. 2020; Gupta et al. 2023).

Additionally, the presence of impurities or natural contaminants can significantly affect the properties and effectiveness of photocatalysts.

It is often necessary to combine multiple analytical techniques to obtain comprehensive information about green photocatalysts. For example, the combination of scanning electron microscopy with UV-visible absorption spectroscopy and surface area analysis can provide insights into the morphology, composition, optical, and surface properties of the materials (Li et al. 2015; Sakir et al. 2020; Galeas et al. 2023; Bôlla de Menezes et al. 2024). Additionally, in situ studies or operating under reaction conditions can help elucidate the mechanisms involved in green photocatalysis.

Determining the cost of a green nanophotocatalyst is complex, as it depends on the availability of the biomaterial, the cost of the metallic salt, the possibility of reuse, energy consumption and effluents, and emissions generated from the process, among other factors. According to the literature, the overall cost of the plant-mediated nanophotocatalyst synthesis process depends mostly on the cost of the metal salt (Joseph and Vijayanandan 2021).

Economic viability also depends on the plant or crop used. The use of local crops is recommended; however, it must be taken into account that environmental conditions such as climate and flowering periods can make it difficult to maintain the commercial availability of biomaterials. Therefore, crops that are minimally subject to the effects of seasonality should be preferred. Furthermore, crops that are easy to harvest should also be preferred to minimize costs (Paradva et al. 2023).

An alternative to overcome the unavailability of cultivated plants due to seasonal conditions is the use of biomass wastes, especially agro-industrial wastes. The residues have biomolecules of interest in their composition, such as polyphenols, flavonoids, and vitamins (Mohamad Rodzi et al. 2023), which can act as reducing and stabilizing agents in the synthesis of nanophotocatalysts. The use of waste in the synthesis of nanophotocatalysts is advantageous because, in addition to contributing as mediators in the synthesis process, they also contribute to minimizing the environmental impact arising from the inadequate disposal of these wastes in the environment. It is also a way of valuing waste as raw material (Paradva et al. 2023).

For the green nanophotocatalyst to be commercially and industrially competitive, its reusability is highly desirable. Reusability determines the overall stability, robustness, and efficiency of the nanophotocatalyst. Therefore, catalysts that are reused or regenerated several times without losing their catalytic activity are considered highly stable from an economic point of view (Paradva et al. 2023).

References

Abduljabbar B, El-Zayat M, El-Amier Y, El-Halawany ES (2024) Biosynthesis, characterization, and cytotoxic activities of zinc nanoparticles (ZnNPs) using Euphorbia retusa extract. Kuwait J Sci 51:100108. https://doi.org/10.1016/j.kjs.2023.08.001

AbuKhadra MR, Mohamed AS, El-Sherbeeny AM, Elmeligy MA (2020) Enhanced photocatalytic degradation of acephate pesticide over MCM-41/Co3O4 nanocomposite synthesized

from rice husk silica gel and Peach leaves. J Hazard Mater 389. https://doi.org/10.1016/j.jhazmat.2020.122129
Achudhan D, Vijayakumar S, Malaikozhundan B et al (2020) The antibacterial, antibiofilm, antifogging and mosquitocidal activities of titanium dioxide (TiO2) nanoparticles green-synthesized using multiple plants extracts. J Environ Chem Eng 8. https://doi.org/10.1016/j.jece.2020.104521
Adamski J, Lopes AA, Lopez AB et al (2024) Development of composite films of bismuthates immobilized in alginate biopolymers for application in photocatalysis under visible light. J Photochem Photobiol A Chem 452:115575. https://doi.org/10.1016/j.jphotochem.2024.115575
Ahmad I, Akhtar MS, Ahmed E, Ahmad M (2020) Facile synthesis of Pr-doped ZnO photocatalyst using sol–gel method and its visible light photocatalytic activity. J Mater Sci Mater Electron 31:1084–1093. https://doi.org/10.1007/s10854-019-02620-2
Ahmed MA, Mohamed AA (2024) Advances in ultrasound-assisted synthesis of photocatalysts and sonophotocatalytic processes: a review. iScience 27:108583. https://doi.org/10.1016/j.isci.2023.108583
Alamro FS, Toghan A, Ahmed HA et al (2022) Multifunctional leather surface embedded with zinc oxide nanoparticles by pulsed laser ablation method. Microsc Res Tech 85:1611–1617. https://doi.org/10.1002/jemt.24022
Alfaro Cruz MR, Saldaña-Ramírez A, Juárez-Ramírez I, Torres-Martínez LM (2023) Development of SnO2–ZnO thin films as a photocatalyst for obtaining alternative fuels through photocatalytic reactions. Solid State Sci 137. https://doi.org/10.1016/j.solidstatesciences.2023.107112
Al-Hartomy OA, Ubaidullah M, Khatoon S et al (2012) Synthesis, characterization, and dielectric properties of nanocrystalline Ba1-xPbxZrO3 ($0 \leq x \leq 0.75$) by polymeric citrate precursor route. J Mater Res 27:2479–2488. https://doi.org/10.1557/jmr.2012.242
Almatroudi A (2020) Silver nanoparticles: synthesis, characterization and biomedical applications. Open Life Sci 15:819–839. https://doi.org/10.1515/biol-2020-0094
Ashraf A, Wahab R, Al-Khedhairy AA et al (2024) Magnetically separable and visible light-driven photocatalytic activity of graphene oxide based α-Fe2O3 nanocomposite. Mater Chem Phys 316. https://doi.org/10.1016/j.matchemphys.2024.129111
Aygün A, Özdemir S, Gülcan M et al (2020) Synthesis and characterization of Reishi mushroom-mediated green synthesis of silver nanoparticles for the biochemical applications. J Pharm Biomed Anal 178. https://doi.org/10.1016/j.jpba.2019.112970
Bairamis F, Konstantinou I (2021) Wo3 fibers/g-c3n4 z-scheme heterostructure photocatalysts for simultaneous oxidation/reduction of phenol/cr (Vi) in aquatic media. Catalysts 11. https://doi.org/10.3390/catal11070792
Balakrishnan A, Jacob MM, Chinthala M et al (2024) Photocatalytic sponges for wastewater treatment, carbon dioxide reduction, and hydrogen production: a review. Environ Chem Lett 22:635–656. https://doi.org/10.1007/s10311-024-01696-5
Balavandy SK, Shameli K, Biak DRBA, Abidin ZZ (2014) Stirring time effect of silver nanoparticles prepared in glutathione mediated by green method. Chem Cent J 8:11
Bayahia H (2023) The highly efficient green synthesis of nanostructured ZnFe2O4 photocatalysts by using Ziziphus mauritiana and Salvadora persica extracts for the photocatalytic degradation of crystal violet in sunlight. J King Saud Univ Sci 35:102584. https://doi.org/10.1016/j.jksus.2023.102584
Bishnoi S, Kumar A, Selvaraj R (2018) Facile synthesis of magnetic iron oxide nanoparticles using inedible Cynometra ramiflora fruit extract waste and their photocatalytic degradation of methylene blue dye. Mater Res Bull 97:121–127. https://doi.org/10.1016/j.materresbull.2017.08.040
Bôlla de Menezes L, Cristine Ladwig Muraro P, Moro Druzian D et al (2024) Calcium oxide nanoparticles: biosynthesis, characterization and photocatalytic activity for application in yellow tartrazine dye removal. J Photochem Photobiol A Chem 447. https://doi.org/10.1016/j.jphotochem.2023.115182
Boningari T, Inturi SNR, Suidan M, Smirniotis PG (2018) Novel one-step synthesis of sulfur doped-TiO2 by flame spray pyrolysis for visible light photocatalytic degradation of acetaldehyde. Chem Eng J 339:249–258. https://doi.org/10.1016/j.cej.2018.01.063

Bopape DA, Tetana ZN, Mabuba N et al (2023) Biosynthesis of TiO2 nanoparticles using Commelina benghanlensis for the photodegradation of methylene blue dye and antibiotics: effect of plant concentration. Results Chem 5. https://doi.org/10.1016/j.rechem.2023.100825
Borah D, Saikia P, Sarmah P et al (2022) Composition controllable alga-mediated green synthesis of covellite CuS nanostructure: an efficient photocatalyst for degradation of toxic dye. Inorg Chem Commun 142:109608. https://doi.org/10.1016/j.inoche.2022.109608
Bredar ARC, Chown AL, Burton AR, Farnum BH (2020) Electrochemical impedance spectroscopy of metal oxide electrodes for energy applications. ACS Appl Energy Mater 3:66–98. https://doi.org/10.1021/acsaem.9b01965
Cappelluti MD, Hadzifejzovic E, Foord JS, Gregory DH (2020) Flash microwave-assisted solvothermal (FMS) synthesis of photoactive anatase sub-microspheres with hierarchical porosity. RSC Adv 10:37233–37245. https://doi.org/10.1039/d0ra05796g
Cerón-Urbano L, Aguilar CJ, Diosa JE, Mosquera-Vargas E (2023) Nanoparticles of the perovskite-structure CaTiO3 system: the synthesis, characterization, and evaluation of its photocatalytic capacity to degrade emerging pollutants. Nanomaterials 13. https://doi.org/10.3390/nano13222967
Chang CJ, Teng MC, Chen J et al (2021) Microwave solvothermal synthesis of cubic MnS@Ag2S core-shell photocatalysts with improved charge separation and photocatalytic activity. Appl Surf Sci 558:149875. https://doi.org/10.1016/j.apsusc.2021.149875
Cruz D, Falé PL, Mourato A et al (2010) Preparation and physicochemical characterization of Ag nanoparticles biosynthesized by Lippia citriodora (Lemon Verbena). Colloids Surf B Biointerfaces 81:67–73. https://doi.org/10.1016/j.colsurfb.2010.06.025
Delgado AV, González-Caballero F, Hunter RJ et al (2005) Measurement and interpretation of electrokinetic phenomena: (IUPAC technical report). Pure Appl Chem 77:1753–1805. https://doi.org/10.1351/pac200577101753
Duy Trinh N, Hoang HH, Xuan Linh N et al (2019) A facile synthesis and properties of bismuth vanadate (BiVO4) photocatalyst by hydrothermal method. IOP Conf Ser Mater Sci Eng 542. https://doi.org/10.1088/1757-899X/542/1/012059
Elumalai K, Velmurugan S (2015) Green synthesis, characterization and antimicrobial activities of zinc oxide nanoparticles from the leaf extract of Azadirachta indica (L.). Appl Surf Sci 345:329–336. https://doi.org/10.1016/j.apsusc.2015.03.176
Feng X, Huang G, Qiu J et al (2023) Dynamic light scattering in flowing dispersion. Opt Commun 531:129225. https://doi.org/10.1016/j.optcom.2022.129225
Galeas S, Valdivieso-Ramírez CS, Pontón PI et al (2023) Photocatalytic degradation of phenol under visible light irradiation by using ferrous oxalate synthesized from iron-rich mineral sands via a green hydrothermal route. Environ Technol Innov 32:103325. https://doi.org/10.1016/j.eti.2023.103325
Gaur J, Vikrant K, Kim KH et al (2023) Photocatalytic degradation of Congo red dye using zinc oxide nanoparticles prepared using Carica papaya leaf extract. Mater Today Sustain 22. https://doi.org/10.1016/j.mtsust.2023.100339
Ghodake GS, Deshpande NG, Lee YP, Jin ES (2010) Pear fruit extract-assisted room-temperature biosynthesis of gold nanoplates. Colloids Surf B Biointerfaces 75:584–589. https://doi.org/10.1016/j.colsurfb.2009.09.040
Gopinath K, Gowri S, Arumugam A (2013) Phytosynthesis of silver nanoparticles using Pterocarpus santalinus leaf extract and their antibacterial properties. J Nanostructure Chem 3:68
Goutam SP, Saxena G, Singh V et al (2018) Green synthesis of TiO2 nanoparticles using leaf extract of Jatropha curcas L. for photocatalytic degradation of tannery wastewater. Chem Eng J 336:386–396. https://doi.org/10.1016/j.cej.2017.12.029
Guan X, Shi J, Guo L (2013) Ag3PO4 photocatalyst: hydrothermal preparation and enhanced O2 evolution under visible-light irradiation. Int J Hydrog Energy 38:11870–11877. https://doi.org/10.1016/j.ijhydene.2013.07.017
Gupta K, Chundawat TS (2019) Bio-inspired synthesis of platinum nanoparticles from fungus Fusarium oxysporum: its characteristics, potential antimicrobial, antioxidant and photocatalytic activities. Mater Res Express 6. https://doi.org/10.1088/2053-1591/ab4219

Gupta D, Boora A, Thakur A, Gupta TK (2023) Green and sustainable synthesis of nanomaterials: recent advancements and limitations. Environ Res 231:116316. https://doi.org/10.1016/j.envres.2023.116316

Haseena S, Shanavas S, Duraimurugan J et al (2019) Investigation on photocatalytic and antibacterial ability of green treated copper oxide nanoparticles using Artabotrys Hexapetalus and Bambusa Vulgaris plant extract. Mater Res Express 6:125064

Haverkamp RG, Marshall AT (2009) The mechanism of metal nanoparticle formation in plants: limits on accumulation. J Nanopart Res 11:1453–1463. https://doi.org/10.1007/s11051-008-9533-6

He J (2019) Oxide varistors: from microstruture to macro-characteristics. Wiley-VHC Verlag GmbH & Co. KGaA, Weinheim

Hoag GE, Collins JB, Holcomb JL et al (2009) Degradation of bromothymol blue by "greener" nano-scale zero-valent iron synthesized using tea polyphenols. J Mater Chem 19:8671–8677. https://doi.org/10.1039/b909148c

Horwat D, Zakharov DI, Endrino JL et al (2011) Chemistry, phase formation, and catalytic activity of thin palladium-containing oxide films synthesized by plasma-assisted physical vapor deposition. Surf Coat Technol 205. https://doi.org/10.1016/j.surfcoat.2010.12.021

Huang CW, Nguyen VH, Zhou SR et al (2020) Metal-organic frameworks: preparation and applications in highly efficient heterogeneous photocatalysis. Sustain Energy Fuels 4:504–521. https://doi.org/10.1039/c9se00972h

Huang W, Li Y, Fu Q, Chen M (2022) Fabrication of a novel biochar decorated nano-flower-like MoS2 nanomaterial for the enhanced photodegradation activity of ciprofloxacin: performance and mechanism. Mater Res Bull 147. https://doi.org/10.1016/j.materresbull.2021.111650

Hussain I, Singh NB, Singh A et al (2016) Green synthesis of nanoparticles and its potential application. Biotechnol Lett 38:545–560. https://doi.org/10.1007/s10529-015-2026-7

Iqbal W, Tian B, Anpo M, Zhang J (2017) Single-step solvothermal synthesis of mesoporous anatase TiO2-reduced graphene oxide nanocomposites for the abatement of organic pollutants. Res Chem Intermed 43:5187–5201. https://doi.org/10.1007/s11164-017-3049-6

Iravani S (2011) Green synthesis of metal nanoparticles using plants. Green Chem 13:2638–2650. https://doi.org/10.1039/c1gc15386b

Isaacs MA, Davies-Jones J, Davies PR et al (2021) Advanced XPS characterization: XPS-based multi-technique analyses for comprehensive understanding of functional materials. Mater Chem Front 5:7931–7963. https://doi.org/10.1039/d1qm00969a

Jain S, Mehata MS (2017) Medicinal plant leaf extract and pure flavonoid mediated green synthesis of silver nanoparticles and their enhanced antibacterial property. Sci Rep 7:15867

Jamdar M, Goudarzi M, Dawi EA et al (2024) Synthesis of SmMnO3/Sm2O3 nanocomposites as efficient photocatalysts for organic dye degradation by sol gel pechini method. Results Eng 21:101650. https://doi.org/10.1016/j.rineng.2023.101650

Jasbi NE, Solati E, Dorranian D (2021) Role of laser fluence in decoration of graphene nanosheets with TiO2 nanoparticles by pulsed laser ablation method. J Alloy Compd 861:157956

Jigyasa, Singh H, Rajput JK (2021) Graphitic carbon nitride nanosheets (g-C3N4 NS) as dual responsive template for fluorescent sensing as well as degradation of food colorants. Food Chem 343:128451. https://doi.org/10.1016/j.foodchem.2020.128451

Joseph A, Vijayanandan A (2021) Photocatalysts synthesized via plant mediated extracts for degradation of organic compounds: a review of formation mechanisms and application in wastewater treatment. Sustain Chem Pharm 22. https://doi.org/10.1016/j.scp.2021.100453

Kaminska A, Koronski K, Strak P et al (2020) Defect-related photoluminescence and photoluminescence excitation as a method to study the excitonic bandgap of AlN epitaxial layers: experimental and ab initio analysis. Appl Phys Lett 117. https://doi.org/10.1063/5.0027743

Kato K, Shirai T (2022) Highly efficient water purification by WO3-based homo/heterojunction photocatalyst under visible light. J Alloys Compd 901:163434. https://doi.org/10.1016/j.jallcom.2021.163434

Khan MSJ, Sidek LM, Kamal T et al (2024) An efficient wastewater treatment through reduction of organic dyes using Ag nanoparticles supported on cellulose gum beads. Int J Biol Macromol 257. https://doi.org/10.1016/j.ijbiomac.2023.128544

Khatami M, Iravani S (2021) Green and eco-friendly synthesis of nanophotocatalysts: an overview. Comments Inorg Chem 41:133–187. https://doi.org/10.1080/02603594.2021.1895127

Khazaei M, Khazaei A, Nasrollahzadeh M, Tahsili MR (2017) Highly efficient reusable Pd nanoparticles based on eggshell: green synthesis, characterization and their application in catalytic reduction of variety of organic dyes and ligand-free oxidative hydroxylation of phenylboronic acid at room temperature. Tetrahedron 73:5613–5623. https://doi.org/10.1016/j.tet.2017.04.016

Krishna DNG, Philip J (2022) Review on surface-characterization applications of X-ray photoelectron spectroscopy (XPS): recent developments and challenges. Appl Surf Sci Adv 12:100332. https://doi.org/10.1016/j.apsadv.2022.100332

Król A, Pomastowski P, Rafińska K et al (2017) Zinc oxide nanoparticles: synthesis, antiseptic activity and toxicity mechanism. Adv Colloid Interf Sci 249:37–52. https://doi.org/10.1016/j.cis.2017.07.033

Kumar G (2023) Sunlight responsive 2D/2D SnS2/BiVO4 nanocomposite for photocatalytic removal of ciprofloxacin antibiotic from aqueous medium. J Inorg Organomet Polym Mater 33:2710–2720. https://doi.org/10.1007/s10904-023-02711-y

Langa C, Hintsho-Mbita NC (2023) Plant and bacteria mediated synthesis of TiO2 NPs for dye degradation in water. A review. Chem Phys Impact 7:100293

Lelis M, Tuckute S, Urbonavicius M et al (2023) C-TiO2+Ni and ZnO+Ni magnetic photocatalyst powder synthesis by reactive magnetron sputtering technique and their application for bacteria inactivation. Inorganics 11. https://doi.org/10.3390/inorganics11020059

Li X, Xu H, Chen ZS, Chen G (2011) Biosynthesis of nanoparticles by microorganisms and their applications. J Nanomater 2011:270974

Li D, Cheng X, Yu X, Xing Z (2015) Preparation and characterization of TiO2-based nanosheets for photocatalytic degradation of acetylsalicylic acid: influence of calcination temperature. Chem Eng J 279:994–1003. https://doi.org/10.1016/j.cej.2015.05.102

Li JF, Rupa EJ, Hurh J et al (2019) Cordyceps militaris fungus mediated zinc oxide nanoparticles for the photocatalytic degradation of methylene blue dye. Optik (Stuttg) 183:691–697. https://doi.org/10.1016/j.ijleo.2019.02.081

Li Z, Zhong W, Gao D et al (2023) CdIn2S4-xSex solid-solution nanocrystal photocatalyst: one-step hydrothermal synthesis, controllable band structure, and improved H2-evolution activity. Adv Sustainable Syst 2200030:1–11. https://doi.org/10.1002/adsu.202200030

Liang H, Tai X, Du Z (2020) Photocatalytic degradation of nonylphenol ethoxylate and its degradation mechanism. J Mol Liq 302:112567. https://doi.org/10.1016/j.molliq.2020.112567

Lima NA, Mendonc GC, De Lima BS et al (2021) Influence of the synthesis method on CuWO 4 nanoparticles for photocatalytic application. J Mater Sci Mater Electron:1139–1149. https://doi.org/10.1007/s10854-020-04887-2

Ling Y, Dai Y (2020) Direct Z-scheme hierarchical WO3/BiOBr with enhanced photocatalytic degradation performance under visible light. Appl Surf Sci 509:145201. https://doi.org/10.1016/j.apsusc.2019.145201

Liu X, Chen K, Shim JJ, Huang J (2015) Facile synthesis of porous Fe2O3 nanorods and their photocatalytic properties. J Saudi Chem Soc 19:479–484. https://doi.org/10.1016/j.jscs.2015.06.009

Liu H, Zhang H, Wang J, Wei J (2020) Effect of temperature on the size of biosynthesized silver nanoparticle: deep insight into microscopic kinetics analysis. Arab J Chem 13:1011–1019. https://doi.org/10.1016/j.arabjc.2017.09.004

Liu W, Liu N, Song K et al (2024) Fabrication of WO3 photocatalyst by plasma assisted ball milling under different discharge atmospheres. Vacuum 220:112809. https://doi.org/10.1016/j.vacuum.2023.112809

Loddo V, Bellardita M, Camera-Roda G et al (2018) Heterogeneous photocatalysis

Lundin D, Sarakinos K (2012) An introduction to thin film processing using high-power impulse magnetron sputtering. J Mater Res 27:780–792. https://doi.org/10.1557/jmr.2012.8

Ma X, Geiser-Lee J, Deng Y, Kolmakov A (2010) Interactions between engineered nanoparticles (ENPs) and plants: phytotoxicity, uptake and accumulation. Sci Total Environ 408:3053–3061

Mahlaule-Glory LM, Mapetla S, Makofane A et al (2022) Biosynthesis of iron oxide nanoparticles for the degradation of methylene blue dye, sulfisoxazole antibiotic and removal of bacteria from real water. Heliyon 8. https://doi.org/10.1016/j.heliyon.2022.e10536

Masekela D, Hintsho-Mbita NC, Ntsendwana B, Mabuba N (2022) Thin films (FTO/BaTiO3/AgNPs) for enhanced piezo-photocatalytic degradation of methylene blue and ciprofloxacin in wastewater. ACS Omega 7:24329–24343

Meléndrez MF, Cárdenas G, Arbiol J (2010) Synthesis and characterization of gallium colloidal nanoparticles. J Colloid Interface Sci 346:279–287. https://doi.org/10.1016/j.jcis.2009.11.069

Mohamad Rodzi US, Kiew PL, Lam MK, Tan LS (2023) A feasibility study on the green synthesis of iron oxide nanoparticles using Chlorella vulgaris extract for photocatalytic degradation of crystal violet. In: IOP conference series: earth and environmental science. Institute of Physics

Mohammad Sajadi S, Kolo K, Hamad SM et al (2019) Biosynthesis of reusable and recyclable CuO@Magnetite@Hen Bone NCs and its antioxidant and antibacterial activities: a highly stable magnetically nanocatalyst for excellent reduction of organic dyes and adsorption of polycyclic aromatic hydrocarbons. IET Nanobiotechnol 13:124–133. https://doi.org/10.1049/iet-nbt.2018.5014

Mostafa AM, Mwafy EA, Toghan A (2021) ZnO nanoparticles decorated carbon nanotubes via pulsed laser ablation method for degradation of methylene blue dyes. Colloids Surf A Physicochem Eng Asp 627:127204

Mostafazadeh N, Dadashi R, Faraji M et al (2024) Anodizing of commercial galvanized mesh followed by electroless decorating of Ag nanoparticles for application as novel and low-cost photocatalyst for degradation of both dye and microbiological pollutants. J Photochem Photobiol A Chem 447. https://doi.org/10.1016/j.jphotochem.2023.115257

Motakef-Kazemi N, Ataei F, Dorranian D (2023) Synthesis and evaluation of copper–imidazole MOF nanostructures and its graphene nanocomposites by pulsed laser ablation method in liquid. Opt Quant Electron 55:1–17. https://doi.org/10.1007/s11082-023-04775-z

Narayanan KB, Sakthivel N (2010) Phytosynthesis of gold nanoparticles using leaf extract of Coleus amboinicus Lour. Mater Charact 61:1232–1238. https://doi.org/10.1016/j.matchar.2010.08.003

Nasrollahzadeh M, Sajadi SM (2015) Synthesis and characterization of titanium dioxide nanoparticles using Euphorbia heteradena Jaub root extract and evaluation of their stability. Ceram Int 41:14435–14439. https://doi.org/10.1016/j.ceramint.2015.07.079

Nguyen NTT, Nguyen LM, Nguyen TTT et al (2023) Green synthesis of ZnFe2O4@ZnO nanocomposites using Chrysanthemum spp. floral waste for photocatalytic dye degradation. J Environ Manag 326. https://doi.org/10.1016/j.jenvman.2022.116746

Noruzi M (2015) Biosynthesis of gold nanoparticles using plant extracts. Bioprocess Biosyst Eng 38:1–14. https://doi.org/10.1007/s00449-014-1251-0

Osuntokun J, Onwudiwe DC, Ebenso EE (2019) Green synthesis of ZnO nanoparticles using aqueous Brassica oleracea L. var. italica and the photocatalytic activity. Green Chem Lett Rev 12:444–457. https://doi.org/10.1080/17518253.2019.1687761

Pakzad K, Alinezhad H, Nasrollahzadeh M (2019) Green synthesis of Ni@Fe3O4 and CuO nanoparticles using Euphorbia maculata extract as photocatalysts for the degradation of organic pollutants under UV-irradiation. Ceram Int 45:17173–17182. https://doi.org/10.1016/j.ceramint.2019.05.272

Pan Y, Wang J, Lu Z et al (2024) A review on the application of magnetron sputtering technologies for solid oxide fuel cell in reduction of the operating temperature. Int J Hydrog Energy 50:1179–1193. https://doi.org/10.1016/j.ijhydene.2023.10.143

Pantidos N, Horsfall LE (2014) Biological synthesis of metallic nanoparticles by bacteria, fungi and plants. J Nanomed Nanotechnol 5:1

Paradva KC, Jangir R, Kalla S (2023) Green nanomaterials: synthesis and applications in wastewater treatment. Inorg Chem Commun 158:111584

Patil PS (1999) Versatility of chemical spray pyrolysis technique. Mater Chem Phys 59:185–198. https://doi.org/10.1016/S0254-0584(99)00049-8

Pradhan SR, Paszkiewicz-gawron M, Łomot D et al (2022) Bimetallic TiO2 nanoparticles for lignin-based model compounds valorization by integrating an optocatalytic flow-microreactor. Molecules 27:8731

Primo JDO, Bittencourt C, Acosta S et al (2020) Synthesis of zinc oxide nanoparticles by eco-friendly routes: adsorbent for copper removal from wastewater. Front Chem 8:1–13. https://doi.org/10.3389/fchem.2020.571790

Qi Y, Meador WE, Xiong J, Abbaszadeh M (2021) Structural, optical, photocatalytic, and optoelectronic properties of Zn2 SnO4 nanocrystals prepared by hydrothermal method. Nanotecnhnology 32:145702

Qu J, Yuan X, Wang X, Shao P (2011) Zinc accumulation and synthesis of ZnO nanoparticles using Physalis alkekengi L. Environ Pollut 159:1783–1788. https://doi.org/10.1016/j.envpol.2011.04.016

Rafiq U, Mehraj O, Lone S et al (2020) Solvothermal synthesis of Ag2WO4/Sb2WO6 heterostructures for enhanced charge transfer properties and e ffi cient visible-light-driven photocatalytic activity and stability. J Environ Chem Eng 8:104301. https://doi.org/10.1016/j.jece.2020.104301

Rahaman S, Singha MK (2022) Nanoarchitectonics earth-abundant chalcogenide Cu2SnS3 thin film using ultrasonic spray pyrolysis for visible light-driven photocatalysis. Appl Phys A Mater Sci Process 128:1–10. https://doi.org/10.1007/s00339-021-05174-5

Rambabu K, Bharath G, Banat F, Show PL (2021) Green synthesis of zinc oxide nanoparticles using Phoenix dactylifera waste as bioreductant for effective dye degradation and antibacterial performance in wastewater treatment. J Hazard Mater 402. https://doi.org/10.1016/j.jhazmat.2020.123560

Rani N, Singh P, Kumar S et al (2023) Plant-mediated synthesis of nanoparticles and their applications: a review, vol 163. Mater Res Bull

Rao SM, Kiligaridis A, Yangui A et al (2024) Photoluminescence mapping over laser pulse fluence and repetition rate as a fingerprint of charge and defect dynamics in perovskites. Adv Opt Mater 2300996:1–9. https://doi.org/10.1002/adom.202300996

Rasli NI, Basri H, Harun Z (2020) Zinc oxide from aloe vera extract: two-level factorial screening of biosynthesis parameters. Heliyon 6:e03156. https://doi.org/10.1016/j.heliyon.2020.e03156

Ren Y, Zhang N, Arain Z et al (2020) Polymer-induced lattice expansion leads to all-inorganic CsPbBr3 perovskite solar cells with reduced trap density. J Power Sources 475:228676. https://doi.org/10.1016/j.jpowsour.2020.228676

Saad Algarni T, Abduh NAY, Al Kahtani A, Aouissi A (2022) Photocatalytic degradation of some dyes under solar light irradiation using ZnO nanoparticles synthesized from Rosmarinus officinalis extract. Green Chem Lett Rev 15:460–473. https://doi.org/10.1080/17518253.2022.2089059

Sadhukhan P, Sarkar S, Souza SWD et al (2023) Bulk electronic structure of Mn2 NiGa using hard x-ray photoelectron spectroscopy and density functional theory. Phys Scr 98:055912

Sakir M, Salem S, Sanduvac ST et al (2020) Photocatalytic green fabrication of Au nanoparticles on ZnO nanorods modified membrane as flexible and photocatalytic active reusable SERS substrates. Colloids Surf A Physicochem Eng Asp 585:124088. https://doi.org/10.1016/j.colsurfa.2019.124088

Sakka S (2022) History of ferroelectric materials prepared by sol-gel method. J Sol-Gel Sci Technol:140–175. https://doi.org/10.1007/s10971-021-05712-w

Salam MA, AbuKhadra MR, Mohamed AS (2020) Effective oxidation of methyl parathion pesticide in water over recycled glass based-MCM-41 decorated by green Co3O4 nanoparticles. Environ Pollut 259. https://doi.org/10.1016/j.envpol.2019.113874

Samriti, Tyagi R, Ruzimuradov O, Prakash J (2023) Fabrication methods and mechanisms for designing highly-efficient photocatalysts for energy and environmental applications. Mater Chem Phys 307:128108. https://doi.org/10.1016/j.matchemphys.2023.128108

Sang Y, Cao X, Dai G et al (2020) Facile one-pot synthesis of novel hierarchical Bi2O3/Bi2S3 nano flower photocatalyst with intrinsic p-n junction for efficient photocatalytic removals of RhB and Cr (VI). J Hazard Mater 381:120942. https://doi.org/10.1016/j.jhazmat.2019.120942

Sangeetha G, Rajeshwari S, Venckatesh R (2011) Green synthesis of zinc oxide nanoparticles by aloe barbadensis miller leaf extract: structure and optical properties. Mater Res Bull 46:2560–2566. https://doi.org/10.1016/j.materresbull.2011.07.046

Sdobnov A, Piavchenko G, Bykov A, Meglinski I (2024) Advances in dynamic light scattering imaging of blood flow. Laser Photonics Rev 2300494:1–21. https://doi.org/10.1002/lpor.202300494

Seçme A, Bozer BM, Yıldırım Kocaman A et al (2024) Synthesis, characterization, and anticancer properties of Ag nanoparticles derived from walnut leaves tested on cells of L929, MCF-7 and H1299. J Drug Deliv Sci Technol:105478. https://doi.org/10.1016/j.jddst.2024.105478

Selvarajan E, Mohanasrinivasan V (2013) Biosynthesis and characterization of ZnO nanoparticles using Lactobacillus plantarum VITES07. Mater Lett 112:180–182. https://doi.org/10.1016/j.matlet.2013.09.020

Senasu T, Chankhanittha T, Hemavibool K, Nanan S (2022) Solvothermal synthesis of BiOBr photocatalyst with an assistant of PVP for visible-light-driven photocatalytic degradation of fluoroquinolone antibiotics. Catal Today 384–386:209–227. https://doi.org/10.1016/j.cattod.2021.04.008

Seto K (2023) Theoretical description for nonlinear dynamic light scattering based on stimulated Raman effect (NLDLS-SRS). J Raman Spectrosc:719–736. https://doi.org/10.1002/jrs.6534

Shamaila S, Sajjad AKL, Ryma N et al (2016) Advancements in nanoparticle fabrication by hazard free eco-friendly green routes. Appl Mater Today 5:150–199

Shankar PD, Shobana S, Karuppusamy I et al (2016) A review on the biosynthesis of metallic nanoparticles (gold and silver) using bio-components of microalgae: formation mechanism and applications. Enzym Microb Technol 95:28–44

Shanmugaratnam S, Ravirajan P, Yohi S, Velauthapillai D (2024) Well-separated photoinduced charge carriers on hydrogen production using NiS2/TiO2 nanocomposites. ACS Omega. https://doi.org/10.1021/acsomega.3c08194

Sheibani ASMESS (2019) Effect of intermediate ball milling on the synthesis of Cu-doped TiO2 nano-photocatalyst by sol–gel method. J Sol-Gel Sci Technol:173–185. https://doi.org/10.1007/s10971-019-05045-9

Shi Y, Wang K, Zhang S et al (2021) Characterizing the conformational distribution in an amorphous film of an organic emitter and its application in a "self-doping". Organ Light-Emitting Diode Res 0395:25878–25883. https://doi.org/10.1002/anie.202108943

Shinoda T, Yamaguchi Y, Kudo A, Murakami N (2022) Photoacoustic spectroscopic analysis of electron-trapping sites in titanium(IV) oxide photocatalyst powder treated by ball milling. J Phys Chem C 126:20975–20982. https://doi.org/10.1021/acs.jpcc.2c07064

Singaravelu G, Arockiamary JS, Kumar VG, Govindaraju K (2007) A novel extracellular synthesis of monodisperse gold nanoparticles using marine alga, Sargassum wightii Greville. Colloids Surf B Biointerfaces 57:97–101. https://doi.org/10.1016/j.colsurfb.2007.01.010

Singh J, Dutta T, Kim KH et al (2018) "Green" synthesis of metals and their oxide nanoparticles: applications for environmental remediation. J Nanobiotechnol 16:84

Singh J, Kumar V, Kim KH, Rawat M (2019) Biogenic synthesis of copper oxide nanoparticles using plant extract and its prodigious potential for photocatalytic degradation of dyes. Environ Res 177. https://doi.org/10.1016/j.envres.2019.108569

Somanathan T, Abilarasu A, Jermy BR et al (2019) Microwave assisted green synthesis Ce0.2Ni0.8Fe2O4 nanoflakes using calotropis gigantean plant extract and its photocatalytic activity. Ceram Int 45:18091–18098. https://doi.org/10.1016/j.ceramint.2019.06.031

Song S, Xing Z, Zhao H et al (2023) Recent advances in bismuth-based photocatalysts: environment and energy applications. Green Energy Environ. ScienceDirect 8:1232–1264. https://doi.org/10.1016/j.gee.2022.04.004

Srinivasan R, Fasmin F (2021) An introduction to electrochemical impedance spectroscopy. CRC Press

Srivastava N, Mukhopadhyay M (2014) Biosynthesis of SnO2 nanoparticles using bacterium erwinia herbicola and their photocatalytic activity for degradation of dyes. Ind Eng Chem Res 53:13971–13979. https://doi.org/10.1021/ie5020052

Suresh D, Nethravathi PC, Udayabhanu et al (2015) Green synthesis of multifunctional zinc oxide (ZnO) nanoparticles using Cassia fistula plant extract and their photodegradative, antioxidant and antibacterial activities. Mater Sci Semicond Process 31:446–454. https://doi.org/10.1016/j.mssp.2014.12.023

Swain S, Basu S (2021) Visible light-driven photocatalytic reduction of chromium (VI) using TiO2-modified carbon sphere hybrid photocatalyst. Energy Ecol Environ 6:26–34. https://doi.org/10.1007/s40974-020-00157-3

Tănase MA, Marinescu M, Oancea P et al (2021) Antibacterial and photocatalytic properties of ZnO nanoparticles obtained from chemical versus Saponaria officinalis extract-mediated synthesis. Molecules 26. https://doi.org/10.3390/molecules26072072

Tang Z, Yang J, Zhang W (2022) The effects of calcination on sol–gel synthesized Gd2Ti2O7 photocatalyst for degradation of Reactive Brilliant Red X-3B. React Kinet Mech Catal 135:2867–2877. https://doi.org/10.1007/s11144-022-02274-y

Tao CA, Wang JF (2021) Synthesis of metal organic frameworks by ball-milling. Crystals 11:1–20. https://doi.org/10.3390/cryst11010015

Tejashwini DM, Harini HV, Nagaswarupa HP et al (2024) Nanoferrites in photocatalytic wastewater treatment: advancements, characterization, and environmental implications. Results Chem 7. https://doi.org/10.1016/j.rechem.2023.101247

Toledo AC, Assunção DPSF, Boscardin PMD, Paula JFP (2020) Biosynthesis and characterization of silver nanoparticles produced with aqueous extract of Pterodon emarginatus Vogel - Fabaceae seeds associated with gentamicin sulfate and hyaluronic acid with potential antimicrobial activity Brazilian Journal of Development 6(12):100655–100677. 10.34117/bjd 10.34117/bjdv6n12 10.34117/bjdv6n12-526

Usman UL, Allam BK, Sajad S et al (2024) Plant leaves extract assisted eco-friendly fabrication of ZnO-SnO2@Chitosan for UV-induced enhanced photodegradation of single and ternary mixtures of Rhodamine B. Nano-Struct Nano-Objects 37. https://doi.org/10.1016/j.nanoso.2023.101072

Vadlapudi V, Kaladhar DSVGK, Behara M et al (2013) Review: green synthesis of silver and gold nanoparticles. Int J Chem Stud 1:22–31

Varnagiris S, Urbonavicius M, Sakalauskaite S et al (2021) Photocatalytic inactivation of Salmonella typhimurium by floating carbon-doped TiO2 photocatalyst. Materials (Basel) 14:5681

Vijayaraghavan K, Ashokkumar T (2017) Plant-mediated biosynthesis of metallic nanoparticles: a review of literature, factors affecting synthesis, characterization techniques and applications. J Environ Chem Eng 5:4866–4883

Vinodhkumar G, Wilson J, Inbanathan SSR et al (2020) Solvothermal synthesis of magnetically separable reduced graphene oxide/Fe3O4 hybrid nanocomposites with enhanced photocatalytic properties. Phys B Condens Matter 580:411752. https://doi.org/10.1016/j.physb.2019.411752

Wang Y, Wang H, Yang Y, Xin B (2021) Magnetic NiFe2O4 3D nanosphere photocatalyst: glycerol-assisted microwave solvothermal synthesis and photocatalytic activity under microwave electrodeless discharge lamp. Ceram Int 47:14594–14602. https://doi.org/10.1016/j.ceramint.2021.02.041

Wiranwetchayan O, Promnopat S, Thongtem T et al (2020) Effect of polymeric precursors on the properties of TiO2 films prepared by sol-gel method. Mater Chem Phys 240:122219. https://doi.org/10.1016/j.matchemphys.2019.122219

Xie X, Li S, Qi K, Wang Z (2021) Photoinduced synthesis of green photocatalyst Fe3O4/BiOBr/CQDs derived from corncob biomass for carbamazepine degradation: the role of selectively more CQDs decoration and Z-scheme structure. Chem Eng J 420. https://doi.org/10.1016/j.cej.2021.129705

Yakoup AY, Kamel AG, Elbermawy Y et al (2024) Characterization, antibacterial, and cytotoxic activities of silver nanoparticles using the whole biofilm layer as a macromolecule in biosynthesis. Sci Rep 14:1–17. https://doi.org/10.1038/s41598-023-50548-9

Yashni G, Al-Gheethi A, Radin Mohamed RMS et al (2021) Bio-inspired ZnO NPs synthesized from Citrus sinensis peels extract for Congo red removal from textile wastewater via photocatalysis: optimization, mechanisms, techno-economic analysis. Chemosphere 281. https://doi.org/10.1016/j.chemosphere.2021.130661

Yedurkar S, Maurya C, Mahanwar P (2016) Biosynthesis of zinc oxide nanoparticles using ixora coccinea leaf extract—a green approach. Open J Synth Theory Appl 05:1–14. https://doi.org/10.4236/ojsta.2016.51001

Yergaziyeva G, Kuspanov Z, Mambetova M et al (2024) Advancements in catalytic, photocatalytic, and electrocatalytic CO2 conversion processes: current trends and future outlook. J CO2 Util 80:102682. https://doi.org/10.1016/j.jcou.2024.102682

Yeshchenko O, Kutsevol N, Virych P et al (2024) Anti-cancer activity of zinc-tetraphenylporphyrin photosensitizer/dextran-graft-polyacrylamide copolymer/Au(Ag) nanoparticle nanohybrids. RSC Adv 14:5045–5054. https://doi.org/10.1039/d3ra07825f

Yousefi SR, Sobhani A, Alshamsi HA, Salavati-Niasari M (2021) Green sonochemical synthesis of BaDy2NiO5/Dy2O3and BaDy2NiO5/NiO nanocomposites in the presence of core almond as a capping agent and their application as photocatalysts for the removal of organic dyes in water. RSC Adv 11:11500–11512. https://doi.org/10.1039/d0ra10288a

Yu HL, Wu QX, Wang J et al (2020) Simple fabrication of the Ag-Ag2O-TiO2 photocatalyst thin films on polyester fabrics by magnetron sputtering and its photocatalytic activity. Appl Surf Sci 503. https://doi.org/10.1016/j.apsusc.2019.144075

Yue C, Zhu L, Qiu Y et al (2023) Recent advances of plasmonic elemental Bi based photocatalysts in environmental remediation and energy conversion. J Clean Prod 392:136017. https://doi.org/10.1016/j.jclepro.2023.136017

Zhang Y, Sun T, Zhang D et al (2023) Ball-milling preparation of La3+/TiO2 photocatalyst and application in photodegradation of PVC plastics. Coatings 13:1–12. https://doi.org/10.3390/coatings13020317

Zhao C, Pan X, Wang Z, Wang CC (2021) 1 + 1 > 2: a critical review of MOF/bismuth-based semiconductor composites for boosted photocatalysis. Chem Eng J 417:128022. https://doi.org/10.1016/j.cej.2020.128022

Zheng B, Kong T, Jing X et al (2013) Plant-mediated synthesis of platinum nanoparticles and its bioreductive mechanism. J Colloid Interface Sci 396:138–145. https://doi.org/10.1016/j.jcis.2013.01.021

Zuliani A, Cova CM (2021) Green synthesis of heterogeneous visible-light-active photocatalysts: recent advances. Photochem 1:147–166. https://doi.org/10.3390/photochem1020009

Part II
Waste Derived Catalysts for Energy Production

Chapter 3
Earth-Abundant Metal-Based Nanostructured Photocatalysts

Pham Thi Le Na, Minh-Kha Nguyen, Duyen Phuc-Hanh Tran, and Bao-Trong Dang

Abstract Providing functional photocatalysts (PCs) is truly necessary for converting solar energy into sustainable energy sources. The use of noble metals has been a cornerstone in efforts to develop processes due to their renowned capability to create catalysts that are highly active, selective, and durable. However, mounting concerns about their cost and scarcity have led to research efforts toward developing noble metal-free photocatalysts. The development of viable substitute PCs using components that are widely accessible on Earth is, therefore, the subject of increased research and development. There have been far more reports of Earth-abundant PCs lately. Abundant transition metals, such as specific complexes based on nickel (II), iron (II), cobalt (III), copper (I), and zinc (II), have been reported in notable ways. This chapter discusses the traditional and environmentally friendly synthesis of photocatalysts using Earth-abundant metal-based nanostructured photocatalysts, emphasizing the significant advantages in the production of renewable energy, water purification, and improved air quality. The rest of this section presents a summary of the development of nanostructured photocatalysts, serving as a useful reference for researchers and experts in chemistry, materials science, and renewable energy.

Keywords Earth-abundant metal complexes · Nanostructured photocatalysts · Doping · Hydrogen production · CO_2 reduction

P. T. Le Na · M.-K. Nguyen · B.-T. Dang (✉)
Faculty of Chemical Engineering, Ho Chi Minh City University of Technology (HCMUT), Ho Chi Minh City, Vietnam

Vietnam National University Ho Chi Minh City, Linh Trung Ward, Thu Duc City, Ho Chi Minh City, Vietnam
e-mail: dbaotrong@hcmut.edu.vn

D. P.-H. Tran
Department of Civil Engineering, Chung Yuan Christian University, Taoyuan, Taiwan, ROC

C. M. Hussain, G. Hussain (eds.), *Green Nanostructured Photocatalysts for Clean Energy*, Clean Energy Production Technologies,
https://doi.org/10.1007/978-981-95-1618-6_3

3.1 Introduction

Renewable energy resources are under investigation in response to the growing energy demand and environmental issues surrounding fossil fuels. Potential replacements include solar, geothermal, wind, hydropower, and tidal resources. Among these, solar power has widespread support because it is sustainable and cost-effective. Standard methods for harnessing solar energy include photovoltaic cells and photocatalytic reactions (Kahng et al. 2020). The first one may instantly utilize solar radiation to generate power and store it in a secondary device like a battery or super capacitor (Ran et al. 2014). The alternative is photocatalysis, which involves using light to speed up chemical reactions and storing energy in chemical bonds. However, most photocatalytic systems still require rare and costly noble metal photocatalysts to achieve adequate activity, severely limiting their potential for large-scale applications. As a result, the quest to develop photocatalysts that are cost-effective and efficient continues to be a top priority.

Scientists are exploring various strategies to produce nanostructured photocatalysts from free noble metals. Although several wide-bandgap semiconductors, including TiO_2, have shown benchmark results in terms of photocatalytic efficiency, a significant bottleneck still exists in their limited capacity to absorb light of visible region (Singh et al. 2023b). Semiconductors with a narrow bandgap have played a vital role in the effort to capture the full range of visible light (account for 43% of solar energy) with bandgap energies below 3 eV (Chen et al. 2023; Galeas et al. 2023). Earth-abundant metals such as nickel (II), zinc (II), copper (I), iron (II), and cobalt (III) with low bandgaps and affordable prices have been the centerpiece of research into new photocatalysts. The annual growth rate of papers in the Scopus database related to Earth-abundant metals has been approximately 23.11% since 2003 (Fig. 3.1). Bulk forms of these metals have no photocatalytic activity. However, because of their relatively small bandgap, many semiconductors exhibit much more active forms when present as nanoparticles (Table 3.1).

This improvement can be attributed to the intrinsic electrocatalytic activities of Earth-abundant metals being amplified at the nanoscale. Metal oxides and multi-component oxides (binary, ternary quaternary) are among the several varieties of semiconductor photocatalysts produced (Berede et al. 2024; Meky et al. 2024). Incorporating Earth-abundant metals into nanostructures has led to faster charge transfer due to increased light absorption thanks to localized surface plasmon resonance (LSPR) effects and effective charge separation through the formation of Schottky junctions with semiconductors (Ameta and Ameta 2016). Furthermore, these type of metals are at least three orders of magnitude cheaper than Pt, which significantly enhances their practicality in real-world applications. This chapter delves into the latest developments in Earth-abundant photocatalyst materials, focusing on how these materials excel at small scales. Our discussion covers the significant advantages in the production of renewable energy, water purification, and improved air quality.

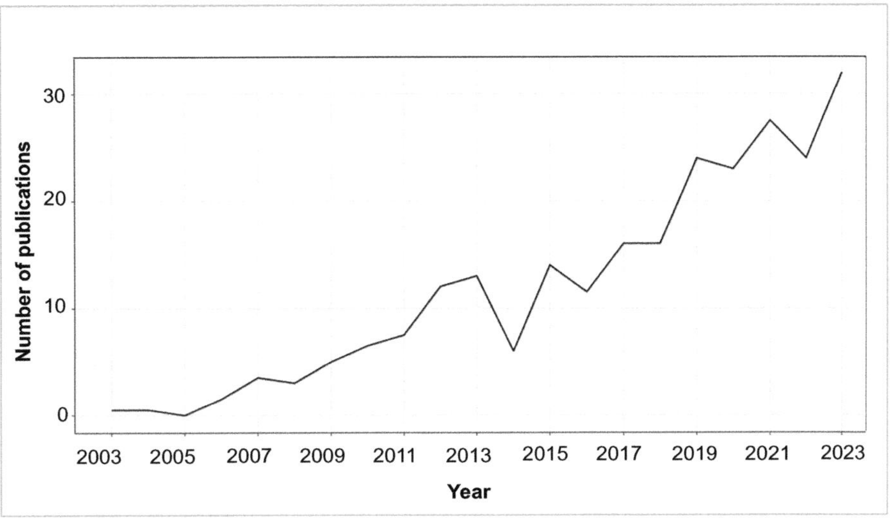

Fig. 3.1 Research trends of Earth-abundant metals related to photocatalysis. Data from research articles (Scopus database, 2003–2023) and analyzed by the author

Table 3.1 The selected bandgaps of different semiconductors derived from Earth-abundant metal

Photocatalysts	Bandgap (eV)	Category
CuO	1.9	Binary
Cu_2O	1.9	Binary
ZnO	3.2	Binary
ZnS	3.7	Binary
ZnSe	2.7	Binary
ZnTe	2.25	Binary
$CuInSe_2$	1.00	Ternary
Cu_2SnS_3	1.16	Ternary
$ZnFe_2O_4$	1.9	Ternary
Cu_2ZnSnS_4	1.5	Quaternary
$Cu_2ZnSnSe_4$	1.00	Quaternary
$FeZn_2Cu_3O_{6.5}$	2.70	Quaternary

3.2 Preparation Methods of Earth-Abundant Metal-Based Photocatalysts

In *Engines of Creation* (1986), Eric Drexler coined the term "nanotechnology" to describe manipulating individual atoms and molecules to build structure specifications. Nanomaterials can be synthesized using two primary strategies: the top-down approach and the bottom-up approach (Fig. 3.2a). Nanomaterial products can be categorized further according to their spatial quantization into several forms:

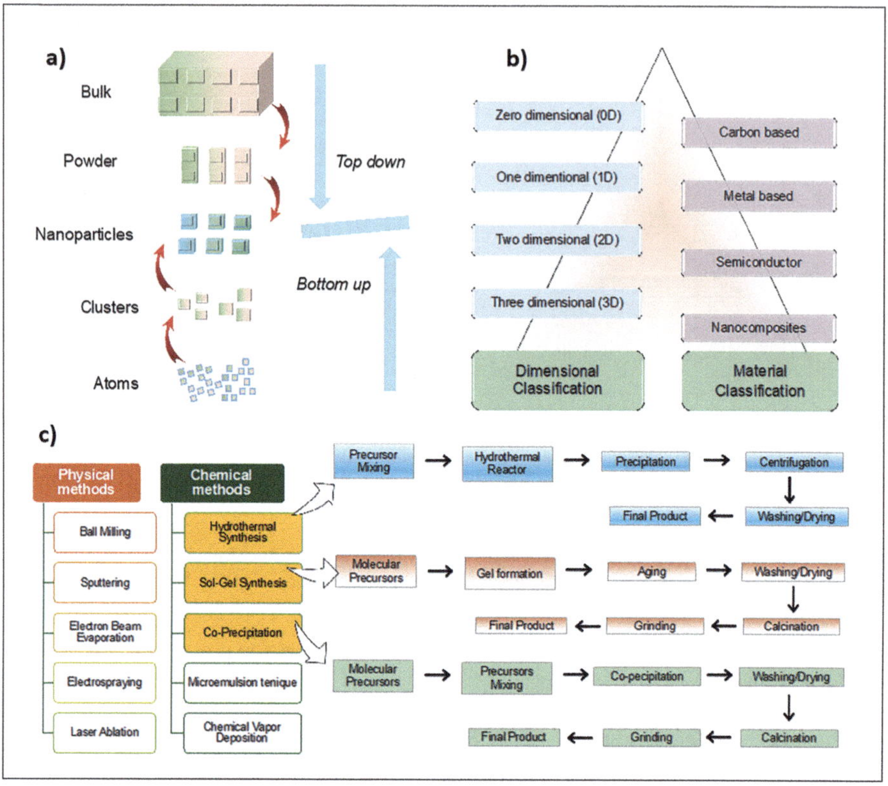

Fig. 3.2 Common techniques for synthesizing and categorizing nanomaterials

nanoparticles (0D); nanorods, nanowires, and nanobelts (1D); atomically thin layered materials (2D); and three-dimensional structures (3D) (Fig. 3.2b). As depicted in Fig. 3.2a, the "top-down" approach involves restructuring bulk materials to nanomaterials such as thermal decomposition, mechanical milling, electro-explosion, laser ablation, and nanolithography. The bottom-up method entails constructing nanomaterials by beginning with fundamental components like atoms or molecules (Abid et al. 2022). The nanoparticles (NPs) can be synthesized by different techniques such as sol-gel process, chemical vapor deposition (CVD), hydrothermal methods, spinning, and soft and hard templating methods. In most cases, the top-down approach is less suitable for creating particles with nonconventional shapes and tiny sizes. Meanwhile, the bottom-up approach such as sol-gel, hydrothermal methods, or co-precipitation are more suitable for nanoparticle synthesis (Fig. 3.2c). They enable the production of particles with particular sizes, dimensions, compositions, and architectures for many purposes (Parashar et al. 2020).

Table 3.2 highlights the study of nanostructured photocatalysts, and the data is derived chiefly from scholarly studies conducted between 2019 and 2024. This focuses on synthesis processes of Earth-abundant metals, primarily employing

Table 3.2 An overview of recent studies on synthesizing nanostructured photocatalysts using metals that are abundant on Earth

No.	Photocatalysts	Synthesis	Enhancement mechanism	Application	References
Sol-gel/green sol-gel synthesis					
1	$C_3N_4/CuWO_4$	Sol-gel	Generate $O_2^{\bullet-}$ or hydroxyl radicals •OH responsible for the inactivation of bacteria	Antibacterial agent	Gupta et al. (2019)
2	Ni/S-TiO_2	Sol-gel technique	Morphology control Generate h^+ and $O_2^{\bullet-}$	Degradation of pollutants (85% MB, 180 mins)	Kunnamareddy et al. (2021)
3	Tb_2ZnMnO_6	Green sol-gel Auto-combustion	Generate superoxide ($O_2^{\bullet-}$) or hydroxyl radicals (•OH)	Degradation of pollutants Methyl orange 35% Methyl violet 55% at neutral pH methyl violet 90% at pH 8	Dara et al. (2021)
4	NiO (*Catharanthus roseus* leaf)	Green sol-gel	Generate superoxide ($O_2^{\bullet-}$) or hydroxyl radicals (•OH)	Degradation of pollutants (Bromothymol blue, 2.76%, 90 mins). Antibacterial activity	Ahmad et al. (2023)
5	ZnO (*Salvia officinalis* and *Abelmoschus esculentus* plants)	Green sol-gel	The electrons transfer from the nanoparticle to oxygen species	Degradation of MB *A. esculentus*, 41.0%, 90 min *S. officinalis*, 86.9%, nanoparticles	Aalami et al. (2024)
6	Hydroxyapatite (using lemon extract)	Green sol-gel	Not specified	Examination of nanoparticle-induced cytotoxicity in human cancer cell lines (T98 and SHSY5)	Baladi et al. (2023a)
7	$ErFeO_3/Fe_3O_4$/rGO nanocomposite	Sol-gel Auto-combustion	Not specified	Cytotoxic activity in human cancer cells (T98 and SHSY5)	Baladi et al. (2023b)

(continued)

Table 3.2 (continued)

No.	Photocatalysts	Synthesis	Enhancement mechanism	Application	References
Hydrothermal/green hydrothermal synthesis					
8	Cu_2S/CD	Hydrothermal	Heterojunction Improve the charge separation	Hydrogen evolution (9.00 mmol·g^{-1}·h^{-1})	Li et al. (2020)
9	CuO-Cu_2O/GO	Hydrothermal	High photogenerated electron-holes transport properties	Degradation of 90% MO, 95% Tetracycline 120 mins)	Zhang et al. (2020c)
10	CuO hierarchical nanostructures	Hydrothermal	Morphology control Small size and large surface area	Degradation of pollutants (83% MB, 150 mins)	Bustos et al. (2021)
11	CuS/GO	Hydrothermal	Heterojunction improve the charge separation	Antibacterial agent (46% *S. aureus*, 56.43% *E. Coli*)	Rui et al. (2022)
12	α-Ferrous oxalate dihydrate (α-FOD)	Green hydrothermal	narrow optical bandgap (2.31–2.49 eV). improve mesoporous structure	Degradation of pollutants (92% Phenol 6 h)	Galeas et al. (2023)
13	$Ni_2(CO_3)(OH)_2·4H_2O$	Hydrothermal	Morphology control	Hydrogen production (10 μmol g^{-1}·h^{-1})	Talebi et al. (2024)
Co-precipitation synthesis					
14	Cu-doped $ZnAl_2O_4$	Co-precipitation method	The increase in surface area increases the photocatalytic activity	75% Cr(VI) reduction, 150 mins by $Zn_{0.2}Cu_{0.8}Al_2O_4$	Akika et al. (2020)
15	$NiAl_2O_4$	Co-precipitation	High surface area Generate superoxide ($O_2^{\bullet -}$) or hydroxyl radicals (•OH)	Degradation of MB 47.1% in 60 min	Chellammal Gayathri et al. (2021)

(continued)

Table 3.2 (continued)

No.	Photocatalysts	Synthesis	Enhancement mechanism	Application	References
16	$CuS/NiMn_2O_4$	Two-step co-precipitation	Effective charge separation in photoinduced carriers and the proficient harnessing of visible-light absorption	Degradation of 84% and 60% for MB and RhB	Rohokale et al. (2021)
17	$Cu_3(PO_4)_2/MgO$ Nanocomposite	Co-precipitation	Optimized electron-hole pair separation and the formation of highly reactive radicals like •OH and $O2^{\bullet -}$	Degradation of pollutants (99% amaranth) Antibacterial activity	Subalakshmi et al. (2024)
18	$Zn_{0.5}Cu_{0.5}Fe_2O_4$	Co-precipitation	Hydroxyl radicals and the holes' oxidative potential	Degradation of MB 94%, 135 min	Abuzeyad et al. (2024a)
19	$Zn_{0.5}Cu_{0.5}Fe_2O_4$	Co-precipitation	Hydroxyl radicals and the holes' oxidative potential	Degradation of methyl orange (MO) azo dye, 96%, 135 min	Abuzeyad et al. (2024b)
20	CuO/ MgAl-LDHs (*Allium cepa* L. peel extract)	Co-precipitation method	The lower electron and hole recombination rates. Generate the •OH, $O_2^{\bullet -}$, h^+, e^- reactive species	Degradation of MB 99.2%, 80 min	Berede et al. (2024)

hydrothermal, sol-gel procedures, and co-precipitation, with some incorporating eco-friendly approaches (green synthesis). Especially, green sol-gel processes have been discovered recently. This approach uses plant extracts containing organic components to synthesize nanoparticles and control their shape. Natural extracts can enhance nanoparticles by providing extra features like biological compatibility or additional reactive sites for catalysis. These materials have applications in treating pollutants and exhibiting antibacterial properties (Baladi et al. 2023a; Ahmad et al. 2023; Aalami et al. 2024). Enhancement strategies frequently focus on charge separation to boost the effectiveness of the photocatalytic process (Li et al. 2020; Rohokale et al. 2021; Xu et al. 2021a, b, Rui et al. 2022; Ücker et al. 2023).

3.3 Photocatalyst Materials Based on Earth-Abundant Metals

3.3.1 Zinc-Based Metal Nanoparticle Photocatalysts

Zinc oxide (ZnO) stands out as a highly promising semiconductor material in photocatalysis (Table 3.3), owing to its distinct properties such as a direct and relatively wide bandgap of 3.37 eV, beneficial band-edge positions for generating oxygen-containing radicals in aqueous systems, high photosensitivity, and broad absorption in the UV/solar spectrum. It possesses a significant exciton binding energy of 60 meV, exhibits excellent electron mobility, and allows for versatile chemical modifications. Additionally, ZnO can be synthesized through cost-effective and scalable methods into diverse morphologies, while maintaining transparency and biocompatibility (Kumar et al. 2020). The photocatalytic efficiency of ZnO is largely determined by its capacity to absorb light, which causes electrons (e^-) to be excited from the valence band (VB) to the conduction band (CB), creating holes (h^+) in the VB that trigger photoredox reactions. However, due to its large bandgap, ZnO mainly responds to UV radiation (~4% of the solar spectrum) and remains inactive under visible light (~43% of the solar spectrum) (Qi et al. 2017). Moreover, the fast recombination of photogenerated e^-/h^+ pairs further restricts its efficiency as a photocatalyst. To address these challenges, ZnO has undergone numerous modifications to enhance its activity under visible light and reduce carrier recombination. These include doping with metals or nonmetals, noble metal deposition, formation of heterojunctions, and integration with carbon-based materials.

The incorporation of metal atoms, including alkali, rare earth, transition, and noble metals, into the ZnO crystal lattice has been widely studied. Introducing metals, particularly transition metals, into ZnO may enhance its photocatalytic properties. Transition metals that have atomic radii similar to that of Zn can be readily integrated into the ZnO lattice. Numerous transition metals, such as Fe (Senthil Kumar and Arunagiri 2021), Co (Abubshait et al. 2023; Heidari-Asil et al. 2022), Ni (Joulaee et al. 2023; Moradi et al. 2023; Sousani et al. 2024), Mn (Dhivya et al. 2019), Cr (Mathai et al. 2023), Ag (Mahrsi et al. 2023), and Cu (Singh et al. 2023a), have been doped to ZnO to narrow the bandgap of this compound from 3.27 eV to approximately 2.2 eV. This bandgap tuning has led to a marked enhancement in the visible-light photocatalytic activity of zinc-based materials. In addition, ZnO nanoparticles, along with engineered metal-based nanoparticles, have shown a spectrum of antibacterial properties, ranging from moderate to excellent, effective against both Gram-negative and Gram-positive bacterial strains (Heinlaan et al. 2008; Jones et al. 2007).

The incorporation of nonmetal dopants into ZnO is particularly effective due to their ability to readily replace oxygen in the lattice or occupy interstitial sites, thanks to their higher solubility in comparison to metallic dopants. When these nonmetal elements take the place of oxygen sites, they disrupt the valence band (VB) and lead to reduced scattering of electrons in the conduction band (CB) (Kumar et al. 2020).

Table 3.3 Several Earth-abundant metal materials for the synthesis of photocatalytic nanostructure

No.	Nanostructured material	Bandgap (eV)	Synthesis method	Key application	Implication	References
	Zinc- and zinc oxide-based nanoparticle photocatalysts					
1	Ce/Zinc oxide nanopetals	2.5–2.8	Green sythesis *Withania coagulans* plant, $Ce(NO_3)_3$ and $Zn(NO_3)_2$	Dye degradation	Not specified	Hassan et al. (2024)
2	ZnO/Carbon Nanotube (CNT) nanocomposites	Not specified	Hydrothermal method	Enhanced photocatalytic activity, removal of dye	Merges ZnO's photocatalytic and CNT's conductive properties	Aljeboree et al. (2024)
3	Cu-doped ZnO nanoparticles	2.85–2.95	Chemical precipitation	Dye degradation, water treatment, bacterial-resistant textile coatings and biomedical applications	Photocatalytic activity depends on Cu-doping concentration, with 5% Cu-doped ZnO exhibiting the highest level of activity inhibition	Singh et al. (2023a)
4	Cr-doped ZnO nanoparticles	3.147–3.17	Hydrothermal	Degradation of organic dye	Degradation of MB dye under natural sunlight	Mathai et al. (2023)
5	ZnO-CdS nanocomposites	ZnO, 3.13; CdS, 2.29	Precipitation, hydrothermal method	Photocatalytic hydrogen production	Utilizes ZnO and CdS's complementary properties	Revathi et al. (2021)
6	Pomegranate peel biochar impregnated with ZnO nano-flowers	2.8	Precipitation	Removal of dye	Not specified	Kaur and Kaur (2022)
7	Au@ZnONPs-MoS_2-rGO nanocomposites	~2.15	Hydrothermal method, mechanical mixing	Photodegradation, hydrogen evolution	Integrates MoS_2 to boost charge separation	Machín et al. (2023)

(continued)

Table 3.3 (continued)

No.	Nanostructured material	Bandgap (eV)	Synthesis method	Key application	Implication	References
	Zirconium-based nanostructure photocatalysts					
8	Zirconium oxide (ZrO_2) nanoparticles	~5.0	Green synthesis	Dye degradation biomedical research	Emphasizes comprehensive toxicity evaluations for safe use	Atalay et al. (2018)
9	Fe-doped zirconium oxide nanoparticles	4.44, 4.29, 4.05 and 3.88 eV for 2 wt%–10 wt% Fe-doped ZrO_2	Co-precipitation	Dye degradation, water purification	Showcases metal-doped ZrO2's potential in environmental cleanup	Hui and Zhang (2021)
10	Strontium-doped zirconium oxide nanoparticles	4.6–4.25	Co-precipitation	Antimicrobial applications, dye degradation	Suggests doped ZrO2 as effective for water treatment and antimicrobial coatings	Jeba et al. (2022a)
11	g-C_3N_4/ZrO_2 composites	ZrO_2, 4.90 eV; g-C_3N_4, 2.73 eV	Ultrasonication and hydrothermal	Organic pollutant degradation	Demonstrates the synergy in composite materials for photocatalysis	Chand and Mondal (2023)
12	Silver-doped zirconium titan	Not specified	Green solution combustion	Photocatalytic degradation, antimicrobial applications	Introduces ag for enhanced photocatalytic activity and antimicrobial efficacy	Akshay et al. (2023)
13	Silver-doped zirconium	ZrO_2, 6 and 6 mol%; Ag doped ZrO_2, 5.6	Single-step solution combustion method	Photocatalytic degradation of dye	Not specified	Dawoud et al. (2020)
	Earth-abundant metal sulfide nanoparticle photocatalyst					

14	Copper sulfide nanoparticles (CuS)	1.98–1.99	Chemical co-precipitation	Photocatalysis, antibacterial films	Demonstrates enhanced photocatalytic and antibacterial performances due to small nanoflower sizes	Jung et al. (2024)
15	Copper-indium sulfide quantum dots	1.5	Hydrothermal, solvothermal, and thermal decomposition	Photocatalysis	Provides quantum confinement effects for improved photocatalytic performance under visible light	Zhang et al. (2023b)
16	Cocatalyst-free nickel sulfide nanostructure	1.83–1.53	Solvothermal method	Photocatalytic degradation of pollutants	Offers efficient photocatalysis without the need for additional cocatalysts	Muniyappa et al. (2022)
17	Zn-in spinel-type sulfides	2.28	Hydrothermal	Photocatalysis	Utilizes unique spinel structures for enhanced photocatalytic degradation	Jia et al. (2022)
	CuS/PbS/ZnO heterojunction photocatalyst	8.84	Hydrothermal, precipitation	Hydrogen production via the photocatalytic dissociation of water molecules	Photocatalysts are applied to organic substrates, including compounds featuring five-membered ring structures	Chiu et al. (2022)

Among nonmetals like fluorine, sulfur, and phosphorus, nitrogen stands out as the favored dopant because of its size similarity to oxygen and the relatively low energy needed for substitution. Additionally, nitrogen's lower electronegativity and ionization energy compared to oxygen decrease the electron density around the nitrogen atoms incorporated into the original oxide (Rajbongshi et al. 2014). Research indicates that nitrogen introduces distinct 2p electronic states just above the maximum of the valence band, thereby enhancing absorption of visible light. Importantly, nitrogen doping has been shown to enhance ZnO's photocatalytic efficiency without disrupting its tetrapod crystal morphology. The development of this hierarchical structure is also significantly affected by the conditions during thermal annealing (Feng et al. 2017).

Various methods have been employed for the synthesis of ZnO nanoparticles, with physical, chemical, and biosynthetic approaches being particularly prevalent. While conventional synthesis techniques tend to be costly and potentially harmful to the environment, green synthesis has emerged as a more economical and environmentally conscious alternative (Jadoun et al. 2021). Many researchers have increasingly focused on developing eco-friendly fabrication methods using principles of green chemistry, which rely on benign biological sources such as bacteria, fungi, and plants (Duan et al. 2015). A number of investigations have demonstrated successful ZnO nanoparticle synthesis utilizing plant-derived extracts, such as from *Allium cepa* L. peel waste (Islam et al. 2024) and *Eucalyptus globulus* leaf (Ansari et al. 2024), for dye degradation. The green synthesis of ZnO marigold-like nanostructures synthesized with *Eucalyptus globulus* extract displayed outstanding photocatalytic efficiency, achieving nearly 98% degradation of organic dyes under natural sunlight (Ansari et al. 2024). According to current studies, zinc-based photocatalysts perform well in applications such as breaking down organic contaminants, purifying water, and producing sustainable energy through water splitting and carbon dioxide reduction. However, challenges such as photocorrosion, recombination of charge carriers, and limited visible-light absorption hinder their practical application, highlighting the need for further research to overcome these limitations.

3.3.2 Zirconium-Based Metal Nanoparticle Photocatalysts

Zirconium is present in the Earth's crust at a quantity of approximately 130 milligrams per kilogram and about 0.026 micrograms per liter in seawater. It is the 18th most abundant element (Perks and Mudd 2019). Due to their advantageous structural and chemical features, zirconium-containing photocatalysts are being widely explored for applications in pollution control and renewable energy processes (Table 3.3). Zirconium dioxide (ZrO_2) is one of the most studied forms of zirconium for photocatalytic applications (Hassan and Jalil 2022; Melchor-lagar et al. 2020).

Zirconium dioxide exhibits excellent stability, chemical inertness, and its effectiveness as a photocatalyst when exposed to both ultraviolet and visible light.

Nonetheless, ZrO2 possesses a low surface area of 23 m2/g and is only effective under ultraviolet (UV) light with a wavelength shorter than 388 nm due to its large bandgap of approximately 5.0 eV. This characteristic limits its ability to utilize the solar energy available on Earth's surface, which is about 4%, thereby reducing its photocatalytic efficiency (Teeparthi et al. 2018). In recent years, nanostructured forms of zirconium dioxide, such as nanoparticles, nanotubes, and nanosheets, have been synthesized to enhance the surface area and facilitate charge carrier separation, leading to improved photocatalytic efficiency. A range of synthesis techniques such as sol-gel processing, hydrothermal treatment, and biological approaches can be employed to fabricate these nanostructures (Vinayagam et al. 2022). For instance, Zhang et al. successfully fabricated mesoporous ZrO_2 nanoparticles using the sol-gel process, revealing their efficiency in organic pollutant degradation under visible light (Zhang et al. 2018).

Researchers are increasingly focusing on the sustainable and cost-efficient production of metal oxide nanoparticles through the use of plant extracts. To summarize, the process involves mixing the plant extract with a Zr aqueous solution, stirring the mixture, and then using a microwave oven at high temperatures to evaporate the solution until a white powder is produced (Shinde et al. 2018). The plant extract functioned as both a reducing and stabilizing agent, effectively limiting nanoparticle agglomeration. The synthesis of ZrO_2 nanoparticles (NPs) using *Ficus benghalensis* leaf extract as a capping agent proved highly effective in the adsorption and breakdown of methylene blue (MB) and methyl orange (MO), achieving removal efficiencies of 91% and 69%, respectively. This approach aligns with the method described by Wu et al., where ZrO_2 NPs were produced using a green method with chitosan as a scaffold. Specifically, an aqueous chitosan solution in acetic acid was mixed into a zirconium solution with continuous stirring until precipitation occurred, followed by drying and calcination. The study found that ZrO_2 powders prepared with chitosan of low-molecular-weight chitosan with a larger surface area in ZrO_2 synthesis enhanced rhodamine B removal, reaching 86.2% within 2 h (Wu et al. 2016).

Two prevalent methods to enhance the spectral responsiveness of photocatalysts are through extra-modification and self-modification (Nasir 2020). Self-modification involves changes to the photocatalyst's own structure and surface properties (Hassan and Jalil 2022), whereas extra-modification entails the integration of additional materials into the photocatalyst (Li et al. 2016). Examination of existing studies reveals numerous research efforts focused on extra-modification to enhance catalytic efficiency. Various modification approaches, including doping, heterojunction formation, morphological tuning, and material coupling, have been demonstrated to effectively address the previously mentioned limitations (Aldeen et al. 2022). Specifically, the incorporation of dopants or cocatalysts including carbon, nitrogen, metals (Fe, Cr, Ag, Sr, Bi, etc.) can further enhance the photocatalytic activity of zirconium-based photocatalysts as the bandgap is narrowed (Alajmi et al. 2023; Hui and Zhang 2021; Li et al. 2023; Naraginti et al. 2015; Sharafinia et al. 2022; Wang et al. 2017a).

Applications of zirconium-based photocatalysts span various environmental remediation processes, including water purification, air purification, pollutant degradation, and hydrogen production. Several studies demonstrated the use of zirconium-based photocatalysts for the removal of organic dyes from wastewater, achieving high degradation efficiency under solar irradiation (Jeba et al. 2022a, b;Albornoz et al. 2021; Bakry et al. 2022; Huang et al. 2023; Hui and Zhang 2021; Sharafinia et al. 2022). Moreover, zirconium-based photocatalysts have shown promise in photocatalytic CO_2 reduction for solar fuel production (Samriti et al. 2023). However, additional studies are essential to refine synthesis approaches, improve material stability, and gain deeper insight into the fundamental photocatalytic mechanisms to enable real-world applications.

3.3.3 Earth-Abundant Metal Sulfides

Recent advancements in nanophotocatalysts have underscored the significance of earth-abundant metal sulfides, such as CuS, NiS, and Zn-In-S, due to their exceptional photocatalytic performance and cost-effectiveness. These materials offer promising applications in environmental pollutant degradation, hydrogen production, and antibacterial activities, addressing critical sustainability and public health challenges (Table 3.3). The synthesis of nanostructured metal sulfides plays a pivotal role in defining their photocatalytic efficiency. Techniques such as chemical co-precipitation, hydrothermal methods, and green synthesis using plant extracts have been explored to tailor nanoparticle morphologies for enhanced photocatalytic activities (Chiu et al. 2022; Kahng et al. 2020; Muniyappa et al. 2022; Zhang et al. 2023a). For example, Jung et al. reported a low-cost approach for synthesizing copper sulfide nanoparticles (CuS NPs) using a chemical co-precipitation technique, which yielded nanoparticles with significantly smaller nanoflower sizes, showcasing the importance of controlling synthesis conditions to control both the morphology and functional properties of these photocatalysts (Jung et al. 2024).

The photocatalytic mechanism of metal sulfide nanoparticles typically involves the absorption of light energy, generation of electron-hole pairs, and subsequent reactions that contribute to the breakdown of pollutants. The photocatalytic performance of metal sulfide nanoparticles can be improved through various strategies such as elemental doping, heterojunction formation, or surface engineering. These modifications enhance light harvesting, facilitate charge separation and migration, and promote the production of reactive oxygen species. However, the exact photocatalytic behavior may vary depending on the specific type of metal sulfide, its band structure, surface characteristics, and how it interacts with both the light source and target pollutants.

Metal sulfide nanostructured photocatalysts have found applications in dye degradation, water treatment, and hydrogen production from water splitting. For instance, the research by (Jung et al. 2024) on CuS NPs highlighted their superior photocatalytic performance in dye degradation and their potential as antibacterial agents. Similarly, studies on Zn-In spinel-type sulfides (Jia et al. 2022) and CuS/PbS/ZnO heterojunction photocatalysts have shown strong performance in both pollutant degradation and hydrogen generation via water splitting, highlighting their multifunctionality in environmental remediation and clean energy technologies (Chiu et al. 2022).

Despite the promising capabilities of earth-abundant metal sulfide nanostructured photocatalysts, several challenges remain, including the need for improved stability, recyclability, and scalability of these materials for broadening their practical applicability. Future research directions may prioritize the development of composite and hybrid structures, surface modification techniques, and the exploration of novel earth-abundant materials to address these challenges and improve overall photocatalytic efficiency.

3.3.4 Copper-Based Materials

Copper, an inexpensive transition metal with multiple oxidation states, exhibits limited photocatalytic activity in its pure form. However, its performance can be significantly improved through integration with other photocatalysts. In composite systems, metallic Cu enhances visible-light absorption, promotes charge separation through Schottky junctions with semiconductors, increases the number of active sites, and facilitates charge transfer. Many Cu-based materials also demonstrate considerable photocatalytic activity as single-phase systems. Nevertheless, their narrow bandgap often leads to rapid electron-hole recombination, while the negative conduction band of Cu_2O favors reduction reactions. To address these limitations, strategies such as morphology and size control, facet engineering, elemental doping, and heterojunction construction have been widely explored to enhance their photocatalytic efficiency.

Single-atom Cu catalysts offer an effective approach for generating reactive oxygen species. Cu-N_3 sites anchored on tubular graphitic carbon nitride (g-C_3N_4) facilitate singlet oxygen formation via enhanced charge transport, electron-hole dynamics, and exciton effects (Luo et al. 2021). Additionally, co-loading Pt-Cu atoms onto N-vacancy-rich crystalline g-C_3N_4 improves photocatalytic performance, where Pt acts as a hydrogen storage site and Cu enhances CO_2 adsorption, collectively contributing to prolonged charge separation (Cheng et al. 2022).

Nanoscale CuO and Cu_2O, with bandgaps ranging from 1.2 to 2.1 eV, exhibit strong visible-light absorption. Engineered CuO nanostructures, such as precipitation-derived nanorods with high surface exposure, demonstrate effective photocatalytic activity under indoor light (Kumar et al. 2019). Similarly, fragmented nanosheets enhance charge separation via quantum confinement (Nazim et al. 2021). Despite its broader bandgap, Cu_2O suffers from low efficiency and poor stability. To address this, heterojunctions like Cu_2O-Au and Cu_2O-TiO_2 have been designed to trap electrons and suppress charge recombination (Lei et al. 2022; Yang et al. 2022a).

CuS is a p-type semiconductor with a bandgap of 1.2–2.2 eV (Zhao et al. 2015), offering abundant electron states and vacancies favorable for photocatalysis. Various CuS-based architectures have been explored to enhance performance (Zhang et al. 2019). For example, a flower-like CuS/graphene oxide composite synthesized via hydrothermal method exhibited improved activity due to its high surface area and efficient charge separation (Lv et al. 2022). Similarly, CuS combined with g-C_3N_4 shows synergistic photothermal and photocatalytic effects. CuS/protonated g-C_3N_4 composites prepared through hydrothermal and electrostatic methods demonstrated strong visible-light activity (Ding et al. 2020). Moreover, ternary ZnO/ZnS/CuS heterojunctions outperformed binary systems, likely due to extended light absorption and enhanced charge carrier separation enabled by their stepwise band structure (Zeng et al. 2022). CuSe, another p-type semiconductor, possesses both direct (2.1–2.39 eV) and indirect (1.2–1.7 eV) bandgaps, enabling photocatalytic activity across UV to near-infrared regions with strong light-harvesting capability. In contrast, $CuWO_4$ is an n-type semiconductor with a bandgap of 2.2–2.4 eV, notable for its low toxicity and stability in neutral and acidic media (Karthika et al. 2019). However, its photocatalytic performance is hindered by low photocurrent and rapid electron-hole recombination. To address these limitations, various $CuWO_4$ morphologies, such as nanofibers, thin-walled nanotubes, AgI/$CuWO_4$ Z-scheme systems, and $CuWO_4$/g-C_3N_4 composites, have been investigated (Chen et al. 2023). These structures enhance surface area, improve charge transport, and suppress recombination, thereby improving overall photocatalytic efficiency.

A wide range of copper-based semiconductors have demonstrated promising photocatalytic properties through tailored structural modifications and hybridization strategies. These materials benefit from visible-light absorption, engineered charge separation pathways, and synergistic effects when combined with other functional components. Continued development of heterostructures and morphological control remains key to overcoming inherent limitations such as charge recombination and stability, thereby enabling more efficient and robust photocatalytic systems.

3.3.5 *Iron-Based Materials*

Iron-based oxides are attractive candidates for photocatalytic and photoelectrocatalytic applications in environmental remediation and solar fuel conversion. Among the six known phases, hematite (α-Fe_2O_3) is the most stable under aerobic conditions and features a narrow bandgap, high chemical stability, low toxicity, and suitable optoelectronic properties, making it a widely studied photocatalyst (Sivula et al. 2011). However, its performance is limited by poor charge transport, short carrier lifetimes and diffusion lengths, and defect-induced electronic traps (Lee and Han 2013). Strategies such as morphology tuning, elemental doping, and surface modification have been employed to address these limitations and enhance catalytic efficiency.

Iron oxyhydroxides (FeOOH) exist in five polymorphs, among which goethite (α-FeOOH) is the most extensively studied due to its diverse morphologies and facet exposures. Its photocatalytic activity is significantly enhanced in the presence of H_2O_2, where surface hydroxyl groups facilitate hydroxyl radical generation by acting as hole acceptors (Belattar et al. 2018). However, compared to hematite, goethite suffers from poor stability in alkaline environments, limiting its applicability in water-splitting reactions. Moreover, the modification strategies for hematite remain constrained by its binary oxide nature.

Ferrites (MFe_2O_4, M = metal ion) are promising alternatives to hematite for visible-light-driven photocatalysis. Typical spinel ferrites such as $NiFe_2O_4$ and $CoFe_2O_4$ possess inverse spinel structures, where the distribution of metal ions between tetrahedral and octahedral sites is influenced by factors like ionic radius, electronic configuration, and electrostatic energy (AlSalka et al. 2019). These materials offer advantages including superparamagnetism, low toxicity, facile synthesis, strong adsorption capacity, and cost-effectiveness. Despite the potential of iron-based core-shell nanostructures for enhanced recovery and reusability, the long-term stability of the magnetic core remains a critical challenge (Ramadan et al. 2019).

3.3.6 Cobalt-Based Materials

Cobalt-based heterogeneous photocatalysts are broadly categorized into NPs and single atoms, metal oxides, carbides, nitrides, and phosphides. Among them, Co NPs are often employed to enhance charge carrier separation and provide abundant active sites (Liu et al. 2019). For example, hierarchical nanocages composed of $ZnIn_2S_4$ supported on Co/N-doped g-C_3N_4 exhibit high H_2 evolution efficiency and stability without cocatalysts, owing to improved charge mobility and surface area (Wang et al. 2019a). Additionally, atomically dispersed Co on $W_{18}O_{49}$ nanowires serves as an active center for CO_2 reduction (Zhang et al. 2021a), while Co-modified TiO_2/MXene composites enable effective N_2 fixation in pure water without sacrificial agents (Gao et al. 2021b). Efforts have also focused on cobalt oxides such as Co_3O_4, $Co(OH)_2$, and $CoO(OH)$, with emphasis on tuning structural features, i.e., crystallinity, morphology, particle size, exposed facets, and oxidation states, to optimize photocatalytic performance (Sun et al. 2022; Guan et al. 2017).

Metal sulfides have gained increasing attention as photocatalysts owing to their distinctive optical and electrical properties. The less positive valence band derived from 3p orbitals, along with low effective mass carriers, enables favorable band alignment, broad light absorption, and rapid charge transport (Wang et al. 2021). Notably, coupling cobalt-based sulfides enhances charge separation and transfer via interfacial potential gradients (Zhang et al. 2021c). Consequently, various cobalt sulfides and their heterostructures have emerged as efficient photocatalysts.

Transition metal phosphides (TMPs) have emerged as efficient and cost-effective cocatalysts for photocatalysis, offering high electrical conductivity and chemical stability. Typically synthesized via gas-solid phosphidation of precursors such as metal oxides or hydroxides using reactive PH_3 gas, TMPs retain the morphology and size of their precursors (Yang et al. 2021). However, the toxicity of PH_3 demands strict handling precautions. Alternatively, hydrothermal and solvothermal methods offer safer, scalable routes for TMP synthesis. Cobalt-based phosphides (CoP), i.e., with a narrow bandgap (~1.7 eV) and suitable band alignment, have been widely explored for enhancing photocatalytic activity by modulating band structure and charge transfer pathways. Various CoP-based composites, including CoP/$Cd_xZn_{1-x}Se$, CoP/BP, CoP/$Zn_{0.5}Cd_{0.5}S$, CoP/$CeVO_4$, and CoP/TiO_2, have demonstrated improved performance (Sun et al. 2022).

3.3.7 Nickel-Based Materials

Nickel and its compound-based cocatalysts, e.g., sulfides, phosphides, carbides, and nitrides, possess narrow bandgaps that enhance photocatalytic light absorption. In contrast, wide-bandgap materials like $Ni(OH)_2$ and NiO improve light harvesting via multiple scattering when structured hierarchically (Meng et al. 2018). Nickel-based cocatalysts form various interfacial junctions with photocatalysts, including Schottky barriers, ohmic contacts, and p-n junctions, which facilitate efficient charge separation and transfer (Wang et al. 2020). Their high surface area and tunable surface structures further increase the number of active sites by introducing surface defects and exposing specific crystal facets with distinct nickel coordination environments (Shen et al. 2019b). These features contribute to reduced overpotential and enhanced activity in hydrogen evolution and CO_2 reduction reactions (Mou et al. 2019). Moreover, nickel-based cocatalysts improve photocatalyst durability by diverting redox reactions away from the host surface, thus mitigating photocorrosion and supporting long-term stability (Shen et al. 2019a, 2020).

Nickel metal and its compounds serve as effective cocatalysts in photocatalysis by enhancing charge separation and reaction kinetics. Metallic Ni acts as an electron trap, as demonstrated in 0D/3D Ni/$NiTe_2$ Schottky heterojunctions, where electron transfer from $NiTe_2$ to Ni forms a space charge layer and band bending at the interface, suppressing charge recombination via the Schottky barrier (Zhang et al. 2020a). Ni nanoparticles also provide active sites for hydrogen evolution. Similarly, Ni@NiOx cocatalysts deposited on $SrTiO_3$ enhance photocatalytic activity by promoting charge separation, with NiOx functioning as a hole trap (Han et al. 2017).

$Ni(OH)_2$, when coupled with g-C_3N_4 and Ni_3B, lowers the bandgap and improves H_2 evolution performance, with optimal activity observed for g-C_3N_4–3%Ni_3B-2% $Ni(OH)_2$ (Lu et al. 2018). Effective charge separation was also achieved in B-$Ni(OH)_2$/$Mn_{0.3}Cd_{0.7}S$ systems (Zhang et al. 2021b). NiS, owing to its metallic conductivity, facilitates electron transfer under UV light, enhancing separation and reducing recombination in NiS/Sm_2MoO_6 (Xu et al. 2021a). Nickel-based systems can suffer from hydrogen poisoning due to strong Ni-H bonding. However, phosphide-based materials like Ni-P@C and $Ni_2P/g-C_3N_4$ mitigate this by forming Ni-P bridges favorable for hydrogen evolution (Ran et al. 2021). Other Ni-based cocatalysts, such as $NiSe_{12}/TiO_2$ (Gao et al. 2021a, b), Ni_3C/CdS (Ma et al. 2018), and nickel(II) phthalocyanine/$BiVO_4$ (Feng et al. 2021), have shown improvements in light absorption, charge mobility, and interfacial kinetics. To further optimize performance, diverse interfacial engineering strategies, e.g., Z-scheme systems, p-n junctions, type I/II heterojunctions, S-scheme junctions, Schottky barriers, and homojunctions, are being actively explored in nickel-based photocatalysis (Krishnan et al. 2023).

3.4 Photocatalyst Applications

3.4.1 Water Purification

In an effort to solve energy and environmental issues, photocatalyst technologies that make use of earth-abundant metals are gaining popularity in the scientific community. Figure 3.3 illustrates their potential uses including photoactivated CO_2 reduction, water purification, air purification, hydrogen production, bacteria inactivation, and nitrogen oxide degradation. For example, the effectiveness of photocatalysis systems in removing different contaminants from water is the main focus of research at the moment (Table 3.2), with the end objective being water purification. In such case, treatment often employs heterogeneous catalysis, where the catalyst is in a distinct phase from the reactants, due to its enhanced performance. This separation facilitates the recovery and reuse of the catalyst, making heterogeneous photocatalysis generally preferred for clean water.

Metal oxides and semiconductors are the most prevalent heterogeneous photocatalysts which is mainly tested by degrading various pollutants under visible light, such as methylene blue (MB) (Bustos et al. 2021), amaranth dye (Subalakshmi et al. 2024),and methyl orange (MO) (Sajid and Alomayri 2022), antibiotics (Meky et al. 2024), and phenol (Lone et al. 2023). For example, tetracycline (TC) and methyl orange (MO) can be rapidly catalysed oxidized by the nanocomposite Cu_2O-CuO/graphite oxide (GO) (Zhang et al. 2020c) (Table 3.2). After 120 min of visible-light exposure, the degradation ratios of TC and MO are 90% and 95%, respectively. Charge separation prevents the rapid recombination of electron-hole pairs produced by light absorption, enabling them to engage in redox processes on the catalyst's surface. For instance, materials like Cu_2S/CD (Li et al. 2020) and Au-FeOOH (Singh et al. 2023b) are noted for their significant degradation of pollutants over certain time frames. However, under light exposure, some photocatalysts degrade and eventually lose their activity. To meet this challenge, materials that retain their structural and functional integrity in operating environments must be developed.

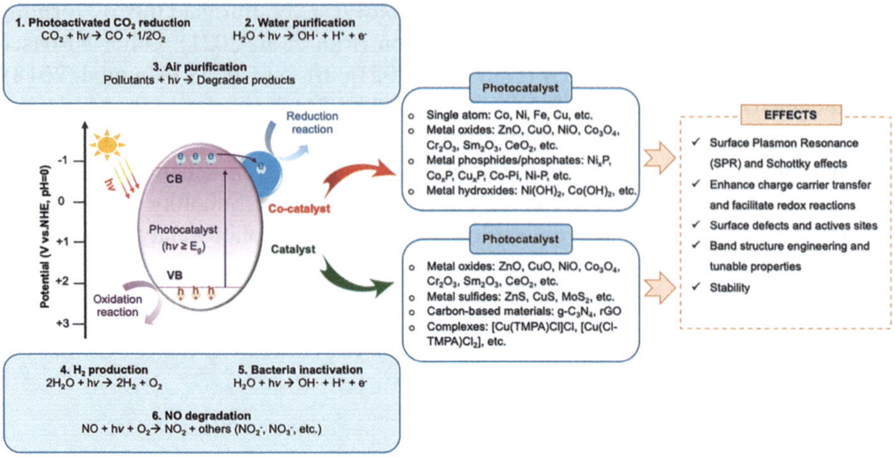

Fig. 3.3 Photocatalyst applications based on earth-abundant metals The primary advantage of photocatalysis over other methods of water purification is that it eliminates the need for further disposal procedures by transforming organic pollutants into inorganic ions, water, carbon dioxide, and other molecules. Other methods of removing the contaminants include activated carbon adsorption and air stripping, which just concentrate the chemicals and transfer them to the adsorbent or air (Dang et al. 2023; Gotore et al. 2022; Le Na et al. 2024). They do not convert them into nontoxic wastes, unlike in the case of photocatalysis

3.4.2 Photocatalytic Hydrogen Production

Earth-abundant metals offer intrinsic catalytic activity and can directly facilitate the hydrogen evolution reaction (HER) without the need for supplementary cocatalysts. For example, copper-based materials, such as CuO, Cu_2O, and CuS, have shown promising photocatalytic activity for water splitting under visible-light irradiation due to their high optical absorption, nontoxicity, good charge separation, and easy interface bonding (Zhang et al. 2024). In addition, carbon-based materials, typically graphite carbon nitride (g-C_3N_4), have been widely studied in H_2 photocatalysis due to their excellent electron transfer ability and suitable band position (−1.1 eV-CB and + 1.6 eV-VB vs NHE) (Mishra et al. 2019; Wang et al. 2009). Despite its potential, the disadvantages are inevitable, leading many researchers to improve the photocatalytic activity of g-C_3N_4 using various techniques, including doping, heterostructures, etc. Kong and co-workers reported metallic Ni doping accelerated H_2 production by creating a Schottky barrier between Ni and g-C_3N_4 for excellent photogenerated electron transfer (Kong et al. 2016).

Furthermore, using abundant metals as cocatalysts is widely investigated in sustainable and cost-effective photocatalytic systems for H_2 production (Ran et al. 2014). Examples of earth-abundant metals used as cocatalysts include Fe, Co, Ni, and Cu. These metals possess favorable catalytic properties, which act as electron traps and promote the hydrogen evolution reaction (HER) by facilitating the transfer of photogenerated electrons and protons. Many different types of cocatalysts have been synthesized from cheap and abundant elements, such as single atoms (Co, Ni, Fe, etc.) (Ejsmont et al. 2023; Lee et al. 2014; Xu and Xu 2015; Yang et al. 2022b), metal oxides (ZnO, CuO, NiO, etc.) (Huerta-Flores et al. 2016; Li et al. 2019; Wang et al. 2019b), metal phosphides/phosphates (NiP, CoP, CuP, Co-Pi, etc.) (Shen et al. 2018; Wang et al. 2017b; Zhao et al. 2023), and others (Ma et al. 2016, 2019; Yu et al. 2013).

Single-atom catalysts derived from earth-abundant metals have low cost, high selectivity, and high active sites. Inclusion of these atoms in the photocatalyst support provides better efficiency and selectivity through interface and LSPR effects, reducing the activation energy or overvoltage of surface H_2 and O_2 evolution reactions, thereby improving the H_2 photocatalytic evolution efficiency (Huang et al. 2022; Meng et al. 2019; Xiao et al. 2023). In addition, incorporating earth-abundant cocatalysts into the semiconductor structure increases the number of active sites and enlarges the contact area, thereby enhancing the rate of surface hydrogen and oxygen evolution reactions (Dong et al. 2017; Du et al. 2018). Li and co-workers developed a low-cost Cu ion-based cocatalyst derived from MOFs, which significantly enhanced photocatalytic hydrogen evolution performance. Transferring the binding metal charge to the binding ligand from the amino group (-NH_2) in the MOF structure enhances the carrier transfer and separation rate, enhancing overall photocatalytic efficiency (Li et al. 2022).

3.4.3 Photocatalytic Inactivation of Microorganisms in Water

Waterborne pathogens pose significant health risks and are responsible for various waterborne diseases. Earth-abundant metal-based photocatalysts also offer a promising solution for effectively inactivating microorganisms such as bacteria, viruses, and algae. Cu-based nanostructures, including copper oxide (CuO) and copper sulfide (CuS), have demonstrated potent photocatalytic activity against various microorganisms. Upon light irradiation, these materials generate reactive oxygen species (ROS) such as hydroxyl radicals (•OH), which induce oxidative damage/rupture to the cell membranes and biomolecules of microorganisms and electrostatic interaction with various biological macromolecules, leading to their inactivation (Alavi and Moradi 2022). Pugazhendhi and co-workers reported that CuO in nanoparticle form is a potential candidate against bacterial (*Staphylococcus aureus* and *Staphylococcus epidermidis*) and fungal (*Candida albicans*) pathogens (Pugazhendhi et al. 2018). A novel Z-scheme $C_3N_4/CuWO_4$ photocatalyst was successfully synthesized using the sol-gel technique (Gupta et al. 2019). As a result of their longer charge carrier lifetime and increased absorption of visible light, photocatalysts are more effective against *E. coli* and *S. aureus*.

Furthermore, CuO inorganic semiconductor nanoparticles incorporated into the conjugated polymer matrix showed inhibition of *E. coli* growth after 2 h of photocatalytic treatment under UV-vis light mainly through the formation of ROS such as $\cdot O_2^-$ which is effective in degrading bacteria (Djellabi et al. 2020). Zhang and colleagues synthesized CuS ultrathin nanosheets featuring a narrow bandgap and abundant vacancies, facilitating strong interaction with *E. coli*; the narrow band structure promotes increased heat dissipation, leading to elevated surface temperatures that disrupt the bacteria structure, rendering them inactive (Zhang et al. 2023b).

Similar to Cu-based photocatalysts, ZnO has also demonstrated its potential to disinfect bacteria in water treatment applications. Photogenerated holes during photocatalytic degradation in ZnO interact with ·OH groups in water, generating ROS, which can inhibit bacterial growth. Mondal and colleagues revealed outstanding antibacterial properties against *Staphylococcus aureus* (Gram-positive bacteria) and *Pseudomonas aeruginosa* (Gram-negative bacteria) (Mondal et al. 2023). In addition, NiO NPs showed excellent antibacterial activity against a variety of microorganisms, such as the Gram-positive bacteria *Bacillus subtilis (B. subtilis)* and the Gram-negative bacteria *Escherichia coli (E. coli)* (Singh et al. 2022). NiO NPs penetrate inside microbial cells, damaging the bacterial cell membrane through electrostatic interactions with DNA and breaking phosphate and hydrogen bonds, which then leads to protein denaturation, thus changing the structure of bacteria.

3.4.4 Photocatalytic Treatment of NO_x

Earth-abundant metal-based photocatalysts offer a promising solution for treating nitrogen oxides (NO_x) in polluted air. Nitrogen oxides (NO_x) in ambient air consist mainly of nitric oxide (NO) and nitrogen dioxide (NO_2), with NO accounting for 90 to 95%, formed from vehicle traffic and fuel combustion, contributing to air pollution and thus adversely affecting human health and the environment. Photocatalytic NO degradation using earth-abundant metal-based nanostructures provides a sustainable approach to mitigating air pollution.

The ability to degrade NO pollution is demonstrated through the reaction between ROS when exposed to light, promoting the oxidation of NO to NO_2 and further to nitrate ions (NO_3^-) or nitrogen gas (N_2). Geng and co-workers reported the synthesized 2D/2D heterojunction Z-scheme α-Fe_2O_3/g-C_3N_4 improved broad visible-light absorption and promoted charge transfer (Geng et al. 2021). The ZnO/TiO_2 composite synthesized through a simple ball milling method showed significantly higher NO removal photocatalytic efficiency than commercial ZnO nanoparticles (Van Viet et al. 2021). Besides, the metal-based photocatalyst provides active sites to adsorb and activate NO_2 molecules, promoting their conversion to less toxic nitrogen species. Zhang and co-workers reported that the hydrogen atoms in α-$Ni(OH)_2$ enhance the adsorption and activation of oxygen atoms to promote NO and NO_2 to be oxidized to nitrate, thereby no toxic NO_2 produce and has great potential in air purification (Zhang et al. 2020b).

3.5 Conclusions and Perspectives

Extensive research is being conducted to improve the synthesis, stability, and understanding of earth-abundant metal-based nanoparticle photocatalysts. Metal-based materials like zirconium, zinc, copper, iron, cobalt, and nickel and metal sulfides have shown positive results. In this brief perspective, we shall look over some of the recently investigated material candidates, methodologies, and possible applications. According to recent research, the efficiency of photocatalyst materials and engineering strategies is exhibited in the following:

- For zirconium, nanostructured ZrO_2 has shown improved photocatalytic performance, and green synthesis, utilizing plant extracts, is a viable approach.
- For zinc, by incorporating heterojunction construction and doping into ZnO, its visible-light absorption is enhanced, while charge carrier recombination is reduced.
- For earth-abundant metal sulfides, CuS and NiS are cost-effective photocatalysts and efficient in degrading pollutants and producing hydrogen.
- For copper, improved photocatalytic efficiency is being investigated by modifying copper-based catalysts; copper sulfides, selenides, and tungstates have demonstrated encouraging outcomes.
- For iron, hematite (α-Fe_2O_3) and other iron oxides possess excellent photocatalytic properties; nevertheless, they demand modifications to improve the rate of charge transfer.
- For cobalt, cobalt oxide and cobalt sulfide catalysts have been shown to speed up photocatalysis.
- For nickel, nickel cocatalysts contribute to improved photocatalytic activity by enhancing the light absorption and charge separation capabilities of photocatalysts, which, in turn, increases the number of active sites available.

3.5.1 The Efficiency of Photocatalysts in Several Environmental Applications

- For water treatment, metal oxides and semiconductors are the most prevalent and effective heterogeneous photocatalysts.
- In the realm of disinfection, copper-based nanostructures such as CuO and CuS, along with ZnO and NiO nanoparticles, have shown potent activity against waterborne pathogens. When exposed to light, they produce reactive oxygen species that cause oxidative damage to cell membranes and internal biomolecules, effectively neutralizing harmful microorganisms.
- Addressing air quality, photocatalysts like the α-Fe_2O_3/g-C_3N_4 heterojunction and ZnO/TiO_2 composites offer promising strategies for the reduction of NO_x pollutants. These materials effectively facilitate the conversion of toxic NO_x to

harmless nitrates or nitrogen gas, aiding in air purification and mitigating health and environmental impacts.

- In the field of hydrogen production, which is vital for advancing renewable energy, copper-based materials and graphite carbon nitride are notable for their capacity to facilitate water splitting under visible light. Earth-abundant metals like iron, cobalt, nickel, and copper are employed as cocatalysts in this process, acting as electron traps to boost the hydrogen evolution reaction and support the production of clean fuel. These insights emphasize the potential of designing photocatalytic materials for targeted environmental applications, thereby enabling their use in practical settings to address pollution and energy challenges.

Acknowledgments We acknowledge Ho Chi Minh City University of Technology (HCMUT), VNU-HCM, for supporting this study.

References

Aalami Z, Hoseinzadeh M, Hosseini Manesh P, Aalami AH, Es'haghi Z, Darroudi M, Sahebkar A, Hosseini HA (2024) Synthesis, characterization, and photocatalytic activities of green sol-gel ZnO nanoparticles using Abelmoschus esculentus and Salvia officinalis: a comparative study versus co-precipitation-synthesized nanoparticles. Heliyon 10:e24212. https://doi.org/10.1016/j.heliyon.2024.e24212

Abid N, Khan AM, Shujait S, Chaudhary K, Ikram M, Imran M, Haider J, Khan M, Khan Q, Maqbool M (2022) Synthesis of nanomaterials using various top-down and bottom-up approaches, influencing factors, advantages, and disadvantages: a review. Adv Colloid Interf Sci 300:102597. https://doi.org/10.1016/j.cis.2021.102597

Abubshait HA, Saad M, Iqbal S, Abubshait SA, Bahadur A, Raheel M, Alshammari FH, Alwadai N, Alrbyawi H, Abourehab MAS, Elkaeed EB, Qayyum MA, Somaily HH (2023) Co-doped zinc oxide nanoparticles embedded in Polyvinylalcohol hydrogel as solar light derived photocatalyst disinfection and removal of coloured pollutants. J Mol Struct 1271:134100. https://doi.org/10.1016/j.molstruc.2022.134100

Abuzeyad OH, El-Khawaga AM, Tantawy H, Gobara M, Elsayed MA (2024a) Photocatalytic degradation of methylene blue dye by promising zinc copper ferrite nanoparticles for wastewater treatment. J Inorg Organomet Polym Mater 34:2705–2715. https://doi.org/10.1007/s10904-024-03006-6

Abuzeyad OH, El Khawaga AM, Elsayed MA, Gobara M (2024b) Chemical routes to materials merits of photocatalytic activity of synthesized for wastewater treatment. J Mater Sci 59:4152–4166. https://doi.org/10.1007/s10853-024-09485-9

Ahmad W, Bhatt SC, Kaur N (2023) Plant extract mediated approach towards the synthesis of NiO nanoparticles: evaluation of its antibacterial, antioxidant and photocatalytic activity. Vietnam J Chem 61:445–454. https://doi.org/10.1002/vjch.202200156

Akika FZ, Benamira M, Lahmar H, Trari M, Avramova I, Suzer Ş (2020) Structural and optical properties of cu-doped ZnAl2O4 and its application as photocatalyst for Cr(VI) reduction under sunlight. Surf Interfaces 18:100406. https://doi.org/10.1016/j.surfin.2019.100406

Akshay S, Vidya YS, Manjunatha HC, Munirathnam R, Gurushantha K, Sridhar KN, Shivanna M, Kumar S (2023) High energy devices and display technology applications of silver doped zirconium titanate nanoparticles. Mater Chem Phys 308:128228. https://doi.org/10.1016/j.matchemphys.2023.128228

Alajmi BM, Basaleh AS, Ismail AA, Mohamed RM (2023) Hierarchical mesoporous CuO/ZrO2 nanocomposite photocatalyst for highly stable photoinduced desulfurization of thiophene. Surf Interfaces 39:102899. https://doi.org/10.1016/j.surfin.2023.102899

Alavi M, Moradi M (2022) Different antibacterial and photocatalyst functions for herbal and bacterial synthesized silver and copper/copper oxide nanoparticles/nanocomposites: a review. Inorg Chem Commun 142:109590. https://doi.org/10.1016/j.inoche.2022.109590

Albornoz LL, Bortolozzi JP, Banús ED, Brussino P, da Silva SW, Bernardes AM, Ulla MA (2021) Synthesis and characterization of immobilized titanium-zirconium Sn-doped oxides onto metallic meshes and their photocatalytic activity for erythromycin mineralization. Chem Eng J 414:128891. https://doi.org/10.1016/j.cej.2021.128891

Aldeen EMS, Jalil AA, Mim RS, Alhebshi A, Hassan NS, Saravanan R (2022) Altered zirconium dioxide based photocatalyst for enhancement of organic pollutants degradation: a review. Chemosphere 304:135349. https://doi.org/10.1016/j.chemosphere.2022.135349

Aljeboree AM, Abed Hussein S, Abed Jawad M, Alkaim AF (2024) Hydrothermal synthesis of eco-friendly ZnO/CNT nanocomposite and efficient removal of brilliant green cationic dye. Results Chem 7:101364. https://doi.org/10.1016/j.rechem.2024.101364

AlSalka Y, Granone LI, Ramadan W, Hakki A, Dillert R, Bahnemann DW (2019) Iron-based photocatalytic and photoelectrocatalytic nano-structures: facts, perspectives, and expectations. Appl Catal B Environ 244:1065–1095. https://doi.org/10.1016/j.apcatb.2018.12.014

Ameta R, Ameta SC (2016) Photocatalysis. CRC Press, Boca Raton. https://doi.org/10.1201/9781315372396

Ansari N, Ali A, Akhtar MS, Hasan S, Khatoon T, Rahman A, Mohammad S, Inamur Q (2024) Materials science in semiconductor processing green synthesis of zinc oxide marigold shaped clusters using Eucalyptus globulus leaf extract as robust photocatalyst for dyes degradation under sunlight. Mater Sci Semicond Process 173:108087. https://doi.org/10.1016/j.mssp.2023.108087

Atalay H, Çeli̇k A, Ayaz F (2018) Investigation of genotoxic and apoptotic effects of zirconium oxide nanoparticles (20 nm) on L929 mouse fibroblast cell line. Chem Biol Interact 296:98–104. https://doi.org/10.1016/j.cbi.2018.09.017

Bakry AM, Alamier WM, El-Shall MS, Awad FS (2022) Facile synthesis of amorphous zirconium phosphate graphitic carbon nitride composite and its high performance for photocatalytic degradation of indigo carmine dye in water. J Mater Res Technol 20:1456–1469. https://doi.org/10.1016/j.jmrt.2022.07.161

Baladi M, Amiri M, Mohammadi P, Salih Mahdi K, Golshani Z, Razavi R, Salavati-Niasari M (2023a) Green sol–gel synthesis of hydroxyapatite nanoparticles using lemon extract as capping agent and investigation of its anticancer activity against human cancer cell lines (T98, and SHSY5). Arab J Chem 16:104646. https://doi.org/10.1016/j.arabjc.2023.104646

Baladi M, Amiri M, Salavati-Niasari M (2023b) Green sol–gel auto-combustion synthesis, characterization and study of cytotoxicity and anticancer activity of ErFeO3/Fe3O4/rGO nanocomposite. Arab J Chem 16:104575. https://doi.org/10.1016/j.arabjc.2023.104575

Belattar S, Debbache N, Ghoul I, Sehili T, Abdessemed A (2018) Photodegradation of phenol red in the presence of oxyhydroxide of Fe(III) (Goethite) under artificial and a natural light. Water Environ J 32:358–365. https://doi.org/10.1111/wej.12333

Berede HT, Andoshe DM, Gultom NS, Kuo DH, Chen X, Abdullah H, Wondimu TH, Wu Y-n, Wu Y, Zelekew OA (2024) Photocatalytic activity of the biogenic mediated green synthesized CuO nanoparticles confined into MgAl LDH matrix. Sci Rep 14:1–13. https://doi.org/10.1038/s41598-024-52547-w

Bustos ME, Hevia F, Fuentes S, Martínez AI, Zárate RA (2021) A novel route for the preparation of self-assembled CuO hierarchical nanostructures by hydrothermal processing at high pressure. Mater Lett 297:129936. https://doi.org/10.1016/j.matlet.2021.129936

Chand S, Mondal A (2023) G-C3N4/ZrO2 composite material: a pre-eminent visible light-mediated photocatalyst for rhodamine B degradation in the presence of natural sunlight. Ceram Int 49:5419–5430. https://doi.org/10.1016/j.ceramint.2022.10.065

Chellammal Gayathri R, Elakkiya V, Sumathi S (2021) Effect of method of preparation on the photocatalytic activity of $NiAl_2O_4$. Inorg Chem Commun 129:108634. https://doi.org/10.1016/j.inoche.2021.108634
Chen P, Zhang P, Cui Y, Fu X, Wang Y (2023) Recent progress in copper-based inorganic nanostructure photocatalysts: properties, synthesis and photocatalysis applications. Mater Today Sustain 21:100276. https://doi.org/10.1016/j.mtsust.2022.100276
Cheng L, Zhang P, Wen Q, Fan J, Xiang Q (2022) Copper and platinum dual-single-atoms supported on crystalline graphitic carbon nitride for enhanced photocatalytic CO_2 reduction. Chinese J Catal 43:451–460. https://doi.org/10.1016/S1872-2067(21)63879-2
Chiu M-H, Kuo C-C, Huang C-W, Yang W-D (2022) Preparation of CuS/PbS/ZnO heterojunction Photocatalyst for application in hydrogen production. Catalysts 12:1677. https://doi.org/10.3390/catal12121677
Dang BT, Ramaraj R, Huynh KPH, Le MV, Tomoaki I, Pham TT, Hoang Luan V, Thi Le Na P, Tran DPH (2023) Current application of seaweed waste for composting and biochar: a review. Bioresour Technol 375:128830. https://doi.org/10.1016/j.biortech.2023.128830
Dara M, Hassanpour M, Alshamsi HA, Baladi M, Salavati-Niasari M (2021) Green sol-gel auto combustion synthesis and characterization of double perovskite Tb_2ZnMnO_6nanoparticles and a brief study of photocatalytic activity. RSC Adv 11:8228–8238. https://doi.org/10.1039/d0ra10400k
Dawoud TMS, Pavitra V, Ahmad P, Syed A, Nagaraju G (2020) Photocatalytic degradation of an organic dye using ag doped ZrO_2 nanoparticles: Milk powder facilitated eco-friendly synthesis. J King Saud Univ–Sci 32:1872–1878. https://doi.org/10.1016/j.jksus.2020.01.040
Dhivya A, Yadav R, Stella Packiam C (2019) An eco-approach synthesis of undoped and Mn doped ZnO nano-photocatalyst for prompt decoloration of methylene blue dye. In: Materials today: proceedings. Elsevier, pp 494–501. https://doi.org/10.1016/j.matpr.2021.02.751
Ding H, Han D, Han Y, Liang Y, Liu X, Li Z, Zhu S, Wu S (2020) Visible light responsive CuS/protonated g-C_3N_4 heterostructure for rapid sterilization. J Hazard Mater 393:122423. https://doi.org/10.1016/j.jhazmat.2020.122423
Djellabi R, Ali J, Zhao X, Saber AN, Yang B (2020) CuO NPs incorporated into electron-rich TCTA@PVP photoactive polymer for the photocatalytic oxidation of dyes and bacteria inactivation. J Water Process Eng 36:101238. https://doi.org/10.1016/j.jwpe.2020.101238
Dong Y, Kong L, Wang G, Jiang P, Zhao N, Zhang H (2017) Photochemical synthesis of Co_xP as cocatalyst for boosting photocatalytic H_2 production via spatial charge separation. Appl Catal B Environ 211:245–251. https://doi.org/10.1016/j.apcatb.2017.03.076
Du F, Lu H, Lu S, Wang J, Xiao Y, Xue W, Cao S (2018) Photodeposition of amorphous MoS_x cocatalyst on TiO_2 nanosheets with {001} facets exposed for highly efficient photocatalytic hydrogen evolution. Int J Hydrog Energy 43:3223–3234. https://doi.org/10.1016/j.ijhydene.2017.12.181
Duan H, Wang D, Li Y (2015) Green chemistry for nanoparticle synthesis. Chem Soc Rev 44:5778. https://doi.org/10.1039/C4CS00363B
Ejsmont A, Lewandowska-Andralojc A, Hauza K, Goscianska J (2023) In situ modification of co-MOF with graphene oxide for enhanced photocatalytic hydrogen production. Int J Hydrog Energy 48:8965–8970. https://doi.org/10.1016/j.ijhydene.2022.12.084
Feng Y, Lu H, Gu X, Qiu J, Jia M, Huang C, Yao J (2017) ZIF-8 derived porous N-doped ZnO with enhanced visible light-driven photocatalytic activity. J Phys Chem Solids 102:110–114. https://doi.org/10.1016/j.jpcs.2016.11.022
Feng J, Bian J, Bai L, Xi S, Wang Y, Chen C, Jing L (2021) Efficient wide-spectrum photocatalytic overall water splitting over ultrathin molecular nickel phthalocyanine/$BiVO_4$ Z-scheme heterojunctions without noble metals. Appl Catal B Environ 295:120260. https://doi.org/10.1016/j.apcatb.2021.120260
Galeas S, Valdivieso-Ramírez CS, Pontón PI, Guerrero VH, Goetz V (2023) Photocatalytic degradation of phenol under visible light irradiation by using ferrous oxalate synthesized from

iron-rich mineral sands via a green hydrothermal route. Environ Technol Innov 32:103325. https://doi.org/10.1016/j.eti.2023.103325

Gao D, Wu X, Wang P, Yu H, Zhu B, Fan J, Yu J (2021a) Selenium-enriched amorphous NiSe1+x nanoclusters as a highly efficient cocatalyst for photocatalytic H2 evolution. Chem Eng J 408:127230. https://doi.org/10.1016/j.cej.2020.127230

Gao W, Li X, Luo S, Luo Z, Zhang X, Huang R, Luo M (2021b) In situ modification of cobalt on MXene/TiO2 as composite photocatalyst for efficient nitrogen fixation. J Colloid Interface Sci 585:20–29. https://doi.org/10.1016/j.jcis.2020.11.064

Geng Y, Chen D, Li N, Xu Q, Li H, He J, Lu J (2021) Z-Scheme 2D/2D α-Fe2O3/g-C3N4 heterojunction for photocatalytic oxidation of nitric oxide. Appl Catal B Environ 280:119409. https://doi.org/10.1016/j.apcatb.2020.119409

Gotore O, Itayama T, Dang BT, Nguyen TD, Ramaraj R, Osamu N, Shuji T, Maseda H (2022) Adsorption analysis of ciprofloxacin and delafloxacin onto the corn cob derived-biochar under different pyrolysis conditions. Biomass Convers Biorefinery 14:10373–10388. https://doi.org/10.1007/s13399-022-03156-y

Guan J, Ding C, Chen R, Huang B, Zhang X, Fan F, Zhang F, Li C (2017) CoOX nanoparticle anchored on sulfonated-graphite as efficient water oxidation catalyst. Chem Sci 8:6111–6116. https://doi.org/10.1039/c7sc01756a

Gupta R, Boruah B, Modak JM, Madras G (2019) Kinetic study of Z-scheme C3N4/CuWO4 photocatalyst towards solar light inactivation of mixed populated bacteria. J Photochem Photobiol A Chem 372:108–121. https://doi.org/10.1016/j.jphotochem.2018.08.035

Han K, Kreuger T, Mei B, Mul G (2017) Transient behavior of Ni@NiOx functionalized SrTiO3 in overall water splitting. ACS Catal 7:1610–1614. https://doi.org/10.1021/acscatal.6b03662

Hassan NS, Jalil AA (2022) A review on self-modification of zirconium dioxide nanocatalysts with enhanced visible-light-driven photodegradation of organic pollutants. J Hazard Mater 423:126996. https://doi.org/10.1016/j.jhazmat.2021.126996

Hassan A, Jalil A, Ilyas SZ, Iqbal MF, Ali Shah SZ, Baqir Y (2024) Green-route synthesis and ab-initio studies of a highly efficient nano photocatalyst:Ce/zinc-oxide nanopetals. Heliyon 10:e25581. https://doi.org/10.1016/j.heliyon.2024.e25581

Heidari-Asil SA, Zinatloo-Ajabshir S, Alshamsi HA, Al-Nayili A, Yousif QA, Salavati-Niasari M (2022) Magnetically recyclable ZnCo2O4/Co3O4 nano-photocatalyst: green combustion preparation, characterization and its application for enhanced degradation of contaminated water under sunlight. Int J Hydrog Energy 47:16852–16861. https://doi.org/10.1016/j.ijhydene.2022.03.157

Heinlaan M, Ivask A, Blinova I, Dubourguier H, Kahru A (2008) Toxicity of nanosized and bulk ZnO , CuO and TiO 2 to bacteria Vibrio fischeri and crustaceans Daphnia magna and Thamnocephalus platyurus. Chemosphere 71:1308–1316. https://doi.org/10.1016/j.chemosphere.2007.11.047

Huang G, Lin G, Niu Q, Bi J, Wu L (2022) Covalent triazine-based frameworks confining cobalt single atoms for photocatalytic CO2 reduction and hydrogen production. J Mater Sci Technol 116:41–49. https://doi.org/10.1016/j.jmst.2021.11.035

Huang H, Xu TT, Wang BL, Lv N, Zhang YT (2023) Construction of zirconium metal–organic polyhedron/silver chloride heterojunction with boosted photocatalytic activity for the degradation of broad-spectrum antibiotics and mechanism insight. Sep Purif Technol 307:122848. https://doi.org/10.1016/j.seppur.2022.122848

Huerta-Flores AM, Torres-Martínez LM, Moctezuma E, Ceballos-Sanchez O (2016) Enhanced photocatalytic activity for hydrogen evolution of SrZrO3 modified with earth abundant metal oxides (MO, M=Cu, Ni, Fe, Co). Fuel 181:670–679. https://doi.org/10.1016/j.fuel.2016.05.025

Hui Y, Zhang S (2021) A facile synthesis of Fe-doped zirconium oxide nanoparticles for enhancement of Rhodamine B dye degradation. Int J Electrochem Sci 16:1–10. https://doi.org/10.20964/2021.06.53

Islam MF, Islam S, Miah MAS, Huq AKO, Saha AK, Mou ZJ, Mondol MMH, Bhuiyan MNI (2024) Green synthesis of zinc oxide nano particles using Allium cepa L. waste peel extracts

and its antioxidant and antibacterial activities. Heliyon 10:e25430. https://doi.org/10.1016/j.heliyon.2024.e25430
Jadoun S, Arif R, Jangid NK, Meena RK (2021) Green synthesis of nanoparticles using plant extracts: a review. Environ Chem Lett 19:355–374. https://doi.org/10.1007/s10311-020-01074-x
Jeba R, Radhika S, Padma CM, Ascar Davix X (2022a) Strontium doped zirconia for enhanced antimicrobial activity. Mater Today Proc 58:927–931. https://doi.org/10.1016/j.matpr.2021.12.226
Jeba R, Radhika S, Padma CM, Ascar Davix X (2022b) Antimicrobial and photocatalytic activities of strontium doped zirconium oxide nanoparticles. Mater Lett 311:131559. https://doi.org/10.1016/j.matlet.2021.131559
Jia T, Luo F, Wu J, Chu F, Xiao Y, Liu Q, Pan W, Li F (2022) Nanosized Zn-In spinel-type sulfides loaded on facet-oriented CeO2 nanorods heterostructures as Z-scheme photocatalysts for efficient elemental mercury removal. Sci Total Environ 813:151865. https://doi.org/10.1016/j.scitotenv.2021.151865
Jones N, Ray B, Ranjit KT, Manna AC (2007) Antibacterial activity of ZnO nanoparticle suspensions on a broad spectrum of microorganisms. FEMS Microbiol Lett 279(1):71–76. https://doi.org/10.1111/j.1574-6968.2007.01012.x
Joulaee S, Mirzaei M, Hassanpour A, Safardoust-Hojaghan H, Khani A (2023) Efficient removal of anionic and cationic dyes from waste water using green ZnO/NiO/graphene quantum dots nano photocatalyst. Optik (Stuttg) 290:171324. https://doi.org/10.1016/j.ijleo.2023.171324
Jung J, Jeon HJ, Yang SW, Choi MG, Vidyasagar D, Kim JH, Shim RB, Yun Y, Han S, Cho IS, Han JE, Lee S (2024) Cost-effective synthesis of copper sulfide nanoparticles and flexible films for photocatalytic and antibiotic applications. J Mater Res Technol 28:1875–1882. https://doi.org/10.1016/j.jmrt.2023.12.100
Kahng S, Yoo H, Kim JH (2020) Recent advances in earth-abundant photocatalyst materials for solar H2 production. Adv Powder Technol 31:11–28. https://doi.org/10.1016/j.apt.2019.08.035
Karthika A, Karuppasamy P, Selvarajan S, Suganthi A, Rajarajan M (2019) Electrochemical sensing of nicotine using CuWO4 decorated reduced graphene oxide immobilized glassy carbon electrode. Ultrason Sonochem 55:196–206. https://doi.org/10.1016/j.ultsonch.2019.01.038
Kaur M, Kaur H (2022) Synergistic effect of biochar impregnated with ZnO nano-flowers for effective removal of organic pollutants from wastewater. Appl Surf Sci Adv 12:100339. https://doi.org/10.1016/j.apsadv.2022.100339
Kong L, Dong Y, Jiang P, Wang G, Zhang H, Zhao N (2016) Light-assisted rapid preparation of a Ni/g-C3N4 magnetic composite for robust photocatalytic H2 evolution from water. J Mater Chem A 4:9998–10007. https://doi.org/10.1039/C6TA03178A
Krishnan A, Ajith A, Krishnan AV, Saji RE, Syamli S, Shibli SMA (2023) Ni-based electro/photo-catalysts in HER—a review. Surf Interfaces 36:102619. https://doi.org/10.1016/j.surfin.2022.102619
Kumar K, Priya A, Arun A, Hait S, Chowdhury A (2019) Antibacterial and natural room-light driven photocatalytic activities of CuO nanorods. Mater Chem Phys 226:106–112. https://doi.org/10.1016/j.matchemphys.2019.01.020
Kumar SG, Kavitha R, Sushma C (2020) Doped zinc oxide nanomaterials: structure–electronic properties and photocatalytic applications. Interface Sci Technol 31:285–312. https://doi.org/10.1016/B978-0-08-102890-2.00009-9
Kunnamareddy M, Rajendran R, Sivagnanam M, Rajendran R, Diravidamani B (2021) Nickel and sulfur codoped TiO2 nanoparticles for efficient visible light photocatalytic activity. J Inorg Organomet Polym Mater 31:2615–2626. https://doi.org/10.1007/s10904-021-01914-5
Le Na PT, Tuyen NDK, Dang BT (2024) Sorption of four antibiotics onto pristine biochar derived from macadamia nutshell. Bioresour Technol 394:130281. https://doi.org/10.1016/j.biortech.2023.130281
Lee J, Han S (2013) Thermodynamics of native point defects in α-Fe2O 3: An ab initio study. Phys Chem Chem Phys 15:18906–18914. https://doi.org/10.1039/c3cp53311e

Lee G-J, Anandan S, Masten SJ, Wu JJ (2014) Sonochemical synthesis of hollow copper doped zinc sulfide nanostructures: optical and catalytic properties for visible light assisted Photosplitting of water. Ind Eng Chem Res 53:8766–8772. https://doi.org/10.1021/ie500663n

Lei W, Wang H, Zhang X, Yang Z, Kong C (2022) Cu2O-based binary and ternary photocatalysts for the degradation of organic dyes under visible light. Ceram Int 48:1757–1764. https://doi.org/10.1016/j.ceramint.2021.09.255

Li G, Guo C, Yan M, Liu S (2016) Applied catalysis B : environmental Cs x WO 3 nanorods : realization of full-spectrum-responsive photocatalytic activities from UV, visible to near-infrared region. Applied Catal. B, Environ 183:142–148. https://doi.org/10.1016/j.apcatb.2015.10.039

Li L, Yu H, Xu J, Zhao S, Liu Z, Li Y (2019) Rare earth element, Sm, modified graphite phase carbon nitride heterostructure for photocatalytic hydrogen production. New J Chem 43:1716–1724. https://doi.org/10.1039/C8NJ05619F

Li X, Dai K, Pan C, Zhang J (2020) Diethylenetriamine-functionalized CdS nanoparticles decorated on Cu2S snowflake microparticles for photocatalytic hydrogen production. ACS Appl Nano Mater 3:11517–11526. https://doi.org/10.1021/acsanm.0c02616

Li H, Liu X, Song M, Feng H, Yu J, Liu B, Ren Q, Yang Z, Xu M, He Y (2022) Non-noble copper ion anchored on NH2-MIL-101(Fe) as a novel cocatalyst with transient metal centers for efficient photocatalytic water splitting. J Alloys Compd 905:164153. https://doi.org/10.1016/j.jallcom.2022.164153

Li X, He J, Zhang W, You L, Li J (2023) Ag nanoparticles interlayered Fe3O4/Ag/m(TiO2-ZrO2) magnetic photocatalysts with enhanced stability and photocatalytic performance for Cr(VI) reduction. Appl Surf Sci 607:155076. https://doi.org/10.1016/j.apsusc.2022.155076

Liu Q, Zeng C, Xie Z, Ai L, Liu Y, Zhou Q, Jiang J, Sun H, Wang S (2019) Cobalt@nitrogen-doped bamboo-structured carbon nanotube to boost photocatalytic hydrogen evolution on carbon nitride. Appl Catal B Environ 254:443–451. https://doi.org/10.1016/j.apcatb.2019.04.098

Lone MA, Shrotriya V, Lone TA, Alshahrani T, Hossain MA, Adam M, Zaman MB (2023) Rapid hydrothermal synthesis of mesoporous NiS2 and NiSe2 nanostructures for wastewater treatment. Heliyon 9:e22308. https://doi.org/10.1016/j.heliyon.2023.e22308

Lu X, Xie J, Liu SY, Adamski A, Chen X, Li X (2018) Low-cost Ni3B/Ni(OH)2 as an eco-friendly hybrid Cocatalyst for remarkably boosting photocatalytic H2 production over g-C3N4 Nanosheets. ACS Sustain Chem Eng 6:13140–13150. https://doi.org/10.1021/acssuschemeng.8b02653

Luo L, Xiao X, Li Q, Wang S, Li Y, Hou J, Jiang B (2021) Engineering of single atomic Cu-N3Active sites for efficient singlet oxygen production in photocatalysis. ACS Appl Mater Interfaces 13:58596–58604. https://doi.org/10.1021/acsami.1c17782

Lv R, Liang YQ, Li ZY, Zhu SL, Cui ZD, Wu SL (2022) Flower-like CuS/graphene oxide with photothermal and enhanced photocatalytic effect for rapid bacteria-killing using visible light. Rare Metals 41:639–649. https://doi.org/10.1007/s12598-021-01759-4

Ma B, Liu Y, Li J, Lin K, Liu W, Zhan H (2016) Mo2N: An efficient non-noble metal cocatalyst on CdS for enhanced photocatalytic H2 evolution under visible light irradiation. Int J Hydrog Energy 41:22009–22016. https://doi.org/10.1016/j.ijhydene.2016.08.133

Ma S, Deng Y, Xie J, He K, Liu W, Chen X, Li X (2018) Noble-metal-free Ni3C cocatalysts decorated CdS nanosheets for high-efficiency visible-light-driven photocatalytic H2 evolution. Appl Catal B Environ 227:218–228. https://doi.org/10.1016/j.apcatb.2018.01.031

Ma B, Li X, Li D, Lin K (2019) A difunctional photocatalytic H2 evolution composite co-catalyst tailored by integration with earth-abundant material and ultralow amount of noble metal. Appl Catal B Environ 256:117865. https://doi.org/10.1016/j.apcatb.2019.117865

Machín A, Soto-Vázquez L, García D, Cotto MC, Ortiz D, Berríos-Rolón PJ, Fontánez K, Resto E, Morant C, Petrescu F, Márquez F (2023) Photodegradation of ciprofloxacin and levofloxacin by Au@ZnONPs-MoS2-rGO nanocomposites. Catalysts 13:538. https://doi.org/10.3390/catal13030538

Mahrsi MI, Chouchene B, Gries T, Carré V, Girot E, Medjahdi G, Ayari F, Balan L, Schneider R (2023) Novel ZnO/Ag nanohybrids prepared from Ag+-doped layered zinc hydroxides as highly active photocatalysts for the degradation of dyes and ciprofloxacin. Colloids Surf A Physicochem Eng Asp 671:131643. https://doi.org/10.1016/j.colsurfa.2023.131643

Mathai J, Anjana MP, Aleena PA, Thomas SA, Jose AK, Kunjumon J, Nair SS, Rimal Isaac RS, Sajan D (2023) Cr doped ZnO nanoparticles as photocatalyst for the degradation of methylene blue dye. J Indian Chem Soc 100:101067. https://doi.org/10.1016/j.jics.2023.101067

Meky AI, Hassaan MA, Fetouh HA, Ismail AM, El Nemr A (2024) Hydrothermal fabrication, characterization and RSM optimization of cobalt-doped zinc oxide nanoparticles for antibiotic photodegradation under visible light. Sci Rep 14:1–23. https://doi.org/10.1038/s41598-024-52430-8

Melchor-lagar V, Ramos-ramírez E, Morales-pérez A, Rangel-vázquez I, Del Angel G, De Química D, Metropolitana-iztapalapa UA, San A, Atlixco R, Cp N, De México C (2020) Journal of Photochemistry & Photobiology a : chemistry photocatalytic removal of 4-chlorophenol present in water using ZrO 2 / LDH under UV light source. J Photochem Photobiol A Chem 389:112251. https://doi.org/10.1016/j.jphotochem.2019.112251

Meng A, Wu S, Cheng B, Yu J, Xu J (2018) Hierarchical TiO2/Ni(OH)2 composite fibers with enhanced photocatalytic CO2 reduction performance. J Mater Chem A 6:4729–4736. https://doi.org/10.1039/c7ta10073f

Meng A, Zhang L, Cheng B, Yu J (2019) Dual Cocatalysts in TiO2 Photocatalysis. Adv Mater 31:1807660. https://doi.org/10.1002/adma.201807660

Mishra A, Mehta A, Basu S, Shetti NP, Reddy KR, Aminabhavi TM (2019) Graphitic carbon nitride (g–C3N4)–based metal-free photocatalysts for water splitting: a review. Carbon N Y 149:693–721. https://doi.org/10.1016/j.carbon.2019.04.104

Mondal S, Jamal M, Ayon SA, Anik MJF, Billah MM (2023) Synergistic enhancement of photocatalytic and antimicrobial efficacy of nitrogen and erbium co-doped ZnO nanoparticles. J Rare Earths 42:859–868. https://doi.org/10.1016/j.jre.2023.10.002

Moradi SAH, Ghobadi N, Tabatabaeinejad SM (2023) Facile and rapid preparation of progressive ZnO/NiO/rGO nano-photocatalyst and investigation its mechanism and reaction kinetics while decomposition of pharmaceuticals pollutant. Surf Interfaces 39:102939. https://doi.org/10.1016/j.surfin.2023.102939

Mou K, Chen Z, Zhang X, Jiao M, Zhang X, Ge X, Zhang W, Liu L (2019) Highly efficient Electroreduction of CO2 on nickel single-atom catalysts: atom trapping and nitrogen anchoring. Small 15:1903668. https://doi.org/10.1002/smll.201903668

Muniyappa M, Kalegowda SN, Shetty M, Sriramoju JB, Shastri M, Navakoteswara NR, De D, Shankar MV, Rangappa D (2022) Cocatalyst free nickel sulphide nanostructure for enhanced photocatalytic hydrogen evolution. Int J Hydrog Energy 47:5307–5318. https://doi.org/10.1016/j.ijhydene.2021.11.171

Naraginti S, Stephen FB, Radhakrishnan A, Sivakumar A (2015) Zirconium and silver co-doped TiO2 nanoparticles as visible light catalyst for reduction of 4-nitrophenol, degradation of methyl orange and methylene blue. Spectrochim. Acta – Part A Mol Biomol Spectrosc 135:814–819. https://doi.org/10.1016/j.saa.2014.07.070

Nasir JA (2020) Recent developments and perspectives in CdS-based photocatalysts for water splitting. J Mater Chem A 8(40):20735–21342. https://doi.org/10.1039/D0TA05834C

Nazim M, Khan AAP, Asiri AM, Kim JH (2021) Exploring rapid photocatalytic degradation of organic pollutants with porous CuO Nanosheets: synthesis, dye removal, and kinetic studies at room temperature. ACS Omega 6:2601–2612. https://doi.org/10.1021/acsomega.0c04747

Parashar M, Shukla VK, Singh R (2020) Metal oxides nanoparticles via sol–gel method: a review on synthesis, characterization and applications. J Mater Sci Mater Electron 31:3729–3749. https://doi.org/10.1007/s10854-020-02994-8

Perks C, Mudd G (2019) Titanium, zirconium resources and production: a state of the art literature review. Ore Geol Rev 107:629–646. https://doi.org/10.1016/j.oregeorev.2019.02.025

Pugazhendhi A, Kumar SS, Manikandan M, Saravanan M (2018) Photocatalytic properties and antimicrobial efficacy of Fe doped CuO nanoparticles against the pathogenic bacteria and fungi. Microb Pathog 122:84–89. https://doi.org/10.1016/j.micpath.2018.06.016

Qi K, Cheng B, Yu J, Ho W (2017) Review on the improvement of the photocatalytic and antibacterial activities of ZnO. J Alloys Compd 727:792–820. https://doi.org/10.1016/j.jallcom.2017.08.142

Rajbongshi BM, Ramchiary A, Samdarshi S (2014) Influence of N-doping on photocatalytic activity of ZnO nanoparticles under visible light irradiation. Mater Lett 134:111–114. https://doi.org/10.1016/j.matlet.2014.07.073

Ramadan W, Dillert R, Koch J, Tegenkamp C, Bahnemann DW (2019) Changes in the solid-state properties of bismuth iron oxide during the photocatalytic reformation of formic acid. Catal. Today 326:22–29. https://doi.org/10.1016/j.cattod.2018.09.003

Ran J, Zhang J, Yu J, Jaroniec M, Qiao SZ (2014) Earth-abundant cocatalysts for semiconductor-based photocatalytic water splitting. Chem Soc Rev 43:7787–7812. https://doi.org/10.1039/c3cs60425j

Ran Q, Yu Z, Jiang R, Hou Y, Huang J, Zhu H, Yang F, Li M, Li F, Sun Q (2021) A novel, noble-metal-free core-shell structure Ni–P@C cocatalyst modified sulfur vacancy-rich ZnIn2S4 2D ultrathin sheets for visible light-driven photocatalytic hydrogen evolution. J Alloys Compd 855:157333. https://doi.org/10.1016/j.jallcom.2020.157333

Revathi M, Pricilla Jeyakumari A, Saravanan SP (2021) Design and fabrication of ZnO/CdS heterostructured nanocomposites for enhanced hydrogen evolution from solar water splitting. Inorg Chem Commun 134:109056. https://doi.org/10.1016/j.inoche.2021.109056

Rohokale MS, Dhabliya D, Sathish T, Vijayan V, Senthilkumar N (2021) A novel two-step co-precipitation approach of CuS/NiMn2O4 heterostructured nanocatalyst for enhanced visible light driven photocatalytic activity via efficient photo-induced charge separation properties. Phys B Condens Matter 610:412902. https://doi.org/10.1016/j.physb.2021.412902

Sajid MM, Alomayri T (2022) Synthesis of α-Fe2O3 rhombus nanoplates for photocatalytic investigation of cationic and anionic dyes and antibacterial aspect. J Taibah Univ Sci 16:1192–1201. https://doi.org/10.1080/16583655.2022.2154094

Samriti, Tyagi R, Ruzimuradov O, Prakash J (2023) Fabrication methods and mechanisms for designing highly-efficient photocatalysts for energy and environmental applications. Mater Chem Phys 307:128108. https://doi.org/10.1016/j.matchemphys.2023.128108

Senthil Kumar M, Arunagiri C (2021) Efficient photocatalytic degradation of organic dyes using Fe-doped ZnO nanoparticles. J Mater Sci Mater Electron 32:111464. https://doi.org/10.1007/s10854-021-06328-0

Sharafinia S, Farrokhnia A, Lemraski EG, Rashidi A (2022) Decoration of ZnFe2O4 and UiO-66 over g-C3N4 as magnetically novel reusable visible light photocatalyst for degradation of Rh–B. Opt Mater (Amst) 132:112838. https://doi.org/10.1016/j.optmat.2022.112838

Shen R, Xie J, Lu X, Chen X, Li X (2018) Bifunctional Cu3P decorated g-C3N4 Nanosheets as a highly active and robust visible-light Photocatalyst for H2 production. ACS Sustain Chem Eng 6:4026–4036. https://doi.org/10.1021/acssuschemeng.7b04403

Shen R, Liu W, Ren D, Xie J, Li X (2019a) Co1.4Ni0.6P cocatalysts modified metallic carbon black/g-C3N4 nanosheet Schottky heterojunctions for active and durable photocatalytic H2 production. Appl Surf Sci 466:393–400. https://doi.org/10.1016/j.apsusc.2018.10.033

Shen R, Xie J, Xiang Q, Chen X, Jiang J, Li X (2019b) Ni-based photocatalytic H2-production cocatalysts. Cuihua Xuebao/Chinese J Catal 40:240–288. https://doi.org/10.1016/s1872-2067(19)63294-8

Shen R, Zhang L, Chen X, Jaroniec M, Li N, Li X (2020) Integrating 2D/2D CdS/α-Fe2O3 ultrathin bilayer Z-scheme heterojunction with metallic β-NiS nanosheet-based ohmic-junction for efficient photocatalytic H2 evolution. Appl Catal B Environ 266:118619. https://doi.org/10.1016/j.apcatb.2020.118619

Shinde HM, Bhosale TT, Gavade NL, Babar SB, Kamble RJ, Shirke BS, Garadkar KM (2018) Biosynthesis of ZrO2 nanoparticles from Ficus benghalensis leaf extract for photocatalytic activity. J Mater Sci Mater Electron 29:14055–14064. https://doi.org/10.1007/s10854-018-9537-7

Singh A, Goyal V, Singh J, Kaur H, Kumar S, Batoo KM, Gaur J, Pal M, Rawat M, Hussain S (2022) Structurally and morphologically engineered single-pot biogenic synthesis of

NiO nanoparticles with enhanced photocatalytic and antimicrobial activities. J Clean Prod 343:131026. https://doi.org/10.1016/j.jclepro.2022.131026

Singh A, Wan F, Yadav K, Salvi A, Thakur P, Thakur A (2023a) Synergistic effect of ZnO nanoparticles with Cu2+ doping on antibacterial and photocatalytic activity. Inorg Chem Commun 157:111425. https://doi.org/10.1016/j.inoche.2023.111425

Singh J, Roy S, Tran TT, Akhtar S, Lee E, Kim J (2023b) Photocatalytic thin films based on Au nanoparticles covered by iron oxyhydroxides by hydrothermal process. Surf Interfaces 39:102915. https://doi.org/10.1016/j.surfin.2023.102915

Sivula K, Le Formal F, Grätzel M (2011) Solar water splitting: progress using hematite (α-Fe 2O3) photoelectrodes. ChemSusChem 4:432–449. https://doi.org/10.1002/cssc.201000416

Sousani F, Sadrnezhaad SK, Abachi P, Mahtabpour H (2024) Ni0.8Co0.2/Zn0.8Cd0.2S/g-C3N4 ternary nano-photocatalyst for improved visible-light-driven photocatalytic H2 evolution and pollutants degradation. Int J Hydrog Energy 51:281–299. https://doi.org/10.1016/j.ijhydene.2023.10.001

Subalakshmi A, Kavitha B, Srinivasan N, Rajarajan M, Suganthi A (2024) Evaluation of photocatalytic activity of Cu3(PO4)2/MgO nanocomposite for the efficient removal of amaranth dye under solar light irradiation. Inorg Chem Commun 161:112033. https://doi.org/10.1016/j.inoche.2024.112033

Sun W, Zhu J, Zhang M, Meng X, Chen M, Feng Y, Chen X, Ding Y (2022) Recent advances and perspectives in cobalt-based heterogeneous catalysts for photocatalytic water splitting, CO2 reduction, and N2 fixation. Chin J. Catal. 43:2273–2300. https://doi.org/10.1016/S1872-2067(21)63939-6

Talebi P, Greco R, Yamamoto T, Zeynali M, Asgharizadeh S, Cao W (2024) Materials advances hierarchical nickel carbonate hydroxide nanostructures for photocatalytic hydrogen evolution from water splitting †. https://doi.org/10.1039/d3ma00977g

Teeparthi SR, Awin EW, Kumar R (2018) Dominating role of crystal structure over defect chemistry in black and white zirconia on visible light photocatalytic activity. Sci Rep 1–11:5541. https://doi.org/10.1038/s41598-018-23648-0

Ücker CL, Goetzke V, Riemke FC, Oliveira ME, Carreno NLV, Morisso FDP, Teodoro MD, Mastelaro VR, Moreira ML, Raubach CW, da Silva Cava S (2023) The photocatalytic performance of Fe inserted in Nb2O5 obtained by microwave-assisted hydrothermal synthesis: factorial design of experiments. J Photochem Photobiol A Chem 435:114294. https://doi.org/10.1016/j.jphotochem.2022.114294

Van Viet P, Hoang The Vinh T, Thi Ngoc Dung N, Minh Thi C (2021) Facile ball-milling synthesis of TiO2 modified ZnO for efficient photocatalytic removal of atmospheric nitric oxide gas under solar light irradiation. Chem Phys Lett 775:138642. https://doi.org/10.1016/j.cplett.2021.138642

Vinayagam R, Singhania B, Murugesan G, Kumar PS, Bhole R, Kumar M, Varadavenkatesan T, Selvaraj R (2022) Photocatalytic degradation of methylene blue dye using newly synthesized zirconia nanoparticles. Environ Res 214:113785. https://doi.org/10.1016/j.envres.2022.113785

Wang X, Maeda K, Thomas A, Takanabe K, Xin G, Carlsson JM, Domen K, Antonietti M (2009) A metal-free polymeric photocatalyst for hydrogen production from water under visible light. Nat Mater 8:76–80. https://doi.org/10.1038/nmat2317

Wang F, Chen D, Zhang N, Wang S, Qin L, Sun X, Huang Y (2017a) Oxygen vacancies induced by zirconium doping in bismuth ferrite nanoparticles for enhanced photocatalytic performance. J Colloid Interface Sci 508:237–247. https://doi.org/10.1016/j.jcis.2017.08.056

Wang W, An T, Li G, Xia D, Zhao H, Yu JC, Wong PK (2017b) Earth-abundant Ni2P/g-C3N4 lamellar nanohydrids for enhanced photocatalytic hydrogen evolution and bacterial inactivation under visible light irradiation. Appl Catal B Environ 217:570–580. https://doi.org/10.1016/j.apcatb.2017.06.027

Wang S, Wang Y, Zhang SL, Zang SQ, Lou XW (2019a) Supporting ultrathin ZnIn2S4 Nanosheets on Co/N-doped graphitic carbon Nanocages for efficient photocatalytic H2 generation. Adv Mater 31:1903404. https://doi.org/10.1002/adma.201903404

Wang S, Zhu B, Liu M, Zhang L, Yu J, Zhou M (2019b) Direct Z-scheme ZnO/CdS hierarchical photocatalyst for enhanced photocatalytic H2-production activity. Appl Catal B Environ 243:19–26. https://doi.org/10.1016/j.apcatb.2018.10.019

Wang Z, Fan J, Cheng B, Yu J, Xu J (2020) Nickel-based cocatalysts for photocatalysis: hydrogen evolution, overall water splitting and CO2 reduction. Mater Today Phys 15:100279. https://doi.org/10.1016/j.mtphys.2020.100279

Wang J, Lin S, Tian N, Ma T, Zhang Y, Huang H (2021) Nanostructured metal sulfides: classification, modification strategy, and solar-driven CO2 reduction application. Adv Funct Mater 31:2008008. https://doi.org/10.1002/adfm.202008008

Wu T, Shao Q, Ge S, Zhao W, Liu Q (2016) Influence of molecular weight of chitosan on the microstructures and photocatalytic property of ZrO2 prepared by chitosan templates method. Mater Res Bull 83:657–663. https://doi.org/10.1016/j.materresbull.2016.07.021

Xiao B, Shen C, Luo Z, Li D, Kuang X, Wang D, Zi B, Yan R, Lv T, Zhou T, Zhang J, Liu Q (2023) Cu surface doped TiO2: constructing cu single-atoms active sites and broadening the photoresponse range for efficient photocatalytic hydrogen production. Chem Eng J 468:143650. https://doi.org/10.1016/j.cej.2023.143650

Xu Y, Xu R (2015) Nickel-based cocatalysts for photocatalytic hydrogen production. Appl Surf Sci 351:779–793. https://doi.org/10.1016/j.apsusc.2015.05.171

Xu J, Li X, Liu Y, Liu Z (2021a) A 2D/3D heterojunction Photocatalyst using 2D Nanosheet Sm2MoO6 on 3D spherical NiS: a novel high-efficiency Photocatalyst. Phys. Status Solidi Appl. Mater. Sci. 218:2100026. https://doi.org/10.1002/pssa.202100026

Xu J, Liu Y, Li X, Li Y (2021b) Hydrothermal synthesis of Mn3O4/CoS2 as a promising photocatalytic material for boosting visible-light photocatalytic hydrogen production. Phys Status Solidi Appl Mater Sci 218:1–10. https://doi.org/10.1002/pssa.202100025

Yang Y, Zhou C, Wang W, Xiong W, Zeng G, Huang D, Zhang C, Song B, Xue W, Li X, Wang Z, He D, Luo H, Ouyang Z (2021) Recent advances in application of transition metal phosphides for photocatalytic hydrogen production. Chem Eng J 405:126547. https://doi.org/10.1016/j.cej.2020.126547

Yang G, Qiu P, Xiong J, Zhu X, Cheng G (2022a) Facilely anchoring Cu2O nanoparticles on mesoporous TiO2 nanorods for enhanced photocatalytic CO2 reduction through efficient charge transfer. Chin Chem Lett 33:3709–3712. https://doi.org/10.1016/j.cclet.2021.10.047

Yang R, Liu J, Wang B, Wang R, Song Y, Hua Y, Wang C, She Y, Yuan J, Xu H, Li H (2022b) Synergistic effect of isolated co and Fe dual active sites boosting the photocatalytic hydrogen evolution reaction. J Alloys Compd 895:162290. https://doi.org/10.1016/j.jallcom.2021.162290

Yu J, Wang S, Cheng B, Lin Z, Huang F (2013) Noble metal-free Ni(OH)2–g-C3N4 composite photocatalyst with enhanced visible-light photocatalytic H2-production activity. Cat Sci Technol 3:1782–1789. https://doi.org/10.1039/C3CY20878H

Zeng Z, Yuan Y, Lu H, Gao S, Ren S, Rong P, Jiao S, Wang D, Wang J (2022) Low-temperature synthesis of ZnO/ZnS/CuS heterojunction based on solution-processed nanosheet array with enhanced photocatalytic activity. J Chin Chem Soc 69:249–256. https://doi.org/10.1002/jccs.202100353

Zhang J, Gao Y, Jia X, Wang J, Chen Z, Xu Y (2018) Oxygen vacancy-rich mesoporous ZrO2 with remarkably enhanced visible-light photocatalytic performance. Sol Energy Mater Sol Cells 182:113–120. https://doi.org/10.1016/j.solmat.2018.03.023

Zhang F, Zhuang HQ, Zhang W, Yin J, Cao FH, Pan YX (2019) Noble-metal-free CuS/CdS photocatalyst for efficient visible-light-driven photocatalytic H2 production from water. Catal Today 330:203–208. https://doi.org/10.1016/j.cattod.2018.03.060

Zhang Q, Jia J, Fan J, Hu X, Liu E, Yang D (2020a) Nanoparticle metal Ni cocatalyst on NiTe2 microsphere for improved photocatalytic hydrogen evolution. Int J Hydrog Energy 45:13340–13352. https://doi.org/10.1016/j.ijhydene.2020.03.030

Zhang R, Ran T, Cao Y, Ye L, Dong F, Zhang Q, Yuan L, Zhou Y (2020b) Oxygen activation of noble-metal-free g-C3N4/α-Ni(OH)2 to control the toxic byproduct of photocatalytic nitric oxide removal. Chem Eng J 382:123029. https://doi.org/10.1016/j.cej.2019.123029

Zhang Z, Sun L, Wu Z, Liu Y, Li S (2020c) Facile hydrothermal synthesis of CuO-Cu2O/GO nanocomposites for the photocatalytic degradation of organic dye and tetracycline pollutants. New J Chem 44:6420–6427. https://doi.org/10.1039/d0nj00577k
Zhang H, Wang Y, Zuo S, Zhou W, Zhang J, Lou XWD (2021a) Isolated cobalt centers on W18O49Nanowires perform as a reaction switch for efficient CO2Photoreduction. J Am Chem Soc 143:2173–2177. https://doi.org/10.1021/jacs.0c08409
Zhang J, Cheng C, Xing FS, Chen C, Huang C (2021b) 0D β-Ni(OH)2 nanoparticles/1D Mn0.3Cd0.7S nanorods with rich S vacancies for improved photocatalytic H2 production. Chem Eng J 414:129157. https://doi.org/10.1016/j.cej.2021.129157
Zhang T, Meng F, Cheng Y, Dewangan N, Ho GW, Kawi S (2021c) Z-scheme transition metal bridge of Co9S8/cd/CdS tubular heterostructure for enhanced photocatalytic hydrogen evolution. Appl Catal B Environ 286:119853. https://doi.org/10.1016/j.apcatb.2020.119853
Zhang J, Bifulco A, Amato P, Imparato C, Qi K (2023a) Copper indium sulfide quantum dots in photocatalysis. J Colloid Interface Sci 638:193–219. https://doi.org/10.1016/j.jcis.2023.01.107
Zhang J, Shi R, Zhang Z, Guo D, Zhang Q (2023b) Modulating the large vacancy types of CuS ultrathin nanosheets via defect engineering to improve the photocatalytic antibacterial performance. Appl Surf Sci 639:158269. https://doi.org/10.1016/j.apsusc.2023.158269
Zhang Y, Jin F, Liu H, Jin Z (2024) In-situ XPS monitoring bimetallic Prussian blue analogs and metal oxides heterojunction for photocatalytic hydrogen production. Fuel 361:130655. https://doi.org/10.1016/j.fuel.2023.130655
Zhao B, Shao G, Fan B, Zhao W, Xie Y, Zhang R (2015) Synthesis of flower-like CuS hollow microspheres based on nanoflakes self-assembly and their microwave absorption properties. J Mater Chem A 3:10345–10352. https://doi.org/10.1039/c5ta00086f
Zhao Z, Chen Z, Wang X (2023) Non-noble metal copper phosphide deposited on the surface of cadmium sulfide nanorods can effectively enhance the photocatalytic hydrogen evolution performance. Int J Hydrog Energy 48:964–974. https://doi.org/10.1016/j.ijhydene.2022.09.223

Chapter 4
Green Carbon-Based Nanostructured Photocatalysts

Ababay Ketema Worku, Minbale Admas Teshager, and Delele Worku Ayele

Abstract Due to its potential to address energy and environmental issues, photocatalysis has gained more attention in recent years as a green and sustainable technology. Because of their special qualities and attributes, a variety of metal-free photocatalytic materials have been created and developed in recent years for a range of applications in an effort to provide highly effective and affordable photocatalysts. Although many excellent catalysts have been created, there are still major obstacles in the way of reducing the price and improving the catalytic activity of catalysts. An excellent replacement for outdated catalysts is the creation of carbon-based nanostructured materials (i.e., graphite, fullerenes, carbon nitride, carbon nanotubes, activated carbon, graphene, and diamond). Researchers have worked very hard to improve several metal nanostructures based on carbon as catalysts. Furthermore, remarkable advancements have been made in a number of catalysis-related disciplines, including electrochemical performance, visible-light-induced catalysis, and energy storage and conversion. This chapter provides an overview of the most recent advancements in the design and production of metal nanostructures based on carbon, serving as photocatalysts through a variety of green approaches. Furthermore, in relation to their environmentally benign applications, the photocatalytic properties of the derived carbon-based nanostructure classes are also discussed.

A. K. Worku (✉)
Bahir Dar Energy Center, Bahir Dar Institute of Technology, Bahir Dar University, Bahir Dar, Ethiopia
e-mail: Ababay.Ketema@bdu.edu.et

M. A. Teshager
Department of Chemistry, College of Science, Bahir Dar University, Bahir Dar, Ethiopia

D. W. Ayele
Bahir Dar Energy Center, Bahir Dar Institute of Technology, Bahir Dar University, Bahir Dar, Ethiopia

Department of Chemistry, College of Science, Bahir Dar University, Bahir Dar, Ethiopia

C. M. Hussain, G. Hussain (eds.), *Green Nanostructured Photocatalysts for Clean Energy*, Clean Energy Production Technologies,
https://doi.org/10.1007/978-981-95-1618-6_4

Keywords Carbon-based nanostructured materials · Green technology · Photocatalysis

4.1 Introduction

Lately, due to serious worries about fossil fuels, research in contemporary civilizations has concentrated on using renewable energy sources as alternatives to meet future energy needs (Low et al. 2021). Because solar energy is always present and may be used for a variety of purposes, including heating, water splitting, and the production of electricity through the use of photocatalysts, it is acknowledged as the main renewable energy source (Liu et al. 2012). In response to the energy crisis and environmental concerns, photocatalysts have gained more attention as a potential solution for lowering the production of H_2 and hazardous pollutants. Nonetheless, there are still issues with the photocatalysts' low quantum yield, excessive charge recombination, and restricted light absorption (Lin et al. 2017). In order to address these problems, a number of photocatalysts have been created thus far. Of these, carbon-based photocatalysts have recently generated a great deal of interest because of their wide surface area, good electronic conductivity, low cost of production, and high chemical and thermal durability (Liu et al. 2021a). Carbon nanomaterials (CNMs) are the most promising option for photocatalysts because of these special qualities (Narasimharao and Malkhasian 2022). Due to their high physiochemical stability, earth abundance, and low synthesis cost, the most often employed CNMs for the synthesis of photocatalysts, including graphene, carbon nanotubes (CNTs), carbon quantum dots (CQDs), fullerene, and graphitic carbon nitride (g-C_3N_4), have garnered significant interest. Carbon-based nanostructured materials, or CBNMs,

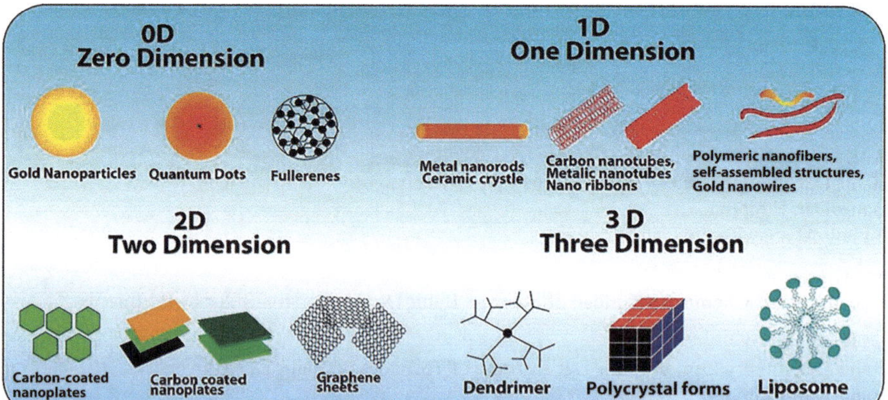

Fig. 4.1 Different carbon-based nanostructured materials. Reproduced under the terms of the CC-BY Creative Commons Attribution 4.0 International license (Abbas et al. 2022). Copyright (2022), Frontiers

have demonstrated significant promise for a range of uses (Lam et al. 2016). Graphite, fullerenes, carbon nitride, carbon nanotubes, activated carbon, graphene, and diamond, among others are a few examples of carbon-based nanostructured materials (Fig. 4.1) (Kim et al. 2015). CBNMs are in high demand for a variety of applications due to their special qualities, which include a high surface area, outstanding chemical stability, good conductivity, exceptional mechanical strength, environmental friendliness, and easy availability (Hafeez et al. 2018). Different morphologies of CBNMs are caused by variations in the arrangement of carbon atoms. There are various allotropic forms of carbon atoms that range from 0D to 3D nanostructures (Feng et al. 2021). The most useful and well-known forms of carbon for a variety of uses, including energy storage, photocatalysis, and medicine, are graphene and its variants. An overview of the most current developments in carbon-based photocatalysts is attempted to be provided in this chapter. There is discussion of carbon's characteristics, capabilities, and ways of development as photocatalysts (Gharib and Arab 2022).

4.2 Nanotechnology in Energy Systems

The study of atom and molecular assemblies is the focus of interdisciplinary collaborations in physics, chemistry, biology, material sciences, and engineering that have led to the development of nanoscience (Gnanasekaran et al. 2022). Nanotechnology can have a significant impact on energy because it can be utilized at every level of the energy value chain, but it also needs the integration of numerous scientific, engineering, and technical disciplines and skills, more so than in other areas (Ensafi and Karimi-Maleh 2010a). The conversion of primary energy sources is more ecologically friendly and efficient thanks to nanotechnology (Chung et al. 2018). Nanostructured materials and nanotechnology are playing a major role in the field of solar photovoltaics, which generates electricity by converting sunlight. Researchers are looking into adding a layer of quantum dots, which are tiny semiconductor blobs formed on the surface of another behind the traditional multilayer composite (Benaissa et al. 2021). Utilizing waste heat is also aided by thermoelectric energy conversion utilizing nanostructured semiconductors. The Carnot efficiency, assuming discharge to a reservoir at 25 °C, is the highest efficiency at which a heat engine may convert some of the flow of heat from a hotter body to a cooler body into meaningful work. Because the Carnot efficiency is dependent on the temperature of both hotter and colder reservoirs, it will be 63.8% at 523 °C (Cheng et al. 2012). The Carnot limit, however, does not hold true in reality or theory. Furthermore, there are a lot of advantages to manufacturing batteries with nanotechnology (Chai et al. 2021a). First, because there is less combustible electrode material available, there is a decreased chance of batteries catching fire (Chai et al. 2021b). Furthermore, nanotechnology primarily has the ability to extend a battery's usable life and shorten its recharge time (Berkani et al. 2022). Nanoparticles are applied to an electrode's surface to provide these advantages. As a result, more current may pass between the electrode and the chemicals inside the battery due to the electrode's increased surface area. This method could improve hybrid car efficiency by lowering the weight of the batteries that are required to generate sufficient power (Balachandar et al. 2020). Additionally, by separating the battery's liquids from its solid electrodes when there is no draw on the battery, the use of nanomaterials can lengthen the battery's lifespan (Ali et al. 2022). The low-level discharge that happens in a typical battery is avoided by this separation, thereby extending the battery's shelf life. Moreover, one type of high-efficiency light bulb uses a nano-engineered polymer matrix (Alamgholiloo et al. 2020). The new bulbs have twice the efficiency of small fluorescence light bulbs and are shatterproof. Other researchers are employing arrays of nanoscale structures called plasmonic cavities to generate high-efficiency LEDs. Modernizing incandescent light bulbs by encircling the traditional filament with crystalline material that partially converts waste infrared radiation into visible light is another concept now under investigation (Wu et al. 2019).

4.3 Green Synthesis of Different Nanostructured Carbon-Based Materials

Nowadays, the principle of nanoparticle synthesis via green chemistry is the main emphasis on the environmentally friendly process, and the final products are free from any hazard. Bio-based raw materials such as plant, algae, bacteria, and yeast are crucial for green synthesis of different nanostructured carbon-based materials (Badr et al. 2022). Synthesis of nanoparticles from bio-based raw materials requires a simple synthesis procedure: simple mixing of plant extract with metal salt (Fig. 4.2). Moreover, properly sized nanoparticles can be synthesized from plant extract and metal salt. Particle size control is possible by controlling some conditions such as pH, temperature, and mixing ratio of reactants (Shi et al. 2020).

Carbon-based nanostructured materials (CBNMs), with their green synthesis, prominent properties, characterizations, and various applications, get great attention in recent advance and state-of-the –art of the area (Fig. 4.3) (Zhang et al. 2018).

4.4 Carbon-Based Photocatalysts

CBNMs are effective photocatalysts for different applications such as hydrogen generation, wastewater treatment, and heavy metal recovery. One of the most vital and significant attributions of carbon-based nanostructured materials (CBNMs) for photocatalytic application is its excellent stability in both acidic and alkaline conditions (Shakeel et al. 2019).

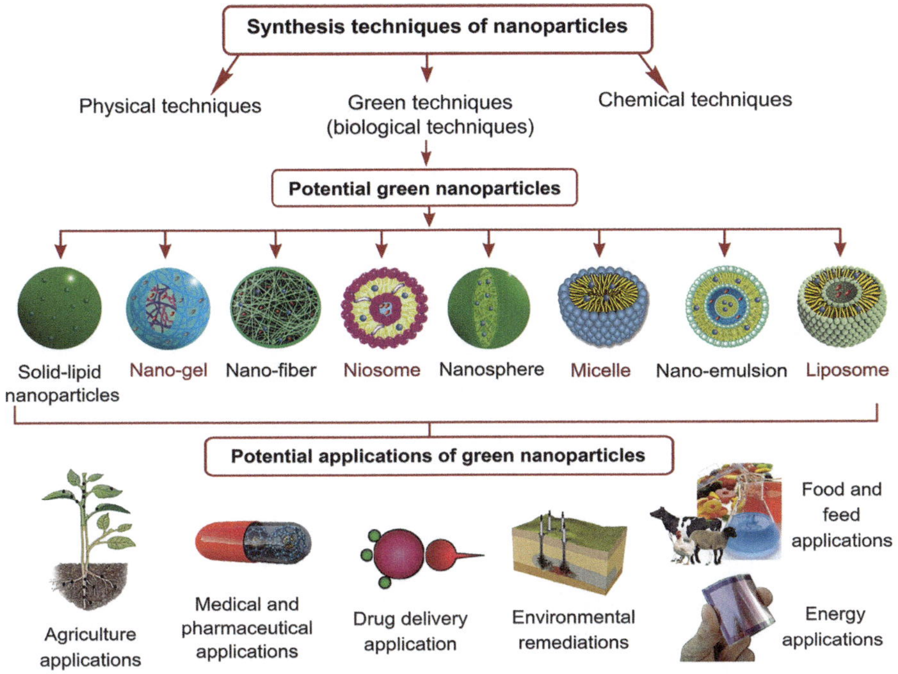

Fig. 4.2 Green synthesis of nanoparticles. Reproduced under the terms of the CC-BY Creative Commons Attribution 4.0 International license (Osman et al. 2024). Copyright (2024), Springer

4.4.1 CNTs as Photocatalyst

Photocatalytic water-splitting technology based on CNT-modified nanomaterials has exhibited great potential for hydrogen production in view of their low cost and high stability (Samanta et al. 2014). For example, it has offered new opportunities for achieving high photocatalytic activity with high stability. Combination of CNTs with graphene not only increases reaction sites but also inhibits the recombination of photoexcited electron-hole pairs. In addition, CNT-based photocatalysts also revealed high activity on the photocatalytic degradation of organic pollutants due to π-system or formation of heterojunction (Qin et al. 2016). For example, CNT-modified hierarchical microspheres ZnO enhanced the visible light adsorption and charge separation process, exhibiting excellent photocatalytic performance much better than the pure ZnO for the reduction of organic molecules in the industrial effluents (Ong et al. 2016). Combination of CNT with other photocatalysts could enhance the conductivity and facilitate the charge transfer process during the photocatalytic reaction (Liu et al. 2021b). To further improve the performance in the future, more efforts should be made to in situ synthesize CNT-based composite in order to strengthen the synergetic interaction between CNT and other nanostructures. When combined with other photocatalysts, carbon nanotubes (CNTs) can

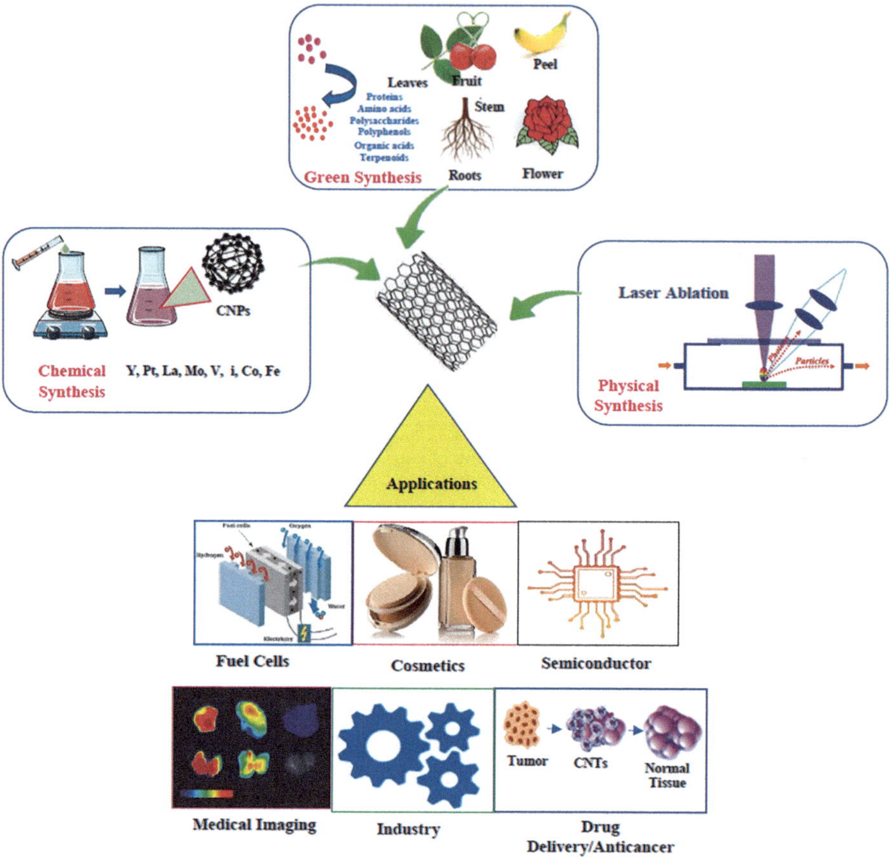

Fig. 4.3 Green synthesis methods and applications of carbon-based nanostructured materials. Reproduced under the terms of the CC-BY Creative Commons Attribution 4.0 International license (Qasim et al. 2023). Copyright (2023), MDPI

improve conductivity and speed up the charge transfer process in photocatalytic reactions. More work should be done to in situ synthesis of CNT-based composites in order to enhance the synergistic interaction between CNT and other nanostructures and further increase performance in the future (Peng et al. 2011).

4.4.2 Fullerene as Photocatalyst

Because of its distinct chemical and physical properties, fullerene (C_{60}), a closed-shell allotrope with 20 hexagons and 12 pentagons that retain 30 orbital bonds with 60 p-electrons, has been acknowledged as the most significant carbon allotrope (Martha et al. 2014). Moreover, fullerene-based carbon materials work better in photocatalytic applications due to C_{60}'s superiority as an electron giver and acceptor (Dai et al. 2009). Fullerene encapsulation into carbon nanotubes (CNTs) is an incredibly efficient method that was developed to create heterojunctions with distinct electrical properties. For water splitting, Song et al. created a new C_{60}/graphene/g-C_3N_4 composite with a high hydrogen production efficiency (Joseph et al. 2020). The ability of g-C_3N_4 to produce hydrogen was significantly increased by the synergistic effect of graphene and C_{60}, which also accelerated the separation of photo-generated electron and hole pairs and increased the efficiency of photo-generated electron transit and use. Although fullerene and its derivatives are widely utilized in organic photovoltaic devices, photocatalysis for the creation of hydrogen and the breakdown of organic contaminants is still in its infancy. Up until now, the majority of studies have only looked at C_{60}; however, future research on photocatalysts should pay more attention to other fullerene family members including the C_{70} and its derivatives (Kumaresan et al. 2019).

4.4.3 CQDs as Photocatalyst

Since CQDs have distinct optical and electrical features in addition to a unique fluorescence emission feature, they represent an emergent and recently created class of CNMs with well-controlled intrinsic characteristics (Chakraborty et al. 2019). Since their 2004 discovery, CQDs have found use in a wide range of applications, such as photocatalysts, chemical sensors, bioimaging, and nanomedicine. CQDs demonstrated the most promising potential for photocatalytic H_2 generation, especially in photocatalytic applications (Lin et al. 2021). Additionally, CQDs may effectively separate electrons and holes by acting as both a donor and an acceptor of electrons. This broadens the photo-absorption range of semiconductor materials with high bandgaps to visible regions. Wang et al. found that the hydrogen production performance of the metal-doped CQDs in conjunction with CdS nanowires as a co-catalyst was significantly higher than that of the undoped CQDs/CdS composite. Owing to its potential as an electron acceptor and donor, CQD needs more research in the area of photocatalytic applications. More focus should be placed on creating new and simple green synthesis techniques to create CQD-based CNMs, particularly metal-free catalysts (Ensafi and Karimi-Maleh 2010b).

4.4.4 g-C_3N_4 as Photocatalyst

Because of its easily tunable electronic band structure, extremely stable physiochemical features, ease of manufacture, and low cost, g-C_3N_4 has recently become a hot topic for photocatalytic activity development. (Jang et al. 2008), Created a 3D flower-shaped TiO_2 hybridized with a 2D g-C_3N_4 nanosheet using calcination and hydrothermal methods. The resulting TiO_2/g-C_3N_4 composite showed a much greater photocatalytic hydrogen generation efficiency, i.e., 7.7 and 1.9 times higher, respectively, than the pure TiO_2 and g-C_3N_4. According to reports, one factor in the enhanced photocatalytic efficacy is the prolonged visible light absorption by g-C_3N_4. Furthermore, doping may potentially increase g-C_3N_4's activity. It has been shown that g-C3N4's NO removal rate may be increased by 1.5 times following Sr doping. The density functional theory (DFT) method is an effective tool for understanding the energy-related mechanism of photocatalytic reactions and for methodically representing the electronic structures. The findings demonstrated that several Sr doping techniques, including intercalation, cavity padding, replacing triazine N, and bridging N, could lower the bandgap of g-C_3N_4, thereby promoting charge transfer (Lv et al. 2019).

4.4.5 Graphene as Photocatalyst

Discovered in 2004, graphene has exceptional physical and chemical capabilities. Its ability to hold tightly integrated sp^2-hybridized atoms into an ordered two-dimensional (2D) honeycomb lattice presents new potential for the development of highly stable and effective photocatalytic materials (Rao et al. 2019). There has been a recent surge in demand for metal-free photocatalyst synthesis. By using an impregnation chemical reduction technique, Gong et al. were able to produce graphene/g-C_3N_4 nanocomposites that acted as active photocatalysts for the generation of H_2 in visible light (Tahir et al. 2020a). Furthermore, because of their special optical, electrical, thermal, and mechanical qualities, 2D graphene and transition metal dichalcogenides (TMDCs) have emerged as adaptable nanomaterials for the quick development of photocatalysts. Unconventional quantum Hall effect, good electrical conductivity (2000 S m^{-1}), strong ambipolar electric field effect, relativistic Dirac fermions, high transparency (97.7%), high thermal conductivity (5000 W m^{-1} K^{-1}), high mechanical strength (with a Young's modulus of 1 TPa), superior carrier mobility at room temperature (200,000 cm^2 V^{-1} s^{-1}), extremely high theoretical specific surface area (2630 m^2 g^{-1}), excellent environmental compatibility, low production cost, and high adsorption capacity for organic and inorganic molecules are some of the apparent benefits of pristine graphene, as shown in Fig. 4.4 (Tahir et al. 2020b). Compared with other carbon allotropes (i.e., 0D fullerenes, 1D nanotubes, and 3D graphite), 2D graphene has more promising applications in optical, electrical, biological, and catalytic sectors because of these superior features (Serafin et al. 2022).

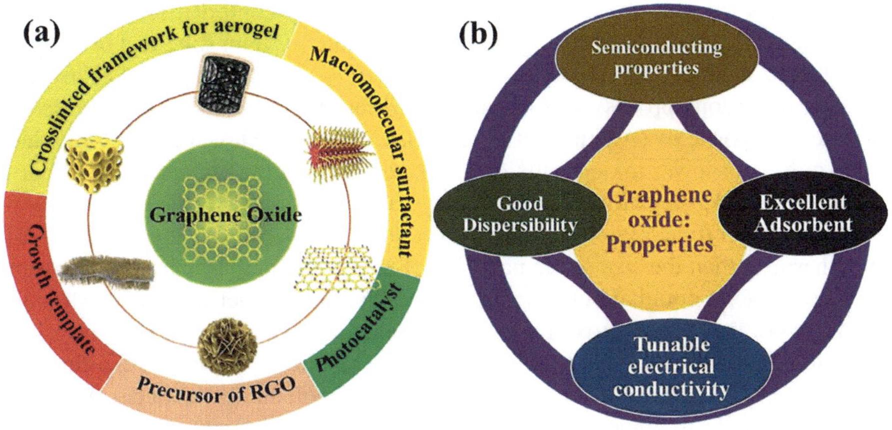

Fig. 4.4 Diagram of (**a**) multifarious roles and (**b**) important properties GO in photocatalysis Reproduced under the terms of the CC-BY Creative Commons Attribution 4.0 International license (Prakash 2022). Copyright (2023), MDPI

4.5 Health and Safety Issues

The electrical, optical, mechanical, magnetic, quantum mechanical, and thermal properties of nanomaterials are exceptional (Mehmandoust et al. 2022). Many nanoscale materials, including nanoparticles, nanofibers, nanocomposites, nanotubes, and nanofilms, all of which are regarded as the next generation of materials, have been used in a range of global sectors due to their special qualities. It has been reported that nanomaterials are already present in over 1500 distinct products and processes, such as sunscreens, car bumpers, wrinkle-resistant clothing, water-splitting reactions, solar cells, concrete, toothpaste, sunscreens, and a variety of optical, diagnostic, and sensing devices (Chang et al. 2023). Photoactive nanomaterials have been the subject of several recent investigations for the production of renewable energy from solar energy. Even in the unlikely event that there are no unexpected release episodes, workers will inevitably be exposed to some degree during the large-scale manufacture of nanomaterials. Determining the level of exposure that can have a significant negative influence on health is essential. Because photoactive nanoparticles have so many uses in the energy and medicinal domains, their toxicities are among the most researched topics in nanotechnology (Wang et al. 2023a). The people who operate in the field of nanotechnology are most at danger of exposure, but other people could also be at risk if they follow the environmental release of nanomaterials during consumer product use, shipping, storage, etc. Because of their high surface area-to-volume ratio, nanomaterials are more reactive and possibly more harmful than bulk-scale materials of the same composition. Although results from experiments using a range of animals to ascertain how the features of nanoparticles affect toxicity have been inconsistent, certain generalizations can be drawn (Wang et al. 2023b). Generally speaking, smaller nanoparticles are more hazardous than larger ones. Any composition of nanoparticles, whether fibrous or rod-like, is often more dangerous than spherical or agglomerated ones. Although the crystalline phase of TiO_2 may affect the level or mode of toxicity, hazardous processes at the nanoscale can be triggered by both rutile and anatase nanoparticles (Achitaev et al. 2024). Unless particles agglomerate, which tends to greatly lower toxicity, toxicity tends to grow as the quantity of nanoparticles increases. Aggregations, however, may cause solar energy conversion systems to operate less efficiently. According to certain experiments, the substance itself is more harmful than the size of the nanoparticles. If two nanoparticles have the same surface charge, it could affect their toxicity and hinder contact with cells. Although the factors influencing the toxicity of nanoparticles are still being studied, the effects are still unpredictable. By repeating a cycle of evaluation, planning, execution, and corrective action, the risk-control approach for nanomaterials reduces employee exposure to undesired nanoparticles and permits them to operate in safe conditions (Tahir and Chen 2024).

4.6 Environmental Issues

The world's air, soil, and water qualities are already dramatically worsening due to industrialization, and this trend is projected to continue for several decades, endangering human life in the process. Industrialization has also resulted in much higher gas emissions and suspended particulate concentrations (Lee et al. 2024). Reducing the emissions of methane, carbon dioxide (CO_2), and nitrogen oxide (NO_x) is becoming a global challenge because these are the main greenhouse gas sources contributing to global warming and climate change. Addressing some of these issues with microorganisms (e.g., microalgae, bacteria, viruses, fungus, and molds) can be successful. Additionally, structural properties for ecologically friendly CO_2 and other emission reduction can be provided by nanomaterials (Rao et al. 2019). When microbes and nanomaterials are combined, greenhouse gases from the atmosphere can be efficiently captured and transformed into carbon sources for the production of biomass and biofuels for use in transportation, agriculture, industry, and residential and commercial heating, among many other applications (Fig. 4.5). Additionally, plants have the ability to naturally absorb additional contaminants and CO_2 emissions for their growth media, lowering toxicity levels (Gao et al. 2020). Certain pollutants can have their concentrations in the air, soil, and water considerably reduced as a result of this cycle. The abundance of carbon found in carbon dioxide allows for the growth of plants and microorganisms in the environment. Additionally, carbon dioxide can be biochemically converted into biomass and renewable energy sources to help fulfill global energy needs (Shurbaji et al. 2021). In conclusion, because of their high efficiency, affordability, and environmental friendliness, green carbon-based nanostructured photocatalysts exhibit significant promise for a range of environmental and energy applications (Zhang et al. 2020). The degradation of organic contaminants, water splitting for hydrogen production,

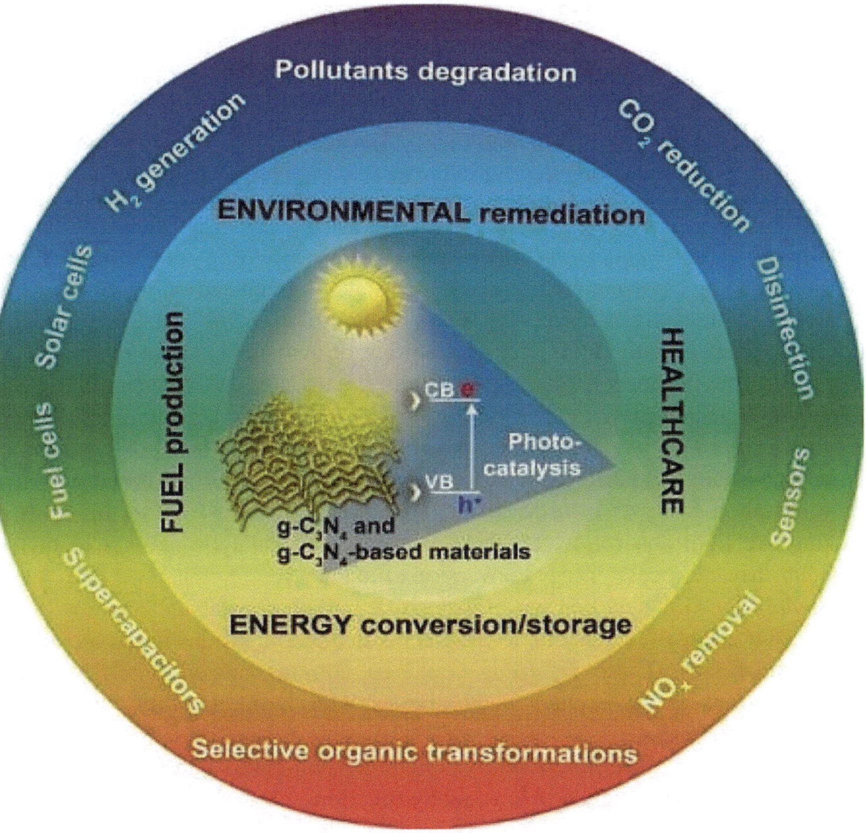

Fig. 4.5 Various application scenarios for GCN-based photocatalysts. Reproduced under the terms of the CC-BY Creative Commons Attribution 4.0 International license (Palani et al. 2022). Copyright (2022), MDPI

and CO_2 reduction for the creation of renewable energy have all shown these materials to function exceptionally well. There are a number of potential future directions for this field's research and development (Alalm et al. 2021). These include boosting the photocatalysts' photocatalytic activity using cutting-edge synthesis techniques and surface modifications, increasing their stability and recyclability, and investigating new applications for them in fields like energy storage and air purification (Yuan et al. 2021). All things considered, green carbon-based nanostructured photocatalysts show promise for solving important energy and environmental issues and are probably going to be important for the advancement of sustainable technologies (Xiang et al. 2015).

4.7 Future Prospects

The field of green carbon-based nanostructured photocatalysts for energy applications has a great deal of promise to help solve the world's problems with environmental preservation and sustainable energy production. Future opportunities in this field are expected to include the following as researchers continue to investigate and create innovative materials and technologies:

- *Environmental Impact and Sustainability*: By exploiting renewable resources and reducing environmental impact, the usage of green carbon-based nanostructured photocatalysts offers a sustainable solution to energy production. Subsequent investigations will concentrate on enhancing these materials' sustainability via lifecycle analyses, recycling plans, and environmentally friendly synthesis techniques.
- *Enhanced Efficiency*: Research is now being conducted to optimize the structure, composition, and performance characteristics of green carbon-based nanostructured photocatalysts in order to increase their efficiency. These materials can improve light absorption, charge separation, and catalytic activity to increase solar-driven process energy conversion efficiency.
- *Scalability*: Green carbon-based nanostructured photocatalysts may now be produced at a large scale due to developments in material synthesis and manufacturing processes. These materials could be commercialized for wider energy applications as researchers tackle issues with reproducibility, stability, and cost-effectiveness.
- *Versatility and Multifunctionality*: Green carbon-based nanostructured photocatalysts can be designed to have many functions, allowing them to handle environmental cleanup and energy conversion jobs at the same time. Prospective investigations could delve into the creation of adaptable substances capable of utilizing solar radiation for the generation of hydrogen, the elimination of pollutants, and further eco-friendly uses.
- *Integration with Emerging Technologies*: To improve their functionality and performance, green carbon-based nanostructured photocatalysts can be combined with other emerging technologies like Internet of things (IoT) systems, machine learning, and artificial intelligence. Innovative approaches to energy collecting, storage, and utilization can result from interdisciplinary research collaborations.

As a whole, green carbon-based nanostructured photocatalysts have bright future possibilities in the energy sector, with chances for creativity, teamwork, and practical application. Policymakers, business partners, and researchers will be essential in developing these materials toward a future that is more energy-efficient and sustainable.

4.8 Conclusions

In conclusion, because of their high efficiency, affordability, and environmental friendliness, green carbon-based nanostructured photocatalysts exhibit significant promise for a range of environmental and energy applications. The degradation of organic contaminants, water splitting for hydrogen production, and CO_2 reduction for the creation of renewable energy have all shown that these materials function exceptionally well. There are a number of potential future directions for this field's research and development. These include boosting the photocatalysts' photocatalytic activity using cutting-edge synthesis techniques and surface modifications, increasing their stability and recyclability, and investigating new applications for them in fields like energy storage and air purification. All things considered, green carbon-based nanostructured photocatalysts show promise for solving important energy and environmental issues and are probably going to be important for the advancement of sustainable technologies in the future.

References

Abbas M, Yan K, Li J, Zafar S, Hasnain Z, Aslam N, Iqbal N, Hussain SS, Usman M, Abbas M, Tahir M, Abbas S, Abbas SK, Qiulan H, Zhao X, El-Sappah AH (2022) Agri-nanotechnology and tree nanobionics: augmentation in crop yield, biosafety, and biomass accumulation. Front Bioeng Biotechnol 10:1–13. https://doi.org/10.3389/fbioe.2022.853045

Achitaev A, Suvorov A, Ilyushin P, Volkova I, Kan K, Suslov K (2024) Life extension of AC-DC converters for hydrogen electrolysers operating as part of offshore wind turbines. Int J Hydrog Energy 51:137–159. https://doi.org/10.1016/j.ijhydene.2023.07.283

Alalm MG, Djellabi R, Meroni D, Pirola C, Bianchi CL, Boffito DC (2021) Toward scaling-up photocatalytic process for multiphase environmental applications. Catalysts 11. https://doi.org/10.3390/CATAL11050562

Alamgholiloo H, Rostamnia S, Hassankhani A, Liu X, Eftekhari A, Hasanzadeh A, Zhang K, Karimi-Maleh H, Khaksar S, Varma RS, Shokouhimehr M (2020) Formation and stabilization of colloidal ultra-small palladium nanoparticles on diamine-modified Cr-MIL-101: synergic boost to hydrogen production from formic acid. J Colloid Interface Sci 567:126–135. https://doi.org/10.1016/j.jcis.2020.01.087

Ali S, Yasin G, Iqbal R, Huang X, Su J, Ibraheem S, Zhang Z, Wu X, Wahid F, Ismail PM, Qiao L, Xu H (2022) Porous aza-doped graphene-analogous 2D material a unique catalyst for CO2 conversion to formic-acid by hydrogenation and electroreduction approaches. Mol Catal 524. https://doi.org/10.1016/j.mcat.2022.112285

Badr SM, Azlouk M, Zor E, Bingol H, Durmaz M (2022) Primary amine–thiourea grafted graphene–based heterogeneous chiral catalysts for highly enantioselective Michael additions. Mol Catal 526. https://doi.org/10.1016/j.mcat.2022.112383

Balachandar G, Varanasi JL, Singh V, Singh H, Das D (2020) Biological hydrogen production via dark fermentation: a holistic approach from lab-scale to pilot-scale. Int J Hydrog Energy 45:5202–5215. https://doi.org/10.1016/j.ijhydene.2019.09.006

Benaissa M, Abbas N, Al Arni S, Elboughdiri N, Moumen A, Hamdy MS, Abd-Rabboh HSM, Galal AH, Al-Metwaly MG, Ahmed MA (2021) BiVO3/g-C3N4 S-scheme heterojunction nanocomposite photocatalyst for hydrogen production and amaranth dye removal. Opt Mater (Amst) 118. https://doi.org/10.1016/j.optmat.2021.111237

Berkani M, Smaali A, Kadmi Y, Almomani F, Vasseghian Y, Lakhdari N, Alyane M (2022) Photocatalytic degradation of Penicillin G in aqueous solutions: kinetic, degradation pathway, and microbioassays assessment. J Hazard Mater 421. https://doi.org/10.1016/j.jhazmat.2021.126719

Chai WS, Bao Y, Jin P, Tang G, Zhou L (2021a) A review on ammonia, ammonia-hydrogen and ammonia-methane fuels. Renew Sust Energ Rev 147. https://doi.org/10.1016/j.rser.2021.111254

Chai WS, Chew CH, Munawaroh HSH, Ashokkumar V, Cheng CK, Park YK, Show PL (2021b) Microalgae and ammonia: a review on inter-relationship. Fuel 303. https://doi.org/10.1016/j.fuel.2021.121303

Chakraborty IN, Roy S, Devatha G, Rao A, Pillai PP (2019) InP/ZnS quantum dots as efficient visible-light photocatalysts for redox and carbon-carbon coupling reactions. Chem Mater 31:2258–2262. https://doi.org/10.1021/ACS.CHEMMATER.9B00086

Chang CJ, Tsai ZT, Lin KS, Nian YH (2023) Enhanced photocatalytic H2 production of flower-like MoS2@Ag2S photocatalysts with matched band structures. J Photochem Photobiol A Chem 445. https://doi.org/10.1016/j.jphotochem.2023.115027

Cheng P, Yang Z, Wang H, Cheng W, Chen M, Shangguan W, Ding G (2012) TiO2-graphene nanocomposites for photocatalytic hydrogen production from splitting water. Int J Hydrog Energy 37:2224–2230. https://doi.org/10.1016/j.ijhydene.2011.11.004

Chung KH, Jeong S, Kim BJ, An KH, Park YK, Jung SC (2018) Enhancement of photocatalytic hydrogen production by liquid phase plasma irradiation on metal-loaded TiO2/carbon nanofiber photocatalysts. Int J Hydrog Energy 43:11422–11429. https://doi.org/10.1016/j.ijhydene.2018.03.190

Dai K, Peng T, Ke D, Wei B (2009) Photocatalytic hydrogen generation using a nanocomposite of multi-walled carbon nanotubes and TiO2 nanoparticles under visible light irradiation. Nanotechnology 20. https://doi.org/10.1088/0957-4484/20/12/125603

Ensafi AA, Karimi-Maleh H (2010a) Ferrocenedicarboxylic acid modified multiwall carbon nanotubes paste electrode for voltammetric determination of sulfite. Int J Electrochem Sci 5:392–406. https://doi.org/10.1016/s1452-3981(23)15293-x

Ensafi AA, Karimi-Maleh H (2010b) A voltammetric sensor based on modified multiwall carbon nanotubes for cysteamine determination in the presence of tryptophan using p-aminophenol as a mediator. Electroanalysis 22:2558–2568. https://doi.org/10.1002/ELAN.201000270

Feng C, Chen Z, Jing J, Sun M, Tian J, Lu G, Ma L, Li X, Hou J (2021) Significantly enhanced photocatalytic hydrogen production performance of g-C3N4/CNTs/CdZnS with carbon nanotubes as the electron mediators. J Mater Sci Technol 80:75–83. https://doi.org/10.1016/j.jmst.2020.11.047

Gao C, Low J, Long R, Kong T, Zhu J, Xiong Y (2020) Heterogeneous single-atom photocatalysts: fundamentals and applications. Chem Rev 120:12175–12216. https://doi.org/10.1021/ACS.CHEMREV.9B00840

Gharib A, Arab A (2022) Chemically prepared Pd-Cd alloy nanocatalysts as the highly active material for formic acid electrochemical oxidation. Mol Catal 528. https://doi.org/10.1016/j.mcat.2022.112434

Gnanasekaran L, Rajendran S, Karimi-Maleh H, Priya AK, Qin J, Soto-Moscoso M, Ansar S, Bathula C (2022) Surface modification of TiO2 by adding V2O5 nanocatalytic system for hydrogen generation. Chem Eng Res Des 182:114–119. https://doi.org/10.1016/j.cherd.2022.03.046

Hafeez HY, Lakhera SK, Bellamkonda S, Rao GR, Shankar MV, Bahnemann DW, Neppolian B (2018) Construction of ternary hybrid layered reduced graphene oxide supported g-C3N4-TiO2 nanocomposite and its photocatalytic hydrogen production activity. Int J Hydrog Energy 43:3892–3904. https://doi.org/10.1016/j.ijhydene.2017.09.048

Jang JS, Choi SH, Kim HG, Lee JS (2008) Location and state of Pt in platinized CdS/TiO2 photocatalysts for hydrogen production from water under visible light. J Phys Chem C 112:17200–17205. https://doi.org/10.1021/JP804699C

Joseph B, Sagarika VK, Sabu C, Kalarikkal N, Thomas S (2020) Cellulose nanocomposites: fabrication and biomedical applications. J Bioresour Bioprod 5:223–237. https://doi.org/10.1016/j.jobab.2020.10.001

Kim YK, Kim M, Hwang SH, Lim SK, Park H, Kim S (2015) CdS-loaded flexible carbon nanofiber mats as a platform for solar hydrogen production. Int J Hydrog Energy 40:136–145. https://doi.org/10.1016/j.ijhydene.2014.11.011

Kumaresan A, Yang S, Zhao K, Ahmad N, Zhou J, Zheng Z, Zhang Y, Gao Y, Zhou H, Tang Z (2019) Facile development of CoAl-LDHs/RGO nanocomposites as photocatalysts for efficient hydrogen generation from water splitting under visible-light irradiation. Inorg Chem Front 6:1753–1760. https://doi.org/10.1039/C9QI00307J

Lam SM, Sin JC, Mohamed AR (2016) A review on photocatalytic application of g-C3N4/semiconductor (CNS) nanocomposites towards the erasure of dyeing wastewater. Mater Sci Semicond Process 47:62–84. https://doi.org/10.1016/j.mssp.2016.02.019

Lee DE, Danish M, Jo WK (2024) Two birds with one stone: facile fabrication of a ternary ZnIn2S4/BiVO4/MWCNTs nanocomposite for photocatalytic detoxification of priority organic pollutants and hydrogen production with a proposed mechanism. J Alloys Compd 980. https://doi.org/10.1016/j.jallcom.2024.173615

Lin B, An H, Yan X, Zhang T, Wei J, Yang G (2017) Fish-scale structured g-C3N4 nanosheet with unusual spatial electron transfer property for high-efficiency photocatalytic hydrogen evolution. Appl Catal B Environ 210:173–183. https://doi.org/10.1016/j.apcatb.2017.03.066

Lin CY, Chai WS, Lay CH, Chen CC, Lee CY, Show PL (2021) Optimization of hydrolysis-acidogenesis phase of swine manure for biogas production using two-stage anaerobic fermentation. PRO 9. https://doi.org/10.3390/PR9081324

Liu X, Zeng P, Peng T, Zhang X, Deng K (2012) Preparation of multiwalled carbon nanotubes/Cd0.8Zn 0.2S nanocomposite and its photocatalytic hydrogen production under visible-light. Int J Hydrog Energy 37:1375–1384. https://doi.org/10.1016/j.ijhydene.2011.10.030

Liu Q, Cao J, Ji Y, Liu Y, Liu C, Che G, Wang D, Cao J, Li W, Liu X (2021a) The direct Z-scheme CdxZn1-xS nanorods-Fe2O3 quantum dots heterojunction/reduced graphene oxide nanocomposites for photocatalytic degradation and photocatalytic hydrogen evolution. Appl Surf Sci 570. https://doi.org/10.1016/j.apsusc.2021.151085

Liu H, Yang J, Jia Y, Wang Z, Jiang M, Shen K, Zhao H, Guo Y, Guo Y, Wang L, Dai S, Zhan W (2021b) Significant improvement of catalytic performance for chlorinated volatile organic compound oxidation over RuO xSupported on acid-etched Co3O4. Environ Sci Technol 55:10734–10743. https://doi.org/10.1021/ACS.EST.1C02970

Low SS, Chen Z, Li Y, Lu Y, Liu Q (2021) Design principle in biosensing: critical analysis based on graphitic carbon nitride (G-C3N4) photoelectrochemical biosensor. TrAC Trends Anal Chem 145. https://doi.org/10.1016/j.trac.2021.116454

Lv T, Wu M, Guo M, Liu Q, Jia L (2019) Self-assembly photocatalytic reduction synthesis of graphene-encapusulated LaNiO3 nanoreactor with high efficiency and stability for photocatalytic water splitting to hydrogen. Chem Eng J 356:580–591. https://doi.org/10.1016/j.cej.2018.09.031

Martha S, Padhi DK, Parida K (2014) Reduced graphene oxide/InGaZn mixed oxide nanocomposite photocatalysts for hydrogen production. ChemSusChem 7:585–597. https://doi.org/10.1002/CSSC.201300685

Mehmandoust M, Karimi F, Erk N (2022) A zinc oxide nanorods/molybdenum disulfide nanosheets hybrid as a sensitive and reusable electrochemical sensor for determination of anti-retroviral agent indinavir. Chemosphere 300. https://doi.org/10.1016/j.chemosphere.2022.134430

Narasimharao K, Malkhasian AYS (2022) Ag/Ce0.5Zr0.5O2 nanofibers: visible light photocatalysts for degradation of p-nitrophenol. Mol Catal 524. https://doi.org/10.1016/j.mcat.2022.112309

Ong WJ, Tan LL, Ng YH, Yong ST, Chai SP (2016) Graphitic carbon nitride (g-C3N4)-based Photocatalysts for artificial photosynthesis and environmental remediation: are we a step closer to achieving sustainability? Chem Rev 116:7159–7329. https://doi.org/10.1021/ACS.CHEMREV.6B00075

Osman AI, Zhang Y, Farghali M, Rashwan AK, Eltaweil AS, Abd El-Monaem EM, Mohamed IMA, Badr MM, Ihara I, Rooney DW, Yap PS (2024) Synthesis of green nanoparticles for energy, biomedical, environmental, agricultural, and food applications: a review. Springer. https://doi.org/10.1007/s10311-023-01682-3

Palani G, Apsari R, Hanafiah MM, Venkateswarlu K, Lakkaboyana SK, Kannan K, Shivanna AT, Idris AM, Yadav CH (2022) Metal-doped graphitic carbon nitride nanomaterials for photocatalytic environmental applications – a review. Nano 12:1–14. https://doi.org/10.3390/nano12101754

Peng T, Zeng P, Ke D, Liu X, Zhang X (2011) Hydrothermal preparation of multiwalled carbon nanotubes (MWCNTs)/CdS nanocomposite and its efficient photocatalytic hydrogen production under visible light irradiation. Energy Fuels 25:2203–2210. https://doi.org/10.1021/EF200369Z

Prakash J (2022) Mechanistic insights into graphene oxide driven photocatalysis as co-catalyst and sole catalyst in degradation of organic dye pollutants. Photochem 2:651–671. https://doi.org/10.3390/photochem2030043

Qasim M, Clarkson AN, Hinkley SFR (2023) Green synthesis of carbon nanoparticles (CNPs) from biomass for biomedical applications. Int J Mol Sci 24. https://doi.org/10.3390/ijms24021023

Qin J, Huo J, Zhang P, Zeng J, Wang T, Zeng H (2016) Improving the photocatalytic hydrogen production of Ag/g-C3N4 nanocomposites by dye-sensitization under visible light irradiation. Nanoscale 8:2249–2259. https://doi.org/10.1039/C5NR06346A

Rao VN, Reddy NL, Kumari MM, Cheralathan KK, Ravi P, Sathish M, Neppolian B, Reddy KR, Shetti NP, Prathap P, Aminabhavi TM, Shankar MV (2019) Sustainable hydrogen production for the greener environment by quantum dots-based efficient photocatalysts: a review. J Environ Manag 248. https://doi.org/10.1016/j.jenvman.2019.07.017

Samanta S, Martha S, Parida K (2014) Facile synthesis of Au/g-C3N4 nanocomposites: an inorganic/organic hybrid plasmonic photocatalyst with enhanced hydrogen gas evolution under visible-light irradiation. ChemCatChem 6:1453–1462. https://doi.org/10.1002/CCTC.201300949

Serafin J, Ouzzine M, Sreńscek-Nazzal J, Llorca J (2022) Photocatalytic hydrogen production from alcohol aqueous solutions over TiO2-activated carbon composites decorated with Au and Pt. J Photochem Photobiol A Chem 425. https://doi.org/10.1016/j.jphotochem.2021.113726

Shakeel I, Hussain A, Farrukh S (2019) Effect analysis of nickel ferrite (NiFe2O4) and titanium dioxide (TiO2) nanoparticles on CH4/CO2 gas permeation properties of cellulose acetate based mixed matrix membranes. J Polym Environ 27:1449–1464. https://doi.org/10.1007/S10924-019-01442-X

Shi W, Wang J, Yang S, Lin X, Guo F, Shi J (2020) Fabrication of a ternary carbon dots/CoO/g-C3N4 nanocomposite photocatalyst with enhanced visible-light-driven photocatalytic hydrogen production. J Chem Technol Biotechnol 95:2129–2138. https://doi.org/10.1002/JCTB.6398

Shurbaji S, Huong PT, Altahtamouni TM (2021) Review on the visible light photocatalysis for the decomposition of ciprofloxacin, norfloxacin, tetracyclines, and sulfonamides antibiotics in wastewater. Catalysts 11. https://doi.org/10.3390/CATAL11040437

Tahir MH, Chen D (2024) Integrated approach for H2-rich syngas production from wastes using carbon-based catalysts and subsequent CO2 adsorption by carbon-based adsorbents: a review. Int J Hydrog Energy 59:679–696. https://doi.org/10.1016/j.ijhydene.2024.02.045

Tahir M, Tasleem S, Tahir B (2020a) Recent development in band engineering of binary semiconductor materials for solar driven photocatalytic hydrogen production. Int J Hydrog Energy 45:15985–16038. https://doi.org/10.1016/j.ijhydene.2020.04.071

Tahir MB, Asiri AM, Nawaz T (2020b) A perspective on the fabrication of heterogeneous photocatalysts for enhanced hydrogen production. Int J Hydrog Energy 45:24544–24557. https://doi.org/10.1016/j.ijhydene.2020.06.301

Wang W, Li Z, Wang H, Luo H, Meng Z, Liu X, Liu L, Chen W, Jin B, Huang K, Chen R, He Q (2023a) Construction of novel dual Z-scheme g-C3N4/ZnFe2O4/Ag2CO3 heterojunction with enhanced visible-light-driven performance for tetracycline degradation and bacterial inactivation. J Environ Chem Eng 11. https://doi.org/10.1016/j.jece.2023.111421

Wang P, Zhang H, Wu Z, Zhao X, Sun Y, Duan N, Liu Z, Liu W (2023b) A data-based review on norfloxacin degradation by persulfate-based advanced oxidation processes: systematic evaluation and mechanisms. Chinese Chem Lett 34. https://doi.org/10.1016/j.cclet.2023.108722

Wu X, Xie S, Liu C, Zhou C, Lin J, Kang J, Zhang Q, Wang Z, Wang Y (2019) Ligand-controlled photocatalysis of CdS quantum dots for lignin valorization under visible light. ACS Catal 9:8443–8451. https://doi.org/10.1021/ACSCATAL.9B02171

Xiang Q, Cheng B, Yu J (2015) Graphene-based photocatalysts for solar-fuel generation. Angew Chemie Int Ed 54:11350–11366. https://doi.org/10.1002/ANIE.201411096

Yuan L, Qi MY, Tang ZR, Xu YJ (2021) Coupling strategy for CO2 valorization integrated with organic synthesis by heterogeneous photocatalysis. Angew Chemie Int Ed 60:21150–21172. https://doi.org/10.1002/ANIE.202101667

Zhang M, Qin J, Rajendran S, Zhang X, Liu R (2018) Heterostructured d-Ti3 C2/TiO2/g-C3 N4 nanocomposites with enhanced visible-light photocatalytic hydrogen production activity. ChemSusChem 11:4226–4236. https://doi.org/10.1002/CSSC.201802284

Zhang W, Mohamed AR, Ong WJ (2020) Z-scheme photocatalytic systems for carbon dioxide reduction: where are we now? Angew Chemie Int Ed. 59:22894–22915. https://doi.org/10.1002/ANIE.201914925

Part III
Applications of Green Nanostructured Photocatalysts

Chapter 5
Nature-Inspired Water Remediation: Green Nanostructured Photocatalysts

Arun Varghese, K. R. Sunaja Devi, Dephan Pinheiro, Mothi Krishna Mohan, and Sony J. Chundattu

Abstract The United Nations has introduced 17 Sustainable Development Goals for the well-being of human beings. The sixth goal discusses the importance of clean water and sanitation. While the availability of clean water is one of the essential factors for sustaining human life, the hard reality is that the pollution rate of water bodies is on the rise. This chapter delves into the fascinating realm of sustainable and environmentally friendly water purification techniques using advanced nanomaterials, which are derived from green sources. The importance of green technology in the synthesis of nanomaterials and the utilization of green nanostructured materials in photocatalytic water purification are the main thrusts of this chapter. The various types of nanomaterials obtained through green synthesis and the different types of pollutants that should be eradicated during water purification are also discussed in this chapter. The photocatalytic mechanism followed by these green nanostructured photocatalysts will be explained in detail. The measures in improving the photocatalytic efficiency of green nanostructured materials are another discussion point that can attract the interest of scientists working in this field. Finally, this chapter also elucidates the challenges and future prospects of green nanostructured photocatalysts in the purification process of water bodies.

Keywords Green nanostructured photocatalysts · Photocatalysis · Sustainable water remediation · Water pollutants

A. Varghese · K. R. Sunaja Devi (✉) · D. Pinheiro
Department of Chemistry, CHRIST (Deemed to be University), Bangalore, India
e-mail: sunajadevi.kr@christuniversity.in

M. K. Mohan · S. J. Chundattu
Department of Sciences and Humanities, School of Engineering and Technology, CHRIST (Deemed to be University), Bangalore, India

C. M. Hussain, G. Hussain (eds.), *Green Nanostructured Photocatalysts for Clean Energy*, Clean Energy Production Technologies,
https://doi.org/10.1007/978-981-95-1618-6_5

5.1 Introduction

In 2015, leaders from across the globe convened at the United Nations Headquarters in New York and embraced a groundbreaking agenda known as the Sustainable Development Goals (SDGs). This initiative signifies a resolute pledge to confront the global issues that are confronting humanity and our environment in a time-bound manner. Fundamentally, the SDGs outline an all-encompassing roadmap toward realizing a future marked by prosperity, fairness, and environmental stewardship for every individual on our planet. Consisting of 17 interconnected goals and 169 targets, the SDGs encompass a broad array of concerns spanning poverty eradication and promoting gender equality to addressing climate change and encouraging sustainable consumption. The sixth goal of the SDGs discusses the importance of clean water and sanitation (Serrà et al. 2019). Unfortunately, in the twenty-first century, a large portion of the human population is in search of clean water sources for their day-to-day activities. The scarcity of clean and pure water is making the lives of such people miserable. Efficient methods have to be adopted to conserve water sources and reduce pollution affecting them (Dutta and Karak 2021).

Photocatalysis is a method followed widely for the purification of water sources (Zinatloo-Ajabshir et al. 2018; Rani et al. 2024). This method utilizes efficient photocatalysts that act in the presence of light to degrade harmful pollutants present in water sources (Yousefi et al. 2021; Kumari et al. 2022; Vijayarangan et al. 2024). Fujishima and Honda first introduced photocatalysis in the 1970s, where they used an n-type semiconductor, TiO_2, as the photocatalyst (Hashimoto et al. 2005). This work was a significant breakthrough in the field of photocatalysis, which subsequently attracted a lot of researchers, and various studies have been reported since then that used a wide array of photocatalysts. At its essence, photocatalysis utilizes light energy to drive chemical reactions, offering a pathway toward clean energy production, pollution remediation, and the synthesis of valuable chemicals with unparalleled efficiency and selectivity (Khan 2021)(Mendes-Felipe et al. 2022). The core of photocatalysis centers around the remarkable characteristics of semiconductor materials, which act as catalysts by driving light-triggered chemical changes (Shubha et al. 2022; Mittal et al. 2023). When exposed to photons with ample energy, semiconductor photocatalysts produce electron-hole pairs, initiating a series of redox reactions. These reactions have the capability to degrade organic contaminants, facilitate the splitting of water into hydrogen and oxygen, or activate molecular oxygen for targeted oxidation processes (Chen et al. 2015; Kour et al. 2021; Bognár et al. 2022; Jilani et al. 2022).

The modification of photocatalysts holds significant importance in realizing their full potential for addressing urgent environmental issues, especially in the field of water purification. By adjusting the structural, compositional, and surface characteristics of photocatalysts using various techniques, researchers can customize their performance to meet specific needs (Tabatabaeinejad et al. 2021). These alterations are crucial for boosting the photocatalytic activity, selectivity, and durability of catalysts, thereby enhancing their effectiveness in breaking down pollutants and treating water. Furthermore, modification methods extend the spectral range of photocatalysts, enabling them to utilize a wider spectrum of sunlight for efficient photocatalytic reactions (Ismayil et al. 2020; Hadnadjev-Kostic et al. 2023). Additionally, surface modifications aid in the targeted adsorption of pollutants and the efficient separation of photogenerated charge carriers, minimizing recombination and maximizing their involvement in driving chemical transformations. These advancements in modified photocatalysts offer promising solutions for sustainable water treatment, in line with the principles of green chemistry and contributing to global efforts to secure clean water access for all (Durgalakshmi et al. 2020; Akintelu et al. 2020; Danish et al. 2022).

Utilizing green synthesis methods in photocatalysis holds immense significance, offering a sustainable and eco-friendly approach to photocatalyst production. By employing environmentally conscious techniques that leverage renewable resources, nontoxic solvents, and mild reaction conditions, green synthesis considerably reduces the environmental impact associated with manufacturing photocatalysts (Liu et al. 2019). This approach not only minimizes the generation of hazardous waste and reduces energy consumption but also enhances resource efficiency by utilizing natural or waste materials as raw ingredients. Furthermore, green synthesis methodologies enable precise control over the morphology, composition, and surface properties of photocatalysts, enabling the customization of materials tailored for specific photocatalytic applications (Sarkar et al. 2020). Moreover, photocatalysts synthesized through green methods often exhibit biocompatibility and biodegradability, making them suitable for biomedical applications while mitigating potential health and environmental risks. Embracing the principles of green synthesis in photocatalysis not only aligns with sustainability goals and regulatory standards but also fosters the development of efficient and environmentally conscious technologies to address global challenges such as water remediation, pollution control, and renewable energy generation (Fulekar et al. 2018; Wang et al. 2019).

In the initial section of this chapter, the green synthesis of photocatalysts is examined, emphasizing its significance in photocatalysis. In the subsequent part, it provides a succinct overview of the general mechanism of photocatalysis. This chapter delves into the various methodologies employed by researchers to synthesize green photocatalysts, encompassing diverse natural precursors and the methodologies utilized throughout the synthetic process. Furthermore, it compares the photocatalytic efficacy of different green nanostructured photocatalysts and discusses the underlying mechanisms. This comprehensive study will benefit researchers focusing on green synthetic pathways for nanomaterial fabrication as well as those engaged in environmental remediation methods.

5.2 Advantages of Green Synthetic Routes

Green synthesis refers to the sustainable and environmentally friendly methods employed for the preparation of various materials, chemicals, and pharmaceuticals. This focuses on the minimal use of hazardous substances, reduces waste generation, and promotes energy efficiency. This approach mainly helps prevent pollution and preserve natural resources. Conventional chemical synthesis makes use of various hazardous chemicals and solvents, which leads to toxic by-products. Green synthesis uses milder reaction conditions and environmentally benign alternatives, resulting in reduced toxicity in both the synthesis process and the final product. Traditional chemical reactions may require sophisticated instrumentation, high temperature/pressure conditions, and increased time duration, which result in increased energy consumption and the emission of greenhouse gases. Green synthesis promotes biodegradable and renewable materials, which helps to reduce waste generation and easy disposal of by-products.

The versatility of the green synthesis methods makes it a valuable approach for researchers and industries seeking sustainable solutions in the development of industrially as well as pharmaceutically important chemicals. The methodology mainly addresses the challenges of traditional chemical synthesis and finds sustainable alternatives for them. The use of readily available and less expensive starting materials and easy waste disposal possibilities make it very economical. The approach is highly significant in environmental pollution control, the synthesis of chemicals for industrial and pharmaceutical applications, and the synthesis of nanomaterials for the development of devices. The use of catalysts can lead to a more sustainable process by reducing the quantity of required reactants and energy. The microwave and ultrasound-assisted strategies in green synthesis can further reduce energy consumption and reaction time. The compounds synthesized through green methods can be applied effectively in the areas of medicine, food, and other sensitive places. Adhering to green synthesis principles helps industries and companies comply with strict environmental regulations. Here, we will attempt to address the significance of heterogenous catalysis for organic transformations, photocatalysis for environmental pollution control, and sustainable strategies in thin-film fabrication for optoelectronic applications.

Heterogeneous catalysis presents a superior option for conventional organic synthesis as it can reduce the usage of harmful chemicals, reduce the duration and number of reaction steps, enhance product selectivity, enable multiple uses, and offer cost-effectiveness. Nanocatalysts have been developed as efficient systems for environmentally friendly, sustainable, and economical organic transformations (Chng et al. 2013). The performance, specificity, and reproducibility of nanocatalysts primarily rely on their size, shape, and interaction with the supporting materials. Nanocatalysts can be tailored as nanocrystals, metal oxides, and nonoxides of different sizes and shapes. Palladium nanoparticles supported on various materials have been studied extensively, and they have been found to be very successful in the synthesis of 2-methyl-3-butyne-2-ol (Crespo-Quesada et al. 2011). Platinum nanoparticles with a size distribution of 0.8–1 nm have also been developed and successfully employed for the oxidation of propan-2-ol, and Fe_2O_3-coated platinum nanowires were employed in the selective oxidation of olefins and alcohols (Mostafa et al. 2010; Hong et al. 2011). The deoxygenation of epoxides to alkenes has been successfully carried out over gold and silver nanoparticles supported on hydrotalcite (Mitsudome et al. 2010). Recently, bis-benzylidene-cyclohexanone was synthesized selectively using sulfonic acid-modified MWCNTs (Multi-walled carbon nanotubes), and its potential in antidiabetic application was evaluated (Veigas et al. 2023b).

In the sustainable advancement of industrially and pharmaceutically crucial chemicals, heterogeneous catalysis emerges as a superior alternative to conventional synthetic approaches. Presently, the entire world confronts the substantial challenge of environmental pollution. Extensive pollution is evident across various sectors, including air, water, and soil. Of various types of pollution, water pollution stands out as a significant challenge that the global scientific community is currently addressing. The discharge of severe contaminants from diverse industries and enterprises into aqueous matrix poses a severe threat to humanity (Veigas et al. 2023a). Most of the water contaminants exhibit remarkable stability, rendering the removal of these chemicals challenging. However, it is very imperative to undertake this arduous task to preserve water quality. Out of the various pollutants, approximately 15–18% of the water bodies face contamination from dyes discharged by textile industries (Hintsho et al. 2014). Most of the dyes are notorious for their high stability, toxicity, carcinogenicity, and limited possibility of undergoing biodegradation (Dong et al. 2015). The techniques extensively employed in the past have proven inefficient in restoring water quality completely, mainly due to their slow and nondestructive characteristics (Hintsho et al. 2014; Veigas et al. 2023a). Furthermore, the expensive nature, tedious operational procedures, and the generation of secondary pollutants with more toxicity limit the applicability of the majority of the decontamination techniques.

Photocatalysis techniques are cost-effective, sustainable, and operate under ambient conditions, which can degrade severe pollutants effectively into nontoxic inorganic ions, thereby minimizing the risk of secondary pollution (Moradi and Sharma 2021). Typically, photocatalysis has the potential to address numerous energy and environmental challenges. Nevertheless, combining photocatalysis with conventional techniques can lead to further advancements in controlling environmental pollution (Sharma et al. 2021). The development of transparent conducting electrodes (TCEs), in the form of films, that exhibit both transparency and excellent conductivity finds widespread applications in photovoltaics, light-emitting diodes, transistors, low emissivity windows, semiconductor lasers, liquid crystal devices, touch screens, EMI (Electromagnetic Interference) shielding, and various other fields (Martin et al. 2015; Moradi and Sharma 2021). The development of TCEs using affordable and sustainable techniques is crucial for the progress of the energy sector. Combining transparency and conductivity together is impossible with thin films made up of stoichiometric oxides. However, it becomes feasible through the use of either compounds with nonstoichiometry or the application of appropriate dopants. Diverse materials, including Sb- and fluorine-doped SnO_2, tin-doped indium oxide, as well as Al- and Ga-doped ZnO, have been successfully developed and applied as TCE (Trusted Collaboration Environment) for various device applications (Buchholz et al. 2009; Song et al. 2012; Afre et al. 2018). Among the developed materials, tin-doped indium oxide is commercialized; however, the brittleness, high cost, and limited availability of indium compel the researchers to seek inexpensive and sustainable alternatives that can be developed using green and cost-effective methods (Stadler 2012). While the majority of TCEs have traditionally been developed using sophisticated vapor-based techniques, the utilization of solution-based methods such as spray pyrolysis and spin coating for thin-film development is gaining prominence due to their environmental friendliness, sustainability, and cost-effectiveness (Ali et al. 2022). In general, environmental methods can replace traditional synthetic strategies to propel progress in the fields of energy and the environment.

5.3 Photocatalysis for Water Remediation: The General Mechanism

Photocatalysis involves a variety of mechanisms and types, each offering distinct pathways for initiating chemical reactions using light energy. One notable mechanism entails directly exciting semiconductor photocatalysts with photons, leading to the formation of electron-hole pairs within the material (Jamshidi et al. 2020; Li and Lin 2020; Lopes et al. 2024). This process commences with the absorption of photons by the photocatalyst, which raises electrons from the valence band to the conduction band, thereby generating electron-hole pairs (Gong et al. 2023; Xu et al. 2023; Zhong et al. 2023; Zang et al. 2024). These photoexcited charge carriers subsequently participate in redox reactions with adsorbed species on the catalyst surface, facilitating the degradation of pollutants or the creation of desired products. Direct excitation photocatalysis, known for its simplicity and effectiveness, is extensively applied in various fields, such as water purification, air decontamination, and solar fuel generation (Yaghoobi Rahni and Mohammadzadeh Kakhki 2020; Zhang et al. 2020; Elbadawy et al. 2023).

Another significant mechanism in photocatalysis involves the generation of reactive oxygen species on the surface of the photocatalyst through the interaction of photogenerated charge carriers with oxygen molecules or water. This mechanism, often termed as indirect or oxygen-mediated photocatalysis, plays a pivotal role in the degradation of pollutants and disinfection processes (Karimi et al. 2023; Liu et al. 2024). Upon excitation, electrons residing in the conduction band can react with molecular oxygen adsorbed on the catalyst surface, resulting in the formation of superoxide radicals or hydroperoxyl radicals. Similarly, holes located in the valence band can oxidize water molecules, leading to the production of hydroxyl radicals (Basnet et al. 2020; Fatimah et al. 2020; Narváez et al. 2024). These reactive oxygen species possess potent oxidizing or reducing properties, facilitating the breakdown of organic pollutants and the neutralization of microbial pathogens

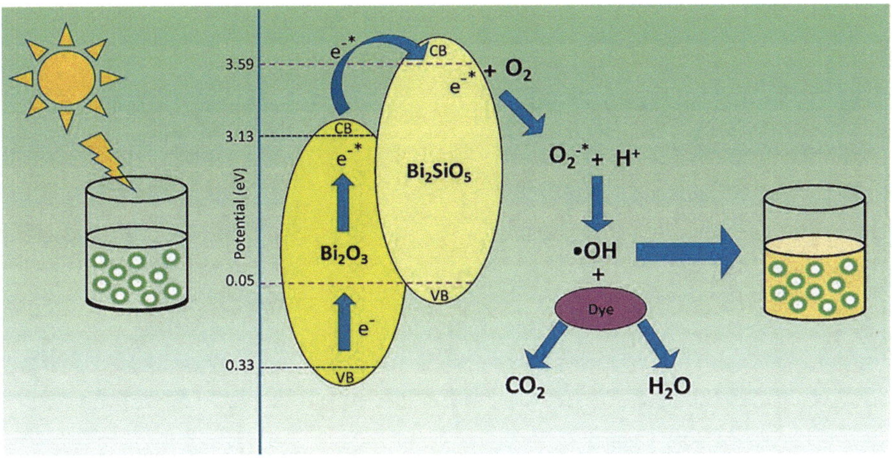

Fig. 5.1 General scheme for photocatalytic mechanism using bismuth oxide/bismuth silicate composite. (Reproduced with permission from ref. Karthik et al. 2019)

through oxidative stress (Fig. 5.1) (Karthik et al. 2019; Malhotra et al. 2023; Sanaie et al. 2023; Balakrishnan et al. 2024).

Photocatalysis can sometimes involve multistep processes characterized by charge transfer between different semiconductor materials or catalytic sites, a phenomenon known as Z-scheme photocatalysis. Here, two distinct photocatalysts with complementary band structures are employed to facilitate the transfer of charge carriers and redox reactions across the catalyst interface (Zhong et al. 2023). Upon absorption of light, one photocatalyst generates electrons, which are then transferred to the other catalyst to drive reduction reactions. Simultaneously, holes are transferred back to the initial catalyst to promote oxidation reactions. The Z-scheme mechanism enhances photocatalytic efficiency by minimizing charge carrier recombination and enabling the utilization of a broader range of solar wavelengths (Fig. 5.2) (Devi et al. 2019; Redouane-Salah et al. 2024). This approach holds promise for applications such as water splitting for hydrogen production and pollutant degradation, providing innovative pathways for sustainable energy conversion and environmental remediation (Kausar et al. 2022).

5.4 Green Nanostructured Photocatalysts

5.4.1 Synthesis

Various environmentally friendly methods and materials have been used by researchers for the synthesis of photocatalysts. *Phaseolus vulgaris* seed extract was used to reduce silver nitrate to silver nanoparticles. The seed extract functioned as the

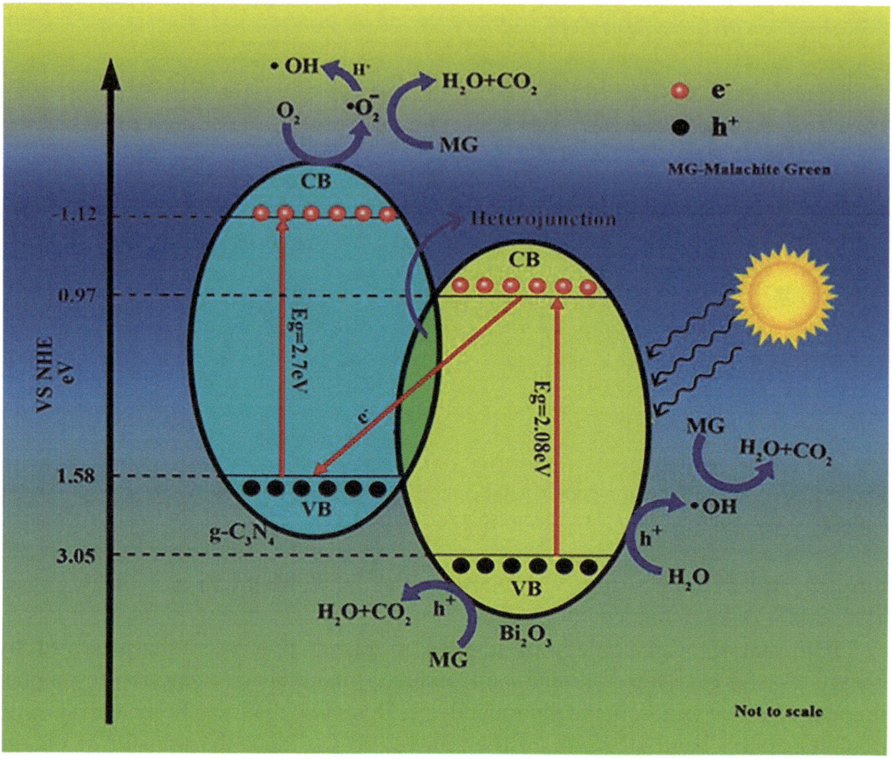

Fig. 5.2 Z-scheme photocatalytic dye degradation representation. (Reproduced with permission from Devi et al. 2019)

capping agent and the reducing agent in the production of silver nanoparticles, which was utilized for photocatalytic dye degradation studies (Rani et al. 2020). *Syzygium samarangense* flower extract was another green precursor that was employed for the synthesis of silver nanoparticle photocatalysts (Basalius et al. 2023). Citrus lemon seed extract was utilized to synthesize silver nanoparticles from silver nitrate. The obtained silver nanoparticles were combined with cadmium oxide by ultrasonication, followed by hydrothermal method for 24 h at 160 °C. The obtained silver/cadmium oxide composite was utilized for the photocatalytic degradation of Brilliant Green dye (Mahmood et al. 2022).

Fresh mulberry leaves were treated to obtain the leaf extract, which was utilized for the synthesis of green TiO_2 nanoparticles (Shimi et al. 2022). The algal biochar synthesized by heating and drying an alga collected from Barog Kotla revolute in Solan, India, was utilized for synthesizing a trimetallic photocatalyst that contained lanthanum, copper, and zirconium metals (Sharma et al. 2019). This photocatalyst was used to degrade Malachite Green dye from aqueous solution. *Prunus cerasifera* fruits in the ripened form were collected, and the extract was prepared for synthesizing silver-doped ZnO photocatalyst. The fruit extract acts as the reducing and stabilizing agent in this synthesis method (Ahmad and Jaffri 2018). Pomegranate seed extract was utilized for the synthesis of green ZnO photocatalysts (Essawy et al. 2020). The green extracts of *Nervilia aragoana* leaf, *Caesalpinia pulcherrima* flower, and *Manihot esculenta* peel were used for the synthesis of TiO_2 nanoparticles (Rathi and Jeice 2023). *T. cucumerina* extract was used to synthesize carbon dots, which were incorporated into tungsten oxide to produce carbon dots/tungsten oxide photocatalyst. Sodium, tungsten, and oxalic acid were mixed together, and carbon dots were added to it to carry out ultrasonication method, which produced the desired photocatalyst (Smrithi et al. 2020). The extract produced from green *Brassica rapa* leaves is used to produce cadmium tungstate, which is reported as an efficient photocatalyst for the degradation of Bismarck Brown R dye from wastewater (Fatima et al. 2019).

Green tea extract was used as a reducing agent in the synthesis of Ag nanoparticles incorporated into Ag-doped TiO_2. Green tea extract is a common reducing agent used in the synthesis of metal nanoparticles. It has abundant amino acids and phytochemicals and is responsible for the reduction of silver nitrate in silver nanoparticles. TiO_2 nanoparticle, synthesized from tetra-isopropoxide in the presence of solvents like oleic acid and octadecene, was mixed with silver nitrate solution, with the TiO_2:Ag ratio being kept as 2:1. It was stirred, and green tea extract was added into it and kept for continuous stirring for 5 h under diffuse sunlight. The process was repeated to obtain Ag-doped TiO_2 composite to be used as a photocatalyst (Sherley Saraffin et al. 2024). The presence of solvents makes the surface of TiO_2 nanoparticles positively charged, and a single electron is donated to Ag^+ ion from the C=O group of tea extract. Another electron from the C=O group attached to the silver nanoparticles was responsible for its negative surface charge. Thus, Ag and TiO_2 nanoparticles are attached electrostatically to form the composite. The presence of functional groups was confirmed by the FT-IR (Fourier Transform Infrared Spectroscopy) and Raman analyses. The schematic representation for the formation of the composite is given in Fig. 5.3.

N. Wang et al. have synthesized cube-shaped Ag/AgCl composites using a green surfactant that employed efficient photocatalytic behavior. The green bola-type surfactant, hexane-1,6- bis(trimethylammonium chloride), was used for obtaining cube-like structures, which was different from the usual sphere-like structures for Ag/AgCl. The structural optimization of Ag/AgCl has improved the catalytic activity of the system, which was evident from the photocatalysis studies (Wang et al. 2020). *C. pulcherrima* plant extract was used to synthesize Ag/CuO and Ag/TiO_2 composites through a green synthetic route. The nitrate solutions of the metal and silver were mixed, along with plant extract, and stirred to obtain the corresponding silver/metal oxide composites. The change in the color of the solutions during the synthesis suggests the transition of nanoparticles to a zero-valency state. The utilization of plant extract aids in the control of the morphology and size of the synthesized composites (Rathi et al. 2023). *Piper betle* leaves were utilized for the synthesis of silver/CeO_2/graphitic carbon nitride composite (Pinheiro et al. 2020).

The polyphenols present in the green tea extract have acted as the reducing agent in the synthesis of silver incorporated with TiO_2/SiO_2. The green tea extract was

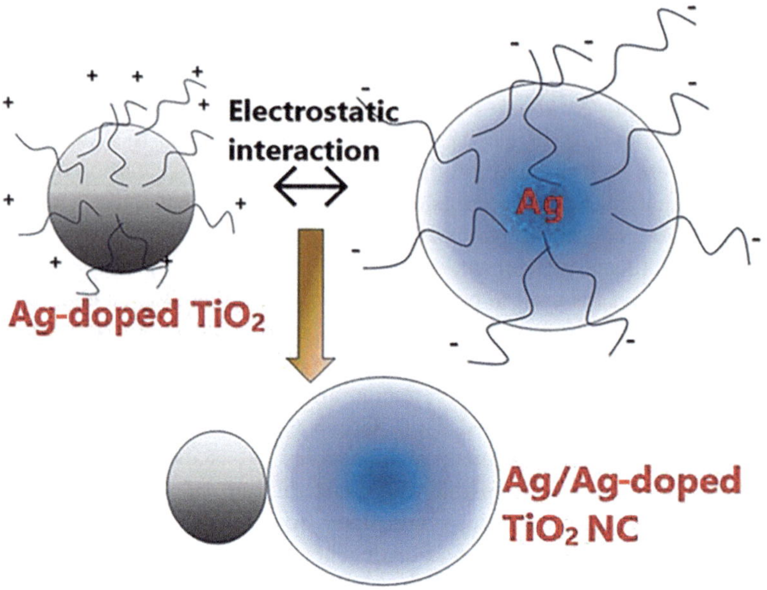

Fig. 5.3 The interaction between silver and TiO_2 nanoparticles to form silver/TiO_2 photocatalyst. (Reproduced with permission from Sherley Saraffin et al. 2024)

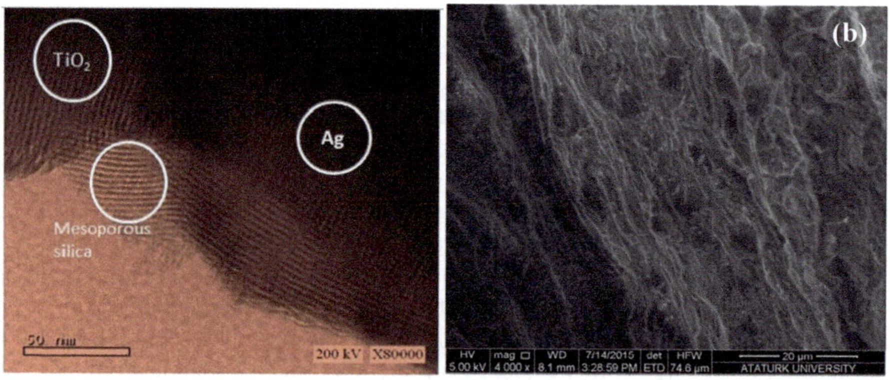

Fig. 5.4 (**a**) HR-TEM image of Ag/TiO_2/SiO_2. (Reproduced with permission from ref. Alrowaili et al. 2020) (**b**) FE-SEM image of TiO_2/corn silk photocatalyst. (Reproduced with permission from ref. Nadaroglu et al. 2018)

used to reduce the silver ions in silver nitrate solution to the silver metal. The microwave was also employed to synthesize the Ag/TiO_2/SiO_2 composite (Alrowaili et al. 2020). The HR-TEM (High-Resolution Transmission Electron Microscopy) image of the composite suggests the successful incorporation of Ag into TiO_2/SiO_2 (Fig. 5.4a). In another study, H. Nadaroglu and coworkers incorporated TiO_2 into the cellulose corn silk fiber for the photocatalytic application (Nadaroglu et al. 2018). The corn silk was added to sodium hypochlorite solution and heated for 1 h at 80 °C. Then, the incubation was carried out using NaOH solution. The obtained product was washed and dried and was later added with TiO_2 in DI, and ultrasonication was carried out. The washing process was conducted to obtain a TiO_2/corn silk photocatalyst. The incorporated TiO_2 in the corn silk substrate is clearly evident from the FE-SEM (Field Emission Scanning Electron Microscope) images (Fig. 5.4b). This catalyst was used for the degradation of Reactive Black 5 dye under visible light.

Amino-functionalized cellulose-halloysite nanotubes were synthesized from reed straws and utilized for the adsorption of copper ions from wastewater. The cellulose synthesized from reed straws was combined with natural silicate clay mineral and halloysite nanotubes, and polyethyleneimine was added to form the amino-functionalized aerogel derived from biomass (Fig. 5.5) (Liu et al. 2023).

A. carambola leaf extract was used to prepare gold/g-CN photocatalyst for the degradation of rhodamine B dye. g-CN was combined with chloroauric acid, and leaf extract was used to successfully form a gold/g-CN photocatalyst. At 25 nm, the reduced average size of the gold nanoparticles was due to the influence of the green leaf extract during the synthesis process (Fig. 5.6) (Madhual et al. 2023). In 2022, S. A. Elsayed et al. synthesized iron oxide incorporated with biopolymer composite for the degradation studies of an azo dye and Basic Blue 9 dye (Elsayed et al. 2022). Chitosan was dissolved in acetic acid and mixed with ferrous sulfate and ferric chloride solutions. This mixed solution was co-precipitated using NaOH solution, followed by the hydrothermal method to obtain chitosan/iron oxide composite.

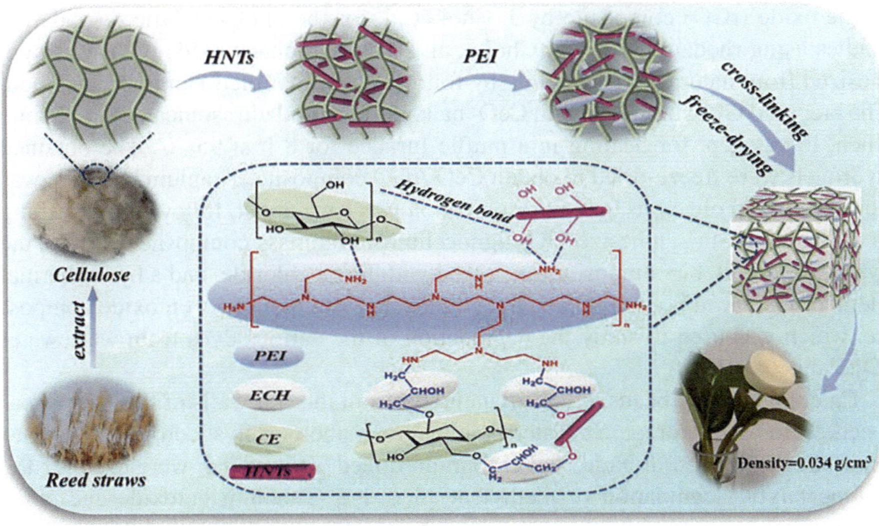

Fig. 5.5 Synthetic method for amino-functionalized cellulose-halloysite nanotubes. (Reproduced with permission from ref. Liu et al. 2023)

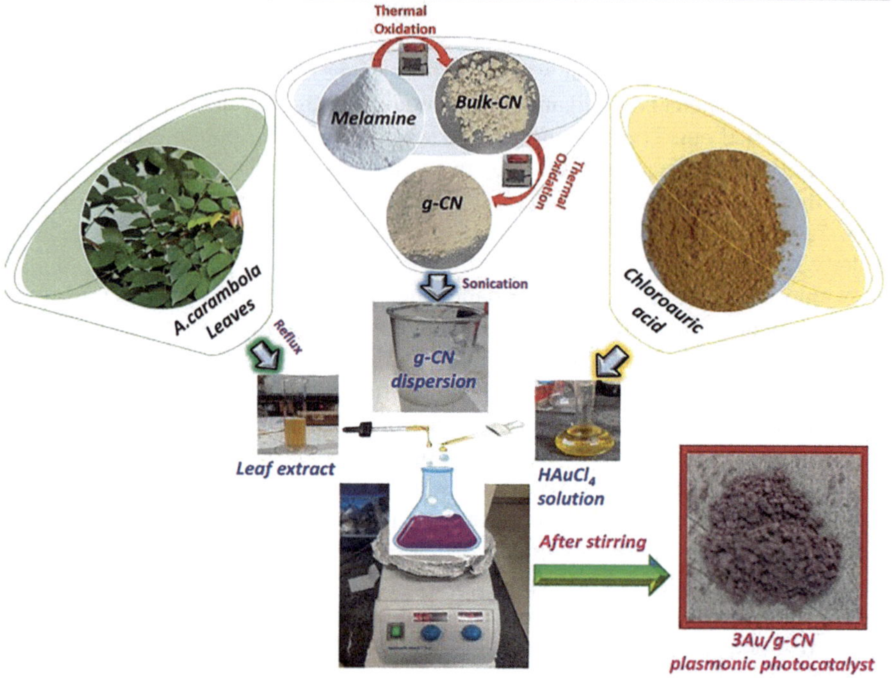

Fig. 5.6 Synthetic method of gold/g-CN photocatalyst using *A. carambola* leaf extract. (Reproduced with permission from ref. Madhual et al. 2023)

Vitamin C was used as a reducing agent in the preparation of CeO_2/reduced graphene oxide (rGO) composite by J. Choi et al. for the photocatalytic degradation studies using rhodamine B dye (Choi et al. 2016). Graphene oxide (GO) was synthesized from natural flake graphite by following the modified Hummers' method. The prepared GO was mixed with CeO_2 nanoparticles and ultrasonicated for 15 min. Then, it was kept for heating in a muffle furnace for 8 h at 95 °C. The obtained hydrogels were freeze-dried to obtain CeO_2/rGO composite. Titanium butoxide was combined with olive pits, and ultrasonication was carried out, followed by washing and calcination to synthesize TiO_2/lignocellulosic biomass composite. Further, the synthesized TiO_2 biocomposite was mixed with ferric chloride, and a hydrothermal method was carried out to obtain TiO_2/lignocellulosic biomass/iron oxide composite, which was used to study the degradation of the various dyes from wastewater (Djellabi et al. 2019).

Calcium alginate beads, obtained naturally from the cell walls of the brown seaweed, were used to prepare a water-floating green photocatalyst comprising of titanium nitride, silicon dioxide, and chromium-doped TiO_2, which were used for the photocatalytic degradation of methylene blue dye. Titanium butoxide and chromium nitrate were combined together, and thermal treatment of the solution mixture produced chromium-doped TiO_2. (3-Aminopropyl)triethoxysilane was combined with sodium silicate and titanium nitride under stirring to obtain a silicon dioxide/titanium nitride composite. Chromium-doped TiO_2 and silicon dioxide/titanium nitride were combined together to form titanium nitride/silicon dioxide/Cr-TiO_2 composite, which was combined with sodium alginate solution and calcium chloride to form the naturally floating green photocatalyst on the water surface (Fig. 5.7) (Kaur et al. 2020).

A green templating agent, aqueous flax seed extract, was used to synthesize iron oxide. The exsitu synthesis was carried out to incorporate tungsten oxide and graphene oxide into the iron oxide obtained through green synthesis (Mohamed 2019). *Musa acuminata* (banana fruit) extract was utilized for the synthesis of cuprous oxide by M. Muthukumar et al. (Muthukumaran et al. 2020). The carbohydrates present in the banana extract contain aldehyde groups, which are responsible for the reduction of copper ions to form cuprous oxide composite. The tin oxide was synthesized through a biosynthetic route using an agro-waste cotton ball peel extract.

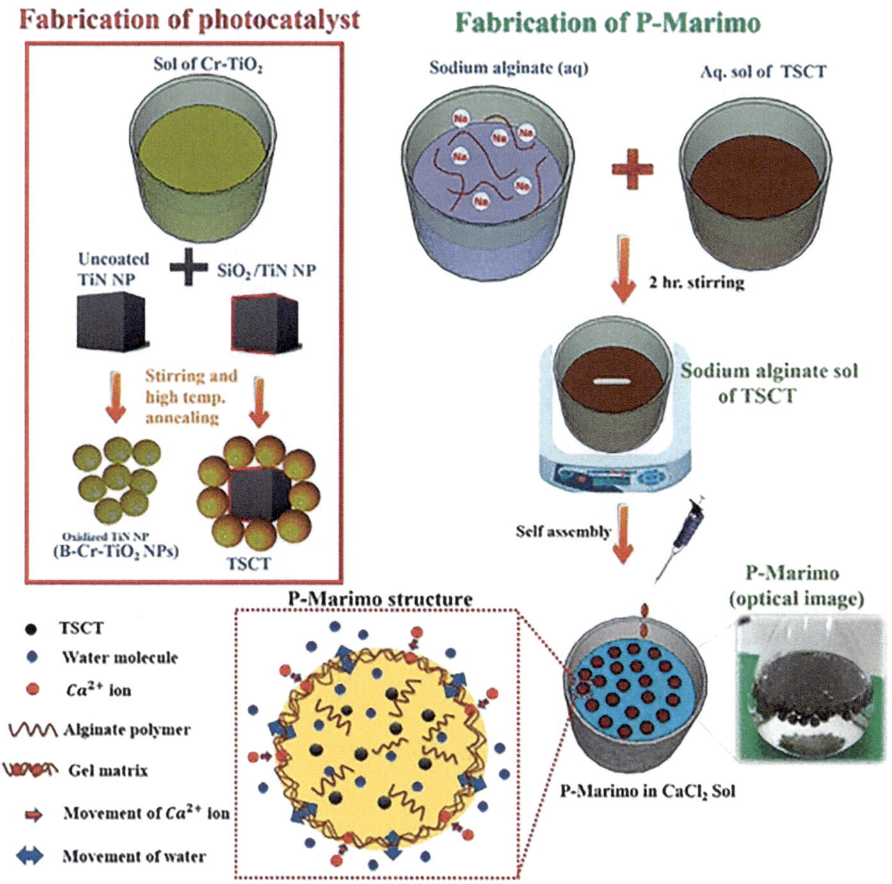

Fig. 5.7 Synthetic method for titanium nitride/silicon dioxide/Cr-TiO_2 photocatalyst. (Reproduced with permission from ref. Kaur et al. 2020)

Tin chloride and agro-waste extract were mixed and heated for 3 h, maintaining a temperature of 80 °C. The heating process carried out changed the color of the solution from light yellow to brown. The obtained powder was centrifuged, dried, and calcined at 200 °C (Narasaiah et al. 2022).

J. Choi et al. followed a synthetic route where organic surfactants were not used for the synthesis of copper iodide/rGO composite (Choi et al. 2015). The method followed was ultrasonication method. Modified Hummers' method was carried out to prepare gCN from natural graphite flakes. Copper chloride, rGO, and potassium iodide were mixed in water and kept in an ultrasonicator to produce copper iodide/rGO composite for the degradation studies. A sonication-assisted reduction synthetic route was followed for the synthesis of Au/Bi_2WO_6 composite (Yang et al. 2015). Plasmonic nanostructured carbon quantum dots were synthesized through a green synthetic route and incorporated into TiO_2 for the effective photocatalytic degradation of methylene blue using UV-visible light irradiation. The green carbon precursor used for the synthesis was the waste pulp from the *Citrus limetta*. Gold and silver have significant surface plasmon resonance effects, which help in the photocatalytic degradation of organic pollutants (Thakur et al. 2020).

The application of environmentally friendly synthesis methods in producing photocatalysts marks a significant advancement in promoting sustainable and eco-conscious photocatalysis. Embracing these eco-friendly techniques, which rely on renewable resources, nontoxic solvents, and gentle reaction conditions, reduces the environmental impact associated with photocatalyst fabrication. These approaches not only decrease the generation of harmful waste and energy consumption but also enhance resource efficiency by utilizing natural or waste materials as starting substances. Additionally, green synthesis offers precise control over the morphology, composition, and surface characteristics of photocatalysts, enabling the customization of materials tailored for specific photocatalytic applications. Furthermore, photocatalysts synthesized through green methods often demonstrate biocompatibility and biodegradability, making them suitable for biomedical uses while mitigating potential health and environmental hazards. By adhering to the principles of green synthesis in photocatalysis, we not only align with sustainability objectives and regulatory standards but also foster the development of efficient and environmentally conscious technologies to address global challenges like water remediation and pollution mitigation. Overall, integrating green synthesis methods holds great promise in propelling innovation in photocatalysis toward a more sustainable and environmentally friendly future.

5.4.2 Photocatalytic Water Remediation

The photocatalysts synthesized through green routes were utilized to degrade harmful organic pollutants in water sources. The various textile dyes and organic molecules present in the water molecules are removed by degradation under the irradiation of light using the photocatalysts. While several single systems for photocatalytic degradation were initially reported, the fast recombination of electrons and holes made the researchers modify the single systems to form semiconductors with heterojunctions, which showed great efficiency.

Green ZnO photocatalysts synthesized from the pomegranate seed extract were utilized for the photocatalytic degradation of flumequine antibiotic. The bandgap of synthesized ZnO was calculated to be 3.07 eV. Under irradiation, the electrons will be transferred from the valence band (VB) to the conduction band (CB) of ZnO, which simultaneously produces holes in the VB. The electrons combine with the oxygen present in the aqueous medium to generate superoxide radicals, which degrade the antibiotics. The holes will generate hydroxide radicals that aid in the degradation process (Fig. 5.8a) (Essawy et al. 2020). TiO_2 nanoparticles synthesized from the *Caesalpinia pulcherrima* flower extract showed a bandgap of 2.89 eV, which was utilized for the photocatalytic degradation of methylene blue (Fig. 5.8b) (Rathi and Jeice 2023).

Cuprous oxide synthesized from banana fruit extract was utilized for the photocatalytic degradation of methylene blue dye. The irradiation of visible light produces photogenerated electrons that move from the VB to the conduction band. The electrons and holes produced will generate hydroxide and superoxide radicals, which degrade the organic pollutants and turn them into harmless products. The catalyst showed a degradation efficiency of 87% at 60 min when methylene blue dye was used (Muthukumaran et al. 2020). The tin oxide nanoparticles synthesized from cotton ball peel extract were utilized for the photodegradation studies using various organic dyes. The excitation of the sample with UV light generates electrons and holes, which, in turn, produce hydroxide and superoxide radicals that are responsible for the photocatalytic degradation of organic dyes. The tin oxide

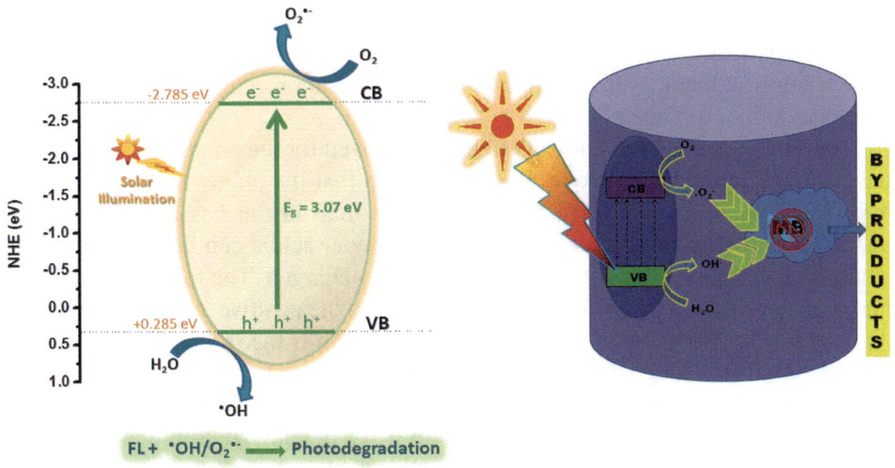

Fig. 5.8 (**a**) Photocatalytic degradation of flumequine antibiotic by ZnO. (Reproduced with permission from ref. Essawy et al. 2020) (**b**) Photocatalytic degradation of methylene blue by TiO_2. (Reproduced with permission from ref. Rathi and Jeice 2023)

catalyst on irradiation with UV light gave a degradation efficiency of 98.5% at 30 min when methylene blue dye was used (Narasaiah et al. 2022).

Ag incorporated with Ag-doped TiO_2 nanocomposite synthesized using green tea extract showed a degradation efficiency of 80% under 50 min on Safranine O dye under illumination with white light, and it was found to be more efficient than TiO_2, pure Ag, and Ag-doped TiO_2. This was due to the transfer of photoinduced electrons from the 4 s orbital of Ag on the surface to the CB of the Ag-doped TiO_2. The holes generated can transfer from the VB of TiO_2 to the 2d orbital of Ag. The ohmic junction formed between the Ag metal and n-type semiconductor TiO_2, and the Schottky junction formed between the Ag metal and the p-type Ag-doped TiO_2, is responsible for the reduction in the recombination of electrons and holes. The photoinduced holes and electrons react with the adsorbed oxygen and water molecules to produce hydroxyl and oxide radicals, which break down the toxic organic dyes into harmless substances (Sherley Saraffin et al. 2024). The cube-like Ag/AgCl composite was utilized for the dye degradation studies using methyl orange as the standard dye. The photocatalytic activity of the composite was studied by observing the UV-visible spectra in the spectrophotometer at 463 nm. The total organic carbon (TOC) value of the initial methyl orange dye was calculated to be 10.4 mg/L, which has significantly reduced to 1.6 mg/L due to the interactions of Ag/AgCl composite under the irradiation of light. The plasmonic absorption behavior of Ag nanoparticles under light irradiation makes it the light-harvesting and electron-generating center, and the transfer of photogenerated electrons would take place toward the CB of AgCl. These photogenerated electrons would react with oxygen to produce superoxide anion radicals, which are responsible for the degradation of dye molecules in water sources (Wang et al. 2020). The Ag/CuO and Ag/TiO_2 composites were evaluated for photocatalytic activity using methylene blue dye (Rathi et al. 2023). The irradiation of visible light ensures 86% degradation of methylene blue at 40 min when Ag/CuO was used. At the same time, Ag/TiO_2 gave a degradation of 94% at a time of 40 min. The addition of Ag nanoparticles, which have a surface plasmon resonance effect, into the metal oxides, reduces the recombination rate, which enhances the photodegradation efficiency.

The TiO_2-incorporated corn silk fiber was utilized for the photodegradation studies using Reactive Black 5 dye. It was observed that the photocatalytic activity of the composite is higher in the acidic medium than in the basic medium. Thus, researchers concluded that the surface of TiO_2 under acidic conditions is positively charged and, under basic conditions, is negatively charged. The four SO_3^- groups in the Reactive Black 5 dye will effectively bind with the positively charged surface of the composite when it is in an acidic medium, which is the reason for the effective photocatalytic dye degradation (Fig. 5.9) (Nadaroglu et al. 2018).

The TiO_2/lignocellulosic biomass/iron oxide composite prepared via the green synthetic route was utilized as a photocatalyst for the degradation of various dyes like methylene blue, rhodamine B, Congo red, and chromium ions from wastewater. The incorporation of olive pits and iron oxide into the TiO_2 has decreased the bandgap of TiO_2, which suggests the higher efficacy of the composite for photocatalytic activity. The hydroxide radicals are mainly responsible for the degradation study.

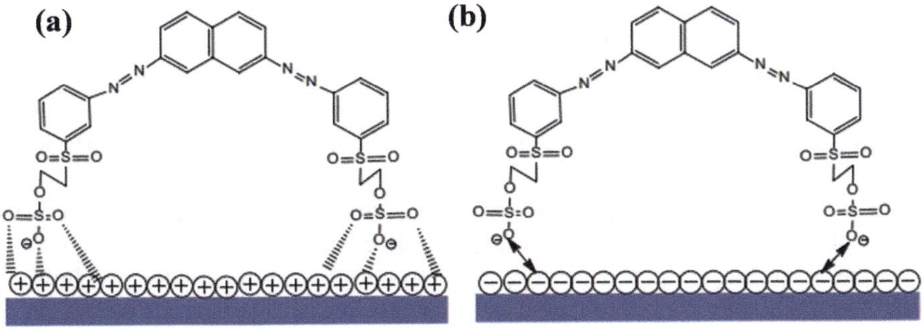

Fig. 5.9 The surface of TiO_2-incorporated corn silk fiber in (**a**) acidic medium and (**b**) basic medium. (Reproduced with permission from ref. Nadaroglu et al. 2018)

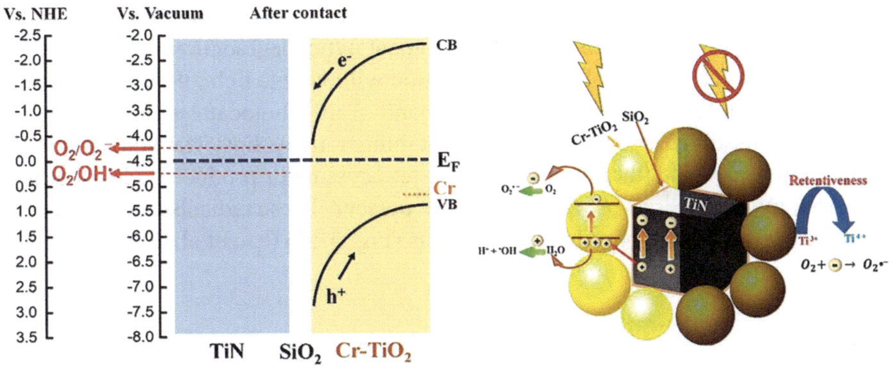

Fig. 5.10 Photocatalytic degradation of titanium nitride/silicon dioxide/Cr-TiO_2 photocatalyst on methylene blue dye. (Reproduced with permission from ref. Kaur et al. 2020)

The presence of olive pits not only functions as the host for holding TiO_2 and iron oxide but also enhances the surface area of the composite. It acts as the photosensitizer for TiO_2, which functions as the electron acceptor that improves photocatalytic behavior (Djellabi et al. 2019).

A naturally water-floating titanium nitride/silicon dioxide/Cr-TiO_2 photocatalyst was synthesized by M. Kaur et al. and studied for the photocatalytic efficiency for degrading methylene blue dye (Kaur et al. 2020). The scavenging studies revealed that the hydroxyl and superoxide radicals are responsible for the degradation of organic dye molecules in the wastewater. Titanium nitride forms a heterojunction with chromium-doped TiO_2, which reduces the electron-hole recombination. The electrons from the chromium-doped TiO_2 degrades the organic pollutants by producing corresponding radicals (Fig. 5.10).

The incorporation of Ag into TiO_2/SiO_2 has decreased the bandgap energy of the TiO_2/SiO_2, which enhances the degradation efficiency when methylene blue dye is used for the analysis. The surface plasmon effect of silver metal produces photogenerated electrons and holes in the form of silver ions, of which electrons move to the conduction band of TiO_2. The electron transfer produces hydroxide and superoxide radicals, which degrade the methylene blue dye. The holes produced also oxidize methylene blue and get reduced to form silver metal, which regenerates the metallic silver, confirming the stability of the composite (Fig. 5.11) (Alrowaili et al. 2020).

In another study, the amino-functionalized cellulose-halloysite nanotube aerogel was synthesized from waste reed straws and utilized initially for the removal of copper ions from wastewater. This copper-adsorbed amino-functionalized cellulose-halloysite nanotube was effectively used to photocatalytically degrade tetracycline. The copper ion-adsorbed amino-functionalized cellulose-halloysite nanotube aerogel was converted into a photocatalyst by reacting with sodium sulfide solution, which gave amino-functionalized cellulose-halloysite nanotube/Cu_2S. For the evaluation of the synthesized photocatalyst, a 340 W Xenon lamp with a filter of 420 nm was used. From the scavenging studies, it was understood that the superoxide radicals were mainly responsible for the photocatalytic degradation of tetracycline using this aerogel photocatalyst. On irradiation with visible light, the electrons from the valence band excite to the conduction band of the photocatalyst and react with the oxygen dissolved in the wastewater containing tetracycline. The holes produced at the same time can either directly oxidize tetracycline or produce hydroxide radicals from hydroxide ions or water molecules present. These radicals can degrade the harmful tetracycline to harmless by-products (Fig. 5.12) (Liu et al. 2023).

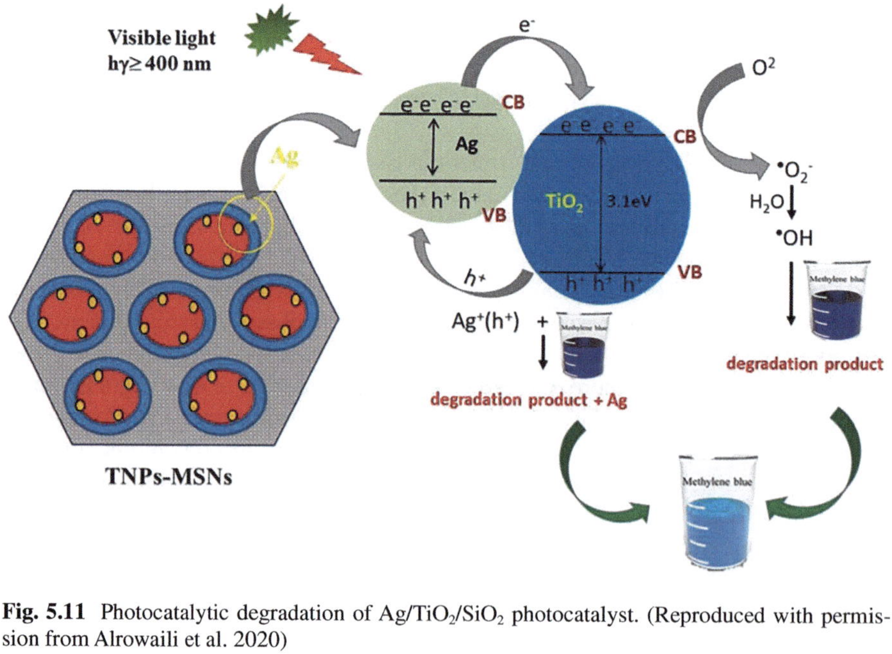

Fig. 5.11 Photocatalytic degradation of $Ag/TiO_2/SiO_2$ photocatalyst. (Reproduced with permission from Alrowaili et al. 2020)

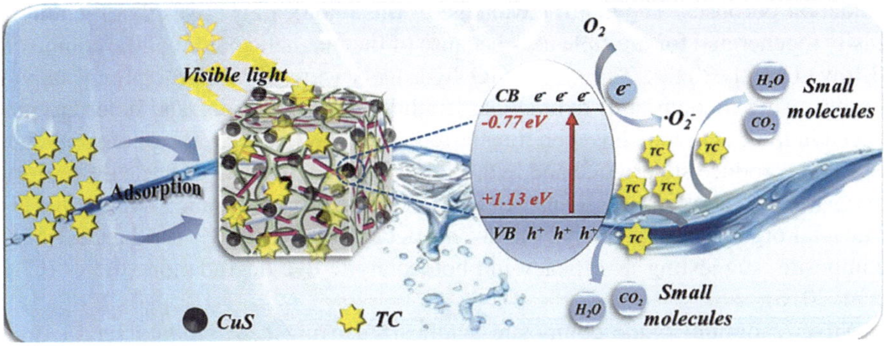

Fig. 5.12 Photocatalytic degradation of amino-functionalized cellulose-halloysite nanotube aerogel photocatalyst. (Reproduced with permission from ref. Liu et al. 2023)

Fig. 5.13 Photocatalytic degradation of rhodamine B using gold/gCN photocatalyst. (Reproduced with permission from ref. Madhual et al. 2023)

Gold was combined with gCN to degrade rhodamine B dye under visible light (Fig. 5.13). The surface plasmon resonance effect of gold nanoparticles enhances the photocatalytic nature of the composite. Under visible light, the electrons move from the valance band of gold nanoparticle to the conduction band, creating a hole in the valence band. The holes produce hydroxide radicals, and electrons generate superoxide radicals. Due to the heterojunction formed between gold nanoparticles and gCN, the photogenerated carriers will easily travel from CB of gCN to gold nanoparticles. Thus, the role of gold is to prevent the recombination rate by acting as an electron pool (Madhual et al. 2023).

The photo-Fenton system under UV light was utilized for oxidizing the Basic Blue 9 dye using the biopolymer chitosan/iron oxide composite. The hydroxyl radicals generated are responsible for the oxidation of pollutants, which has a high rate initially, and as time increases, the rate decreases. The degradation did not happen when dark-Fenton system was used, confirming the necessity of UV light for oxidation purposes. The major advantage of this photocatalyst was that it could be easily regenerated for multiple uses because of the magnetic nature of the composite (Elsayed et al. 2022). The CeO_2/rGO hydrogels were utilized for photocatalytic studies using rhodamine B dye. Under irradiation, the electrons and holes are produced in the CeO_2, which is then transferred to the rGO matrix. The photogenerated electrons produce superoxide and hydroxyl radicals, which can degrade the harmful dye particles. The holes will ensure the direct oxidation of organic dye molecules. The total organic carbon analysis showed a decay value of 74.3% for the CeO_2/rGO composite, suggesting its efficacy in photocatalytic dye degradation studies (Choi et al. 2016).

Silver/cadmium oxide composite synthesized using *Citrus limon* seed extract was utilized for the photocatalytic degradation of Brilliant Green dye. The bandgap of silver nanoparticles reduced significantly from 1.97 eV to 1.23 eV when cadmium oxide was introduced into silver nanoparticles. Under irradiation of visible light, the electrons of both silver and cadmium oxide will transfer from the VB to the CB. This creates holes in both silver nanoparticles and cadmium oxide. The photogenerated electrons in the CB of silver will move toward cadmium oxide, ensuring the inhibition of electron-hole recombination. The photogenerated holes can move from the VB of cadmium oxide to silver nanoparticles. These holes react with the water molecules to produce hydroxide radicals, and electrons react with oxygen to produce superoxide radicals. These radicals react with the Brilliant Green dye and degrade it to carbon dioxide and water (Fig. 5.14) (Mahmood et al. 2022).

Iron oxide/tungsten oxide/GO composite was studied as a photocatalyst for the degradation of methylene blue dye. The characteristic absorption peak at 665 nm in the spectrophotometric analysis was observed to understand the photocatalytic

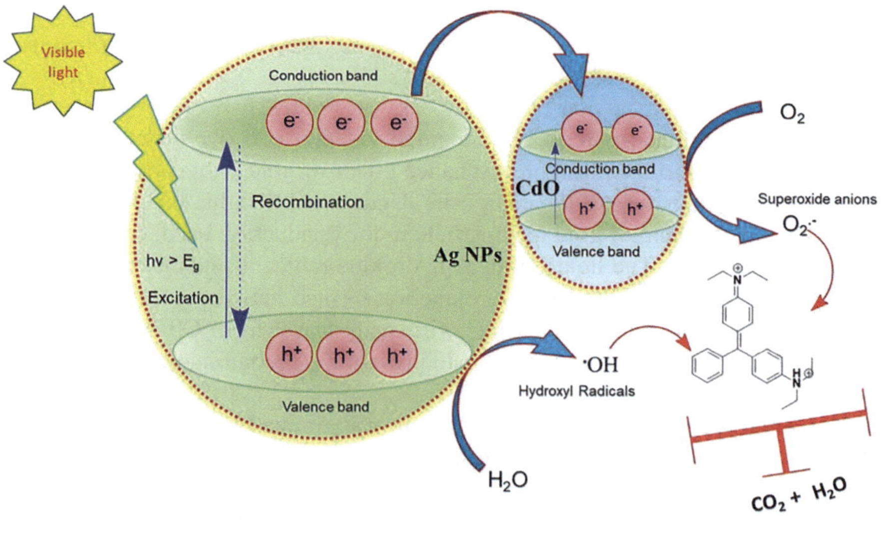

Fig. 5.14 Photocatalytic degradation of Brilliant Green dye by silver/cadmium oxide photocatalyst. (Reproduced with permission from ref. Mahmood et al. 2022)

behavior of the composite. When the composite is exposed to solar light, the electrons and holes will be generated in both iron oxide and tungsten oxide. The heterojunction formed between these two will ensure the transfer of photogenerated electrons from the CB of iron oxide to the CB of tungsten oxide, whereas the holes will move from VB of tungsten oxide to that of iron oxide. This reduces the rate of electron-hole recombination, improving the photocatalytic activity. The hydroxide and superoxide radicals produced by these electrons and holes will degrade the harmful organic pollutants into harmless products (Mohamed 2019).

The copper iodide/rGO composite synthesized through the organic surfactant-free method was utilized for the photocatalytic degradation studies using a solar simulator having AM 1.5G filter and a Xenon lamp of 150 W. Rhodamine B dye was utilized for evaluating the activity of the composite toward degradation. The electrons are photogenerated from the rGO nanosheets, which produce superoxide radicals, degrading the organic pollutants (Choi et al. 2015). Plasmonic gold and silver nanostructured carbon quantum dots were incorporated into TiO_2 nanoparticles and studied for the photocatalytic degradation of methylene blue dye. It was found that the CB of plasmonic structures coincided with that of TiO_2, and thus the light-harvesting effect was enhanced. The photogenerated electrons will move from the VB to the CB of TiO_2 and react with the oxygen present in the water source to produce superoxide radicals, which degrade the pollutants. The holes generated will directly oxidize the methylene blue dye to form harmless products (Thakur et al. 2020). Table 5.1 gives the various nanoparticles synthesized using green precursor and their applications in the removal of pollutants.

The evaluation of the photocatalytic effectiveness of green nanostructured photocatalysts highlights their considerable promise in driving sustainable water remediation and environmental rehabilitation endeavors. Utilizing eco-conscious methodologies and renewable resources in their production represents a significant advancement toward promoting environmentally aware photocatalysis. Through careful design and engineering, these green nanostructured photocatalysts exhibit outstanding photocatalytic performance, effectively decomposing pollutants and purifying water under mild reaction conditions. Their unique nanostructured morphology, composition, and surface properties enable enhanced light absorption, efficient separation of charge carriers, and superior catalytic activity, offering advantages over conventional photocatalysts. Additionally, the inherent biocompatibility and biodegradability of these green photocatalysts make them attractive options for biomedical applications, further broadening their environmental impact. By embracing the principles of green chemistry and sustainable synthesis, researchers can continuously innovate and enhance green nanostructured photocatalysts, laying the groundwork for the development of efficient, eco-friendly technologies to address water pollution, which has become a global challenge. Thus, the integration of green nanostructured photocatalysts signifies a significant advancement in the pursuit of sustainable solutions to urgent environmental issues, providing hope for a cleaner, healthier future for generations to come.

Table 5.1 Various nanoparticles synthesized using green precursor and their applications

Composite	Green precursor	Pollutants removed	References
Silver nanoparticles	*Phaseolus vulgaris* seed extract	Reactive red 141	Rani et al. (2020)
Silver nanoparticles	*Syzygium samarangense* flower extract	Methylene blue	Basalius et al. (2023)
Silver/cadmium oxide	Citrus lemon seed extract	Brilliant green dye	Mahmood et al. (2022)
TiO_2	Fresh mulberry leaves	Methylene blue	Shimi et al. (2022)
TiO_2	*Nervilia aragoana* leaf	Methylene blue	Rathi and Jeice (2023)
Lanthanum/copper/zirconium	Algal biochar	Malachite green	Sharma et al. (2019)
Silver-doped ZnO	*Prunus cerasifera* fruits	Rhodamine B	Ahmad and Jaffri (2018)
ZnO	Pomegranate seed extract	Flumequine antibiotic	Essawy et al. (2020)
Carbon dots/tungsten oxide	*T. cucumerina* extract	Crystal violet	Smrithi et al. (2020)
Cadmium tungstate	*Brassica rapa* leaves	Bismarck Brown R	Fatima et al. (2019)
Silver-doped TiO_2	Green tea extract	Mordant Orange	Sherley Saraffin et al. (2024)
Silver/CuO	*C. pulcherrima* plant extract	Methylene blue	Rathi et al. (2023)
Silver/TiO_2/SiO_2	Green tea extract	Methylene blue	Alrowaili et al. (2020)
Cellulose corn silk fiber/TiO_2	Cellulose corn silk fiber	Reactive black 5	Nadaroglu et al. (2018)
Amino-functionalized cellulose-halloysite nanotubes	Reed straws	Tetracycline	Liu et al. (2023)
Gold/g-CN	*A. carambola* leaf extract	Rhodamine B	Madhual et al. (2023)
Titanium nitride/silicon dioxide/Cr-TiO_2	Cell walls of the brown seaweed	Methylene blue	Kaur et al. (2020)
Iron oxide/GO/WO_3	Aqueous flax seed extract	Crystal violet	Mohamed (2019)
Cuprous oxide	*Musa acuminata* extract	Methylene blue	Muthukumaran et al. (2020)
Tin oxide	Agro-waste cotton ball peel extract	Methyl orange	Narasaiah et al. (2022)

5.5 Conclusions

The utilization of green nanostructured photocatalysts signifies a notable progression in water remediation practices. These environmentally friendly photocatalysts demonstrate outstanding efficacy in decomposing pollutants and cleansing water, presenting a sustainable approach to tackling environmental issues. Through the utilization of natural resources and the application of eco-conscious synthesis techniques, green nanostructured photocatalysts mitigate the environmental repercussions linked with synthesizing photocatalysts in accordance with the views of green chemistry and sustainable growth. The advancement of versatile green nanostructured photocatalysts customized for specific water pollutants and environmental contexts shows significant potential. Integration of additional functionalities, such as enhanced stability, selectivity, and recyclability, can further enhance the applicability and effectiveness of green photocatalysts in water treatment technologies. Furthermore, investigating the synergies arising from the amalgamation of green nanostructured photocatalysts with other cutting-edge water treatment methods, such as membrane filtration or adsorption, has the potential to substitute the creation of integrated and highly effective water remediation systems. The green photocatalysts employed in the water remediation process follow an environmentally friendly method, which aligns with the SDGs put forward by the United Nations. The scarcity of water faced by our society can be tackled to an extent by using these green nanostructured photocatalysts to reuse water sources. The best remedy is not to contaminate or pollute our water sources and to use them wisely so that we can save water for future generations.

References

Afre RA, Sharma N, Sharon M, Sharon M (2018) Transparent conducting oxide films for various applications: a review. Rev Adv Mater Sci 53:79–89. https://doi.org/10.1515/rams-2018-0006

Ahmad KS, Jaffri SB (2018) Phytosynthetic Ag doped ZnO nanoparticles: semiconducting green remediators. Open Chem 16:556–570. https://doi.org/10.1515/chem-2018-0060

Akintelu SA, Folorunso AS, Folorunso FA, Oyebamiji AK (2020) Green synthesis of copper oxide nanoparticles for biomedical application and environmental remediation. Heliyon 6:e04508. https://doi.org/10.1016/j.heliyon.2020.e04508

Ali AY, Holmes NP, Ameri M et al (2022) Low-temperature CVD-grown graphene thin films as transparent electrode for organic photovoltaics. Coatings 12:681. https://doi.org/10.3390/coatings12050681

Alrowaili ZA, Alsohaimi IH, Betiha MA et al (2020) Green fabrication of silver imprinted titania/silica nanospheres as robust visible light-induced photocatalytic wastewater purification. Mater Chem Phys 241:122403. https://doi.org/10.1016/j.matchemphys.2019.122403

Balakrishnan A, Chinthala M, Polagani RK (2024) 3D kaolinite/g-C3N4-alginate beads as an affordable and sustainable photocatalyst for wastewater remediation. Carbohydr Polym 323:121420. https://doi.org/10.1016/j.carbpol.2023.121420

Basalius H, Mani A, Michael A et al (2023) Green synthesis of nano-silver using Syzygium samarangense flower extract for multifaceted applications in biomedical and photocatalytic degradation of methylene blue. Appl Nanosci 13:3735–3747. https://doi.org/10.1007/s13204-022-02523-5

Basnet P, Samanta D, Chanu TI et al (2020) Glycine-A bio-capping agent for the bioinspired synthesis of nano-zinc oxide photocatalyst. J Mater Sci Mater Electron 31:2949–2966. https://doi.org/10.1007/s10854-019-02839-z

Bognár S, Putnik P, Šojić Merkulov D (2022) Sustainable green nanotechnologies for innovative purifications of water: synthesis of the nanoparticles from renewable sources. Nano 12:263. https://doi.org/10.3390/nano12020263

Buchholz DB, Liu J, Marks TJ et al (2009) Control and characterization of the structural, electrical, and optical properties of amorphous zinc−indium−tin oxide thin films. ACS Appl Mater Interfaces 1:2147–2153. https://doi.org/10.1021/am900321f

Chen X, Tan P, Zhou B et al (2015) A green and facile strategy for preparation of novel and stable Cr-doped SrTiO3/g-C3N4 hybrid nanocomposites with enhanced visible light photocatalytic activity. J Alloys Compd 647:456–462. https://doi.org/10.1016/j.jallcom.2015.06.056

Chng LL, Erathodiyil N, Ying JY (2013) Nanostructured catalysts for organic transformations. Acc Chem Res 46:1825–1837. https://doi.org/10.1021/ar300197s

Choi J, Reddy DA, Islam MJ et al (2015) Green synthesis of the reduced graphene oxide–CuI quasi-shell–core nanocomposite: a highly efficient and stable solar-light-induced catalyst for organic dye degradation in water. Appl Surf Sci 358:159–167. https://doi.org/10.1016/j.apsusc.2015.07.170

Choi J, Reddy DA, Islam MJ et al (2016) Self-assembly of CeO2 nanostructures/reduced graphene oxide composite aerogels for efficient photocatalytic degradation of organic pollutants in water. J Alloys Compd 688:527–536. https://doi.org/10.1016/j.jallcom.2016.07.236

Crespo-Quesada M, Yarulin A, Jin M et al (2011) Structure sensitivity of Alkynol hydrogenation on shape- and size-controlled palladium nanocrystals: which sites are most active and selective? J Am Chem Soc 133:12787–12794. https://doi.org/10.1021/ja204557m

Danish MSS, Estrella-Pajulas LL, Alemaida IM et al (2022) Green synthesis of silver oxide nanoparticles for photocatalytic environmental remediation and biomedical applications. Metals (Basel) 12:769. https://doi.org/10.3390/met12050769

Devi KRS, Mathew S, Rajan R et al (2019) Biogenic synthesis of g-C3N4/Bi2O3 heterojunction with enhanced photocatalytic activity and statistical optimization of reaction parameters. Appl Surf Sci 494:465–476. https://doi.org/10.1016/j.apsusc.2019.07.125

Djellabi R, Yang B, Adeel Sharif HM et al (2019) Sustainable and easy recoverable magnetic TiO2-Lignocellulosic Biomass@Fe3O4 for solar photocatalytic water remediation. J Clean Prod 233:841–847. https://doi.org/10.1016/j.jclepro.2019.06.125

Dong S, Feng J, Fan M et al (2015) Recent developments in heterogeneous photocatalytic water treatment using visible light-responsive photocatalysts: a review. RSC Adv 5:14610–14630. https://doi.org/10.1039/C4RA13734E

Durgalakshmi D, Ajay Rakkesh R, Rajendran S, Naushad M (2020) Principles and mechanisms of green Photocatalysis. pp 1–24

Dutta GK, Karak N (2021) Bio-based waterborne polyester supported oxygeneous graphitic carbon nitride nanosheets as a sustainable photocatalyst for aquatic environment remediation. J Clean Prod 285:124906. https://doi.org/10.1016/j.jclepro.2020.124906

Elbadawy HA, Elhusseiny AF, Hussein SM, Sadik WA (2023) Sustainable and energy-efficient photocatalytic degradation of textile dye assisted by ecofriendly synthesized silver nanoparticles. Sci Rep 13:2302. https://doi.org/10.1038/s41598-023-29507-x

Elsayed SA, El-Sayed IET, Tony MA (2022) Impregnated chitin biopolymer with magnetic nanoparticles to immobilize dye from aqueous media as a simple, rapid and efficient composite photocatalyst. Appl Water Sci 12:252. https://doi.org/10.1007/s13201-022-01776-3

Essawy AA, Alsohaimi IH, Alhumaimess MS et al (2020) Green synthesis of spongy Nano-ZnO productive of hydroxyl radicals for unconventional solar-driven photocatalytic remediation of antibiotic enriched wastewater. J Environ Manag 271:110961. https://doi.org/10.1016/j.jenvman.2020.110961

Fatima B, Siddiqui SI, Ahmed R, Chaudhry SA (2019) Green synthesis of f-CdWO4 for photocatalytic degradation and adsorptive removal of Bismarck Brown R dye from water. Water Resour Ind 22:100119. https://doi.org/10.1016/j.wri.2019.100119

Fatimah I, Zunita Pratiwi E, Prio Wicaksono W (2020) Synthesis of magnetic nanoparticles using Parkia speciosa Hassk pod extract and photocatalytic activity for Bromophenol blue degradation. Egypt J Aquat Res 46:35–40. https://doi.org/10.1016/j.ejar.2020.01.001

Fulekar J, Dutta DP, Pathak B, Fulekar MH (2018) Novel microbial and root mediated green synthesis of TiO 2 nanoparticles and its application in wastewater remediation. J Chem Technol Biotechnol 93:736–743. https://doi.org/10.1002/jctb.5423

Gong D, Guo J, Wang F et al (2023) Green construction of metal- and additive-free citrus peel-derived carbon dot/g-C3N4 photocatalysts for the high-performance photocatalytic decomposition of sunset yellow. Food Chem 425:136470. https://doi.org/10.1016/j.foodchem.2023.136470

Hadnadjev-Kostic M, Karanovic D, Vulic T et al (2023) Photocatalytic properties of ZnFe-mixed oxides synthesized via a simple route for water remediation. Green Process Synth 12:1–12. https://doi.org/10.1515/gps-2022-8153

Hashimoto K, Irie H, Fujishima A (2005) TiO 2 Photocatalysis: a historical overview and future prospects. Jpn J Appl Phys 44:8269. https://doi.org/10.1143/JJAP.44.8269

Hintsho N, Petrik L, Nechaev A et al (2014) Photocatalytic activity of titanium dioxide carbon nanotube nanocomposites modified with silver and palladium nanoparticles. Appl Catal B Environ 156–157:273–283. https://doi.org/10.1016/j.apcatb.2014.03.021

Hong H, Hu L, Li M et al (2011) Preparation of Pt@Fe 2 O 3 nanowires and their catalysis of selective oxidation of olefins and alcohols. Chem–A Eur J 17:8726–8730. https://doi.org/10.1002/chem.201003429

Ismayil KM, Varghese A, Antony R (2020) Silver-doped polyaniline–polyvinyl chloride nanocomposite films for photocatalytic and antibacterial applications. J Elastomers Plast 52:103–116. https://doi.org/10.1177/0095244318819238

Jamshidi A, Zonoz FM, Wei Y (2020) A new organic-inorganic nano hybrid based on borotungstic anion and imidazole ring as a two-functional green catalyst. Solid State Sci 100:106093. https://doi.org/10.1016/j.solidstatesciences.2019.106093

Jilani A, Hussain SZ, Melaibari AA, Abu-Hamdeh NH (2022) Development and mechanistic studies of ternary nanocomposites for hydrogen production from water splitting to yield sustainable/green energy and environmental remediation. Polymers (Basel) 14:1290. https://doi.org/10.3390/polym14071290

Karimi F, Zare N, Jahanshahi R et al (2023) Natural waste-derived nano photocatalysts for azo dye degradation. Environ Res 238:117202. https://doi.org/10.1016/j.envres.2023.117202

Karthik K, Devi KRS, Pinheiro D, Sugunan S (2019) Photocatalytic activity of bismuth silicate heterostructures synthesized via surfactant mediated sol-gel method. Mater Sci Semicond Process 102:104589. https://doi.org/10.1016/j.mssp.2019.104589

Kaur M, Shinde SL, Ishii S et al (2020) Marimo-bead-supported core–shell nanocomposites of titanium nitride and chromium-doped titanium dioxide as a highly efficient water-floatable green photocatalyst. ACS Appl Mater Interfaces 12:31327–31339. https://doi.org/10.1021/acsami.0c03781

Kausar F, Varghese A, Pinheiro D, Devi KRS (2022) Recent trends in photocatalytic water splitting using titania based ternary photocatalysts-A review. Int J Hydrog Energy 47:22371–22402. https://doi.org/10.1016/j.ijhydene.2022.05.058

Khan MM (2021) Principles and mechanisms of photocatalysis. In: Photocatalytic Systems by Design. Elsevier, pp 1–22

Kour S, Jasrotia R, Puri P et al (2021) Improving photocatalytic efficiency of MnFe2O4 ferrites via doping with Zn2+/La3+ ions: photocatalytic dye degradation for water remediation. Environ Sci Pollut Res 30:71527–71542. https://doi.org/10.1007/s11356-021-13147-7

Kumari V, Kaushal S, Singh PP (2022) Green synthesis of a CuO/rGO nanocomposite using a Terminalia arjuna bark extract and its catalytic activity for the purification of water. Mater Adv 3:2170–2184. https://doi.org/10.1039/D1MA00993A

Li F, Lin M (2020) Synthesis of biochar-supported K-doped g-C3N4 Photocatalyst for enhancing the polycyclic aromatic hydrocarbon degradation activity. Int J Environ Res Public Health 17:2065. https://doi.org/10.3390/ijerph17062065

Liu Q, Hu S, Yang Z et al (2019) Green synthesis of composite graphene aerogels with robust magnetism for effective water remediation. Materials (Basel) 12:4106. https://doi.org/10.3390/ma12244106

Liu Y, Yang X, Ai J et al (2023) Amino-functionalized biomass-based aerogel derived from waste reed converted into an efficient photocatalyst after adsorption of Cu2+. Chem Eng J 474:145586. https://doi.org/10.1016/j.cej.2023.145586

Liu J, Li J, Yang C, Wang D (2024) Synthesis of g-C3N4/Co3O4@BiVO4 double S-scheme heterojunction photocatalyst for enhanced Rhodamine B degradation with simulated sunlight. Mater Sci Eng B 299:116934. https://doi.org/10.1016/j.mseb.2023.116934

Lopes JL, Estrada AC, Fateixa S et al (2024) Visible-light activation of carbon-supported BiVO4 photocatalysts. J Photochem Photobiol A Chem 447:115174. https://doi.org/10.1016/j.jphotochem.2023.115174

Madhual A, Prabha Devi A, Manjari Mishra P (2023) Bio-based Au/g-CN plasmonic nanophotocatalyst for superior degradation of Rhodamine B under visible light illumination. J Ind Eng Chem 127:567–578. https://doi.org/10.1016/j.jiec.2023.07.042

Mahmood K, Amara U, Siddique S et al (2022) Green synthesis of Ag@CdO nanocomposite and their application towards brilliant green dye degradation from wastewater. J Nanostructure Chem 12:329–341. https://doi.org/10.1007/s40097-021-00418-5

Malhotra M, Sudhaik A, Sonu et al (2023) An overview on cellulose-supported photocatalytic materials for the efficient removal of toxic dyes. Ind Crop Prod 202:117000. https://doi.org/10.1016/j.indcrop.2023.117000

Martin DJ, Liu G, Moniz SJA et al (2015) Efficient visible driven photocatalyst, silver phosphate: performance, understanding and perspective. Chem Soc Rev 44:7808–7828. https://doi.org/10.1039/C5CS00380F

Mendes-Felipe C, Veloso-Fernández A, Vilas-Vilela JL, Ruiz-Rubio L (2022) Hybrid organic–inorganic membranes for photocatalytic water remediation. Catalysts 12:180. https://doi.org/10.3390/catal12020180

Mitsudome T, Noujima A, Mikami Y et al (2010) Supported gold and silver nanoparticles for catalytic deoxygenation of epoxides into alkenes. Angew Chemie Int Ed 49:5545–5548. https://doi.org/10.1002/anie.201001055

Mittal H, Ivaturi A, Khanuja M (2023) MoSe2-modified ZIF-8 novel nanocomposite for photocatalytic remediation of textile dye and antibiotic-contaminated wastewater. Environ Sci Pollut Res 30:4151–4165. https://doi.org/10.1007/s11356-022-22487-x

Mohamed HH (2019) Rationally designed Fe2O3/GO/WO3 Z-scheme photocatalyst for enhanced solar light photocatalytic water remediation. J Photochem Photobiol A Chem 378:74–84. https://doi.org/10.1016/j.jphotochem.2019.04.023

Moradi O, Sharma G (2021) Emerging novel polymeric adsorbents for removing dyes from wastewater: a comprehensive review and comparison with other adsorbents. Environ Res 201:111534. https://doi.org/10.1016/j.envres.2021.111534

Mostafa S, Behafarid F, Croy JR et al (2010) Shape-dependent catalytic properties of Pt nanoparticles. J Am Chem Soc 132:15714–15719. https://doi.org/10.1021/ja106679z

Muthukumaran M, Dhinagaran G, Venkatachalam K et al (2020) Green synthesis of cuprous oxide nanoparticles for environmental remediation and enhanced visible-light photocatalytic activity. Optik (Stuttg) 214:164849. https://doi.org/10.1016/j.ijleo.2020.164849

Nadaroglu H, Lesani A, Soleimani SS, Babagil A (2018) A newly green photocatalyst support for azo dye remediation under UV light irradiation. Appl Water Sci 8:107. https://doi.org/10.1007/s13201-018-0752-4

Narasaiah BP, Banoth P, Sohan A et al (2022) Green biosynthesis of tin oxide nanomaterials mediated by agro-waste cotton Boll Peel extracts for the remediation of environmental pollutant dyes. ACS Omega 7:15423–15438. https://doi.org/10.1021/acsomega.1c07099

Narváez BG, Mendoza-Mendoza E, Peralta-Rodríguez RD et al (2024) Visible-LEDs-induced enhanced photocatalytic and antibacterial activity of BiVO4-based green photocatalysts decorated with silver and graphene. J Photochem Photobiol A Chem 447:115191. https://doi.org/10.1016/j.jphotochem.2023.115191

Pinheiro D, Sunaja Devi KR, Jose A et al (2020) Experimental design for optimization of 4-nitrophenol reduction by green synthesized CeO2/g-C3N4/Ag catalyst using response surface methodology. J Rare Earths 38:1171–1177. https://doi.org/10.1016/j.jre.2019.10.001
Rani P, Kumar V, Singh PP et al (2020) Highly stable AgNPs prepared via a novel green approach for catalytic and photocatalytic removal of biological and non-biological pollutants. Environ Int 143:105924. https://doi.org/10.1016/j.envint.2020.105924
Rani M, Keshu PS et al (2024) Sunlight assisted highly efficient photocatalytic remediation of organic pollutants by green biosynthesized ZnO@WO3 nanocomposite. J Photochem Photobiol A Chem 446:115160. https://doi.org/10.1016/j.jphotochem.2023.115160
Rathi VH, Jeice AR (2023) Green fabrication of titanium dioxide nanoparticles and their applications in photocatalytic dye degradation and microbial activities. Chem Phys Impact 6:100197. https://doi.org/10.1016/j.chphi.2023.100197
Rathi VH, Jeice AR, Jayakumar K (2023) Green synthesis of Ag/CuO and Ag/TiO2 nanoparticles for enhanced photocatalytic dye degradation, antibacterial, and antifungal properties. Appl Surf Sci Adv 18:100476. https://doi.org/10.1016/j.apsadv.2023.100476
Redouane-Salah Z, Boulahbal M, Tuna Ö et al (2024) Construction of Z-scheme FeTiO3/g-C3N4 heterojunction system: characterization and statistical optimization of photocatalytic behavior under visible light irradiation. J Photochem Photobiol A Chem 446:115185. https://doi.org/10.1016/j.jphotochem.2023.115185
Sanaie N, Zhiani R, Hosseiny M et al (2023) Application of green synthesized biogenic Nd2Zr2O7 supported on dendritic fibrous nano-titanium for photocatalytic and antibacterial activities. Results Chem 6:101170. https://doi.org/10.1016/j.rechem.2023.101170
Sarkar S, Ponce NT, Banerjee A et al (2020) Green polymeric nanomaterials for the photocatalytic degradation of dyes: a review. Environ Chem Lett 18:1569–1580. https://doi.org/10.1007/s10311-020-01021-w
Serrà A, Zhang Y, Sepúlveda B et al (2019) Highly active ZnO-based biomimetic fern-like microleaves for photocatalytic water decontamination using sunlight. Appl Catal B Environ 248:129–146. https://doi.org/10.1016/j.apcatb.2019.02.017
Sharma G, Bhogal S, Gupta VK et al (2019) Algal biochar reinforced trimetallic nanocomposite as adsorptional/photocatalyst for remediation of malachite green from aqueous medium. J Mol Liq 275:499–509. https://doi.org/10.1016/j.molliq.2018.11.070
Sharma G, Kumar A, Ghfar AA et al (2021) Fabrication and characterization of xanthan gum-cl-poly (acrylamide-co-alginic acid) hydrogel for adsorption of cadmium ions from aqueous medium. Gels 8:23. https://doi.org/10.3390/gels8010023
Sherley Saraffin R, Bhattacharjee A, Kar B et al (2024) Schottky-junction plasmonic nanocomposite: a non-hazardous excellent visible light photocatalyst for industrial waste management. J Mol Struct 1295:136733. https://doi.org/10.1016/j.molstruc.2023.136733
Shimi AK, Ahmed HM, Wahab M et al (2022) Synthesis and applications of green synthesized TiO2 nanoparticles for photocatalytic dye degradation and antibacterial activity. J Nanomater 2022:1–9. https://doi.org/10.1155/2022/7060388
Shubha JP, Kavalli K, Adil SF et al (2022) Facile green synthesis of semiconductive ZnO nanoparticles for photocatalytic degradation of dyes from the textile industry: a kinetic approach. J King Saud Univ-Sci 34:102047. https://doi.org/10.1016/j.jksus.2022.102047
Smrithi SP, Kottam N, Arpitha V et al (2020) Tungsten oxide modified with carbon nanodots: integrating adsorptive and photocatalytic functionalities for water remediation. J Sci Adv Mater Dev 5:73–83. https://doi.org/10.1016/j.jsamd.2020.02.005
Song J, He Y, Chen J et al (2012) Bicolor light-emitting diode based on zinc oxide Nanorod arrays and poly(2-methoxy, 5-octoxy)-1,4-phenylenevinylene. J Electron Mater 41:431–436. https://doi.org/10.1007/s11664-011-1783-x
Stadler A (2012) Transparent conducting oxides—an up-to-date overview. Materials (Basel) 5:661–683. https://doi.org/10.3390/ma5040661
Tabatabaeinejad SM, Zinatloo-Ajabshir S, Amiri O, Salavati-Niasari M (2021) Magnetic Lu 2 Cu 2 O 5 -based ceramic nanostructured materials fabricated by a simple and green approach for

an effective photocatalytic degradation of organic contamination. RSC Adv 11:40100–40111. https://doi.org/10.1039/D1RA06101A

Thakur A, Kumar P, Kaur D et al (2020) TiO 2 nanofibres decorated with green-synthesized P Au/ Ag @CQDs for the efficient photocatalytic degradation of organic dyes and pharmaceutical drugs. RSC Adv 10:8941–8948. https://doi.org/10.1039/C9RA10804A

Veigas LM, Chandran N, Krishna BM et al (2023a) Heterojunction engineered MWCNT/Ag 3PO4 via organic acid and its natural light-assisted photocatalytic efficiency. Nano-Struct Nano-Obj 34:100975. https://doi.org/10.1016/j.nanoso.2023.100975

Veigas LM, Ravi L, Narasimhaiaha N et al (2023b) In-vitro investigation of the α-amylase inhibition activity of bare Bis-Benzylidene-cyclohexanone synthesized by a highly selective solvent-free route. Chem Select 8:e202301807. https://doi.org/10.1002/slct.202301807

Vijayarangan R, Mohan S, Selvaraj M et al (2024) Surface-engineered anisotropic g-C3N4 photocatalysts via green-exfoliation for visible light-driven water remediation. Surf Interfaces 44:103782. https://doi.org/10.1016/j.surfin.2023.103782

Wang Y, O'Connor D, Shen Z et al (2019) Green synthesis of nanoparticles for the remediation of contaminated waters and soils: constituents, synthesizing methods, and influencing factors. J Clean Prod 226:540–549. https://doi.org/10.1016/j.jclepro.2019.04.128

Wang N, Cheng K, Xu Z-F et al (2020) High-performance natural-sunlight-driven Ag/AgCl photocatalysts with a cube-like morphology and blunt edges via a bola-type surfactant-assisted synthesis. Phys Chem Chem Phys 22:3940–3952. https://doi.org/10.1039/C9CP05273A

Xu H, Zheng X, Li X et al (2023) Persulfate assisted photocatalytic degradation activity and mechanism of tetracycline by Cu enriched reed hydrothermal carbonization carbon: synergistic effect and DFT calculation. Ind Crop Prod 206:117645. https://doi.org/10.1016/j.indcrop.2023.117645

Yaghoobi Rahni S, Mohammadzadeh Kakhki R (2020) Facile and green synthesis of Cu 3 V 2 O 8 nanostructures via Moringa peregrina natural extract as a high performance photo catalyst. Appl Organomet Chem 34:1–8. https://doi.org/10.1002/aoc.5392

Yang J, Wang X, Chen Y et al (2015) Enhanced photocatalytic activities of visible-light driven green synthesis in water and environmental remediation on Au/Bi 2 WO 6 hybrid nanostructures. RSC Adv 5:9771–9782. https://doi.org/10.1039/C4RA15349A

Yousefi SR, Sobhani A, Alshamsi HA, Salavati-Niasari M (2021) Green sonochemical synthesis of BaDy 2 NiO 5 /Dy 2 O 3 and BaDy 2 NiO 5/NiO nanocomposites in the presence of core almond as a capping agent and their application as photocatalysts for the removal of organic dyes in water. RSC Adv 11:11500–11512. https://doi.org/10.1039/D0RA10288A

Zang X, Jiang R, Zhu H-Y et al (2024) A review on the progress of magnetic chitosan-based materials in water purification and solid-phase extraction of contaminants. Sep Purif Technol 330:125521. https://doi.org/10.1016/j.seppur.2023.125521

Zhang X, Huang J, Kang Z et al (2020) Eggshell-templated synthesis of PbS/CaCO3 nanocomposites for CO3− mediated efficient degradation of tetracycline under solar light irradiation. Mol Catal 484:110786. https://doi.org/10.1016/j.mcat.2020.110786

Zhong Y, Zhang X, Wang Y et al (2023) Carbon quantum dots from tea enhance z-type BiOBr/ C3N4 heterojunctions for RhB degradation: catalytic effect, mechanisms, and intermediates. Appl Surf Sci 639:158254. https://doi.org/10.1016/j.apsusc.2023.158254

Zinatloo-Ajabshir S, Salehi Z, Salavati-Niasari M (2018) Green synthesis and characterization of Dy2Ce2O7 ceramic nanostructures with good photocatalytic properties under visible light for removal of organic dyes in water. J Clean Prod 192:678–687. https://doi.org/10.1016/j.jclepro.2018.05.042

Chapter 6
Green Photocatalytic Water Purification

Vaishnavi Sharma and Gururaj Kudur Jayaprakash

Abstract In the era of extreme industrialization, pollution degradation is the biggest issue faced by the world because of its negative impacts on all living beings and the environment. Photocatalysis is one of the most favorable AOPs for pollutant degradation owing to its properties such as simplicity, reproducibility, environmental friendliness, non-toxicity, and technological and economical sustainability. This chapter explains what a photocatalyst is, the process of photocatalytic pollutant degradation, photocatalyst heterostructure modifications, and their application in various pollutant degradation activities like dye degradation, drug degradation, antimicrobial activity, and heavy metal removal.

Keywords Composites · Heterostructure · Pollutants · Metal removal · Semiconductors · Water treatment

6.1 Introduction

Many emerging countries no longer have clean water supplies appropriate even for agriculture due to rapid population increase, harsh weather patterns, and unrestrained industry. This is especially true in dry, drought-prone areas where typical damming techniques for storing water are less successful. As per the 2018 World Water Development Report by the United Nations, the need for access to clean

---

V. Sharma
Laboratory of Quantum Chemistry, School of Advanced Chemical Sciences, Shoolini University, Bajhol, Himachal Pradesh, India

G. K. Jayaprakash (✉)
Laboratory of Quantum Chemistry, School of Advanced Chemical Sciences, Shoolini University, Bajhol, Himachal Pradesh, India

Department of Chemistry, Nitte Meenakshi Institute of Technology, Nitte (Deemed to Be University), Bangalore, Karnataka, India

C. M. Hussain, G. Hussain (eds.), *Green Nanostructured Photocatalysts for Clean Energy*, Clean Energy Production Technologies,
https://doi.org/10.1007/978-981-95-1618-6_6

water is projected to rise by approximately 30% by the year 2050 (Ahmed and Haider 2018). The WHO believes that three out of ten people do not have access to clean water, with the majority depending on contaminated sources, even though access to water has increased since the year 2000. Environmental pollution and energy issues have been made worse by industrial development brought on by the need for dyes, textiles, plastics, fertilizers, and other commodities. Ecosystems and human health are seriously threatened by industrial waste, which contains poisons such as oils, heavy metals, dyes, and other nonbiodegradable compounds (Pérez-Álvarez et al. 2022; Mills and Le Hunte 1997; Gao and Meng 2021). Water that is devoid of biological entities such as pathogens and viruses, as well as low in organic or mineral impurities, is considered desirable treated wastewater. In severely polluted waters, conventional water treatment techniques including chemical and chlorination treatments struggle to remove a variety of contaminants, frequently producing hazardous by-products and having unanticipated negative health effects. Techniques for coagulation and adsorption only concentrate pollutants after treatment. With finer mesh sizes, membrane filtration systems face ongoing problems including fouling and ineffective pollutant removal, which drive up prices (Pérez-Álvarez et al. 2022). So, treatment techniques that are efficient, affordable, and capable of getting rid of a wide range of pollutants are required. Therefore, there is an urgent need for in-depth study and the development of materials that can efficiently capture photonic energy to limit pollution in an environmentally acceptable manner. Utilizing renewable solar energy to drive chemical processes through oxidation and reduction, photocatalysis appears to be a potential option that offers a sustainable way to address environmental issues (Mutamim et al. 2012).

6.2 Photocatalysis

Light-driven water treatment systems have the potential to replace existing ways of treating water and wastewater, providing a more environmentally conscious and sustainable approach (Ahmed and Haider 2018). The utilization of sun energy for the photocatalytic degradation of pollutants is a light-driven catalytic process that includes photoreduction and photo-oxidation processes. Efficient light absorption and a redox reaction that relies on charge carriers generated by photon absorption are the key elements for an effective photocatalytic reaction to take place. Materials with suitable semiconductor band structures are preferred for sunlight absorption and subsequent charge carrier production (Sang et al. 2015).

6.2.1 General Mechanism of Photocatalysis

Photons are absorbed when their energy surpasses or equals the bandgap energy of the semiconductor. This results in an electron moving from the valence band to the conduction band, opening a hole in the valence band. The coupling between this hole and water molecules produces hydroxyl radicals. Concurrently, the photon-excited electrons participate in additional electron transfer events, creating superoxide radical anions. The electron–hole pair travels towards the surface of the photocatalyst, where they react with electron donors or acceptors adsorbed on the surface of the semiconductor, recombine, release energy as heat, or become stuck in metastable surface states. A chain reaction is initiated when these holes catalyze the oxidation of water to create additional hydroxyl radicals. After that, organic materials are oxidized, and electrons can be given to an electron acceptor (Liu et al. 2020; Wenderich and Mul 2016).

6.3 Modifications

So far, numerous structural alterations have been made to improve semiconductor photocatalytic efficacy. These advances are performed by creating vacancies, incorporating impurities, or setting up heterostructures.

6.3.1 Doping

Doping, whether with metals, nonmetals, or both, is a frequently used method that primarily aims to change the band gap of semiconductors. Typically, the dopant acts as a transient charge trapping site, limiting the fast recombination of charges (Sharma et al. 2023). Among the semiconductors investigated for photocatalysis, TiO_2 is the most popular option. Its appeal originates from its durability, affordability, low toxicity, photochemical stability, and superior hydrophilic characteristics. It was shown that carbon-doped TiO_2 had a better photocatalytic efficacy for organic pollutant decomposition than pristine TiO_2. This improvement is due to the introduction of additional states (C 2p) near the valence band edge of TiO_2 (O 2p), which results in a higher energy valence band edge than the reference TiO_2 (Fotiou et al. 2016; Rehman et al. 2009). Nonmetals can interact with photocatalytic oxides in three different ways. Initially, the dopant may create a hybrid structure with the photocatalyst oxide. Alternatively, it may replace oxygen atoms in the structure. Finally, the dopant might be introduced into oxygen-deficient sites to effectively block reoxidation (Ameta et al. 2018). Furthermore, the use of metal doping has been shown to alter semiconductor characteristics. Bansal et al. synthesized

Fe-doped TiO_2 and used it to degrade pentoxifylline. They discovered that the band gap in TiO_2 was reduced from 3.14 to 2.9 eV when Fe was doped.

6.3.2 Co-Doping

Co-doping in photocatalysis is the simultaneous insertion of two or more dopant elements into a semiconductor material to alter its characteristics and increase its photocatalytic activity. Samadi et al. co-doped a TiO_2/SiO_2 nanocomposite with Nd^{3+} and Zr^{4+}, resulting in the active breakdown of methyl orange by decreasing charge recombination (Samadi et al. 2015). Tbessi et al. developed a unique 1% Ce-0.6% Mn/TiO_2 catalyst that achieved 94% diclofenac mineralization, demonstrating impressive photocatalytic performance. This exceptional performance can be ascribed to increased optical activity in the visible region (Tbessi et al. 2019).

6.3.3 Coupling/Heterostructure

The formation of heterostructures is another modification that boosts the photocatalytic activity by facilitating redox reactions occurring simultaneously in different photocatalytic systems. The hydrothermal approach was adopted for developing SnS_2@$ZnIn_2S_4$@kaolinite with remarkable adsorption and photocatalytic degradation capabilities for eliminating tetracycline hydrochloride. The observed reaction rate of the heterostructure under visible light exposure was 0.0231 min^{-1}, which was roughly 2.37 times faster than that of SnS_2 and 20.81 times faster than that of $ZnIn_2S_4$ (Li et al. 2022). Under visible light, Z-scheme CdTe/TiO_2 photocatalysts that were hydrothermally synthesized effectively broke down tetracycline hydrochloride. After 30 minutes of irradiation, the optimized catalyst attained 78% breakdown, as a consequence of the Z-scheme high redox potentials, appropriate charge transfer, effective charge separation, and improved visible-light absorption (Gong et al. 2018).

6.4 Pollutants

6.4.1 Dye Degradation

Industries release unprocessed or inadequately treated wastewater, which leads to water contamination. Sectors such as pharmaceuticals, food production, paper and pulp industries, textiles, and cosmetics all rely extensively on synthetic dyes during the manufacturing process. The textile sector, in particular, uses a substantial amount of water in its processes (Rajkumar and Kim 2006). Dyes are divided into synthetic and natural categories. Synthetic dyes, which are preferred for their wide color range and long-term reliability, are far more widely used than natural alternatives. They are further classed depending on their chemical structures, such as azo, anthraquinone, phthalocyanine, triaryl methane, and sulfur dyes (Popli and Patel 2015). These dyes decrease the penetration of sunlight in water, which harms aquatic life by altering its metabolism. As a result, it is critical to cleanse water-soluble dyes produced in industry before releasing them into the main waters (Ahmed et al. 2020). Photocatalysis has attracted the attention of the global scientific community due to its efficacy in wastewater decontamination and dye degradation. The use of photocatalysts for dye degradation has emerged as a highly effective technology. For instance, Wongaree et al. synthesized Titania-carbon nanotube (CNT) nanocomposite which demonstrated a higher degradation efficiency for methylene blue. These nanotubes inhibit titania agglomeration, increasing surface area and improving electron transfer within composites (Wongaree et al. 2015). The photocatalytic potency of TiO_2 in degrading dyes decreases dramatically due to the high recombination rate of photo-induced charge carriers formed when exposed to UV light. Furthermore, powdered catalysts have the disadvantage of requiring stirring during reactions and making it difficult to separate the catalyst from treated water after each cycle (Han et al. 2009). According to research, dyes with simple structures and low molecular weights decolorize at a faster rate than those with complicated structures, large molecular mass, and prominent substitution (Ahmad et al. 2013). Also, a solvothermal approach was used to create a graphene-Ag/Zn nanocomposite for organic dye degradation via photocatalysis. During synthesis, silver-doped zinc oxide nanoparticles were equally dispersed throughout two-dimensional graphene sheets. This paper describes a unique approach for fabricating graphene-Ag/ZnO nanocomposites on site and demonstrates their extraordinary performance in photocatalytic dye removal (Ahmad et al. 2013).

6.4.2 *Antimicrobial Activity*

Photocatalysis is highly efficient in inhibiting the action of bacterial endospores. Surface damage to bacterial cells is the first step in the procedure, followed by membrane rupture, which causes internal components to leak. Photocatalysis eventually results in the breakdown of cell toxins, which helps with microbial disinfection (Liou and Chang 2012).

Matsunaga et al. were the first to study photocatalytic oxidation as an approach to microbial disinfection. Their findings revealed that a TiO_2-Pt photocatalyst destroyed microbial cells in 1–2 hours. They ascribed the mechanism of action to the photooxidation of Coenzyme A, which inhibits cell respiration and ultimately results in cell death (Matsunaga et al. 1985). Other findings revealed that the attachment of the photocatalyst to the substrate surface promotes the photocatalytic degradation of acid orange-7 (Hajkova et al. 2007). Composites of metal-free graphene/

graphene oxide have lately been widely used as an antibacterial disinfectant. For example, Chella et al. used solvothermal synthesis to create a $MnFe_2O_4$-graphene composite. After 2 hours of exposure to 100 μg ml^{-1} of the ferrite-based graphene composite, 82% of the cells lost viability (Chella et al. 2015). However, Cheng et al. and Nourmohammadi et al. used TiO_2-graphene oxide and ZnO_2-graphene oxide for antibacterial activity against *E. coli*, resulting in 100% bacterial destruction (Chang et al. 2015; Nourmohammadi et al. 2014).

6.4.3 Drug Elimination

Healthcare institutions and industrial pharmaceutical sectors contribute significantly to pharmaceutical pollution as a result of inappropriate disposal of leftover pharmaceuticals (Ruziwa et al. 2023; Shukla et al. 2021). Liu et al. found that adding Bi_2MoO_6 to TiO_2 significantly improves its activity for removing antibiotics from wastewater, particularly under visible light. Using the TiO_2/Bi_2MoO_6 heterojunction, they achieved degradation rates of 78%, 88%, and 78% for oxytetracycline, ciprofloxacin, and tetracycline, respectively, during 1.5 hours of visible light exposure (Liu et al. 2019). Jiang et al. (2018) used Quantum-dot-modified g-C_3N_4 for tetracycline demineralization with a 91% degradation rate (Jiang et al. 2018), whereas Linnick et al. adopted N-doped $Fe_xTi_yO_z$ for tetracycline hydrochloride degradation, which resulted in 32% degradation under UV but 48% under visible light (Linnik et al. 2017).

6.4.4 Removal of Heavy Metal

The buildup of heavy metals such as mercury, lead, copper, nickel, cadmium, and arsenic poses serious health concerns to humans. Furthermore, heavy metal contamination in water bodies and the environment can affect animals and plants, resulting in lower agricultural yields, food and water scarcity, and ecological discrepancies. Mercury, in both elemental (Hg(0)) and ionic (Hg(II)) forms, demonstrates high toxicity and dispersion, necessitating unique photocatalytic approaches when compared to other heavy metals of comparable toxicity (Ynalvez et al. 2016). Xiao et al. created a $CeO_2/BiOIO_3$ heterostructure that includes Ce^{4+}/Ce^{3+} redox centers and oxygen vacancies (Xiao et al. 2020). This new design resulted in a photocatalytic Hg removal efficiency of 86.53% for $CeO_2/BiOIO_3$, which outperformed pure $BiOIO_3$ by 1.3 times.

Moreover, Chen et al. showed that the Z-scheme $ZnTiO_3/Zn_2Ti_3O_8/ZnO$ ternary heterojunction photocatalyst displayed outstanding efficiency in reducing Cr(VI) ions. This improved efficiency is attributable to an effective photo-induced electron–hole pair, as well as the existence of multiple active photocatalytic sites on the surface of the photocatalyst (Chen et al. 2020). CuO/ZnO outperformed pure ZnO for both photocatalytic oxidation of As(III) and concomitant removal by adsorption onto the catalyst surface. Following 12 h of irradiation, the efficiency reached 60% (Samad et al. 2016).

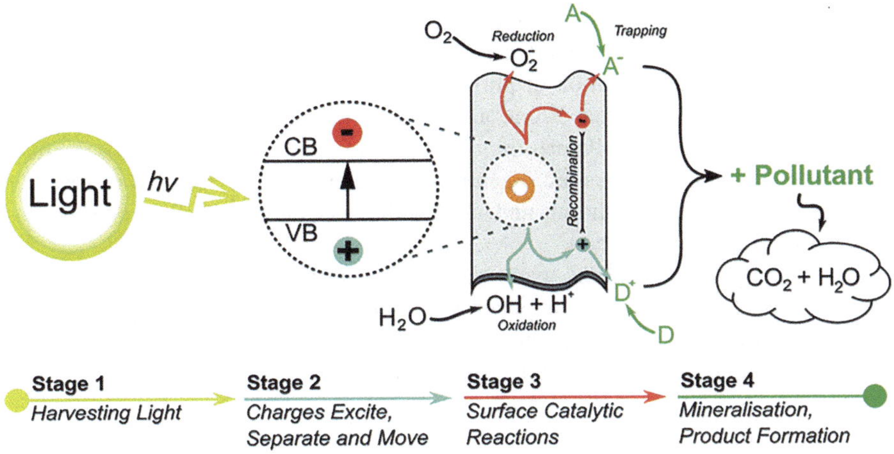

The figure is an illustration of a heterogeneous photocatalytic process schematically laid out across a two-dimensional photocatalyst surface, showing a few of the many reactions to pollutants that can take place when photogenerated charge carriers are employed. A and D The donor species and the surface-adsorbed acceptor species, respectively.

6.5 Conclusion and Future Perspectives

In conclusion, the need for efficient and sustainable water purification technologies has become increasingly urgent due to the rising demand for clean water globally. Conventional methods often struggle to effectively remove a wide range of pollutants, leading to environmental contamination and health risks. Photocatalysis has emerged as a promising solution, harnessing solar energy to drive chemical processes for water treatment. This chapter has provided an overview of the principles of photocatalysis, including its mechanisms and various modifications aimed at enhancing its efficiency.

Through doping, co-doping, and the formation of heterostructures, significant progress has been made in improving the photocatalytic activity of semiconductors. These modifications have enabled the degradation of pollutants such as dyes, antimicrobial agents, pharmaceuticals, and heavy metals. Additionally, advancements in nanotechnology have led to the development of nanocomposites with enhanced properties for water purification.

Looking ahead, further research and development in the field of photocatalysis for water purification are crucial. One avenue for future exploration is the optimization of photocatalyst materials to improve their efficiency, stability, and scalability for practical applications. Additionally, there is a need to explore novel photocatalytic materials and synthesis methods to address specific challenges, such as the degradation of emerging pollutants and the removal of persistent contaminants.

Furthermore, the integration of photocatalysis with other water treatment technologies, such as membrane filtration and adsorption, could lead to synergistic effects and improved overall performance. Collaborative efforts between scientists, engineers, and policymakers are essential to accelerate the translation of research findings into real-world solutions and address the pressing global challenges related to water pollution and scarcity.

In summary, photocatalytic water purification holds great promise for meeting the growing demand for clean water while minimizing environmental impact. Continued innovation and investment in this field are vital to ensure access to safe and sustainable water resources for future generations.

References

Ahmad M, Ahmed E, Hong ZL, Khalid NR, Ahmed W, Elhissi A (2013) Graphene–Ag/ZnO nanocomposites as high performance photocatalysts under visible light irradiation. J Alloys Compd 577:717–727

Ahmed SN, Haider W (2018) Heterogeneous photocatalysis and its potential applications in water and wastewater treatment: a review. Nanotechnology 29(34):342001

Ahmed A, Usman M, Yu B, Ding X, Peng Q, Shen Y, Cong H (2020) Efficient photocatalytic degradation of toxic Alizarin yellow R dye from industrial wastewater using biosynthesized Fe nanoparticle and study of factors affecting the degradation rate. J Photochem Photobiol B Biol 202:111682

Ameta R, Solanki MS, Benjamin S, Ameta SC (2018) Photocatalysis. In: Advanced oxidation processes for waste water treatment. Academic, pp 135–175

Chang YN, Ou XM, Zeng GM, Gong JL, Deng CH, Jiang Y, Liang J, Yuan GQ, Liu HY, He X (2015) Synthesis of magnetic graphene oxide–TiO2 and their antibacterial properties under solar irradiation. Appl Surf Sci 343:1–10

Chella S, Kollu P, Komarala EVP, Doshi S, Saranya M, Felix S, Ramachandran R, Saravanan P, Koneru VL, Venugopal V, Jeong SK (2015) Solvothermal synthesis of MnFe2O4-graphene composite – investigation of its adsorption and antimicrobial properties. Appl Surf Sci 327:27–36

Chen F, Yu C, Wei L, Fan Q, Ma F, Zeng J, Yi J, Yang K, Ji H (2020) Fabrication and characterization of ZnTiO3/Zn2Ti3O8/ZnO ternary photocatalyst for synergetic removal of aqueous organic pollutants and Cr (VI) ions. Sci Total Environ 706:136026

Fotiou T, Triantis TM, Kaloudis T, O'Shea KE, Dionysiou DD, Hiskia A (2016) Assessment of the roles of reactive oxygen species in the UV and visible light photocatalytic degradation of cyanotoxins and water taste and odor compounds using C–TiO2. Water Res 90:52–61

Gao X, Meng X (2021) Photocatalysis for heavy metal treatment: a review. PRO 9(10):1729

Gong Y, Wu Y, Xu Y, Li L, Li C, Liu X, Niu L (2018) All-solid-state Z-scheme CdTe/TiO2 heterostructure photocatalysts with enhanced visible-light photocatalytic degradation of antibiotic waste water. Chem Eng J 350:257–267

Hajkova P, Spatenka P, Horsky J, Horska I, Kolouch A (2007) Photocatalytic effect of TiO2 films on viruses and bacteria. Plasma Process Polym 4(S1):S397–S401

Han F, Kambala VSR, Srinivasan M, Rajarathnam D, Naidu R (2009) Tailored titanium dioxide photocatalysts for the degradation of organic dyes in wastewater treatment: a review. Appl Catal A Gen 359(1–2):25–40

Jiang XH, Wang LC, Yu F, Nie YC, Xing QJ, Liu X, Pei Y, Zou JP, Dai WL (2018) Photodegradation of organic pollutants coupled with simultaneous photocatalytic evolution of hydrogen using quantum-dot-modified g-C3N4 catalysts under visible-light irradiation. ACS Sustain Chem Eng 6(10):12695–12705

Li Y, Yu B, Hu Z, Wang H (2022) Construction of direct Z-scheme SnS2@ ZnIn2S4@ kaolinite heterostructure photocatalyst for efficient photocatalytic degradation of tetracycline hydrochloride. Chem Eng J 429:132105

Linnik O, Chorna N, Smirnova N (2017) Non-porous iron titanate thin films doped with nitrogen: optical, structural, and photocatalytic properties. Nanoscale Res Lett 12:1–10

Liou JW, Chang HH (2012) Bactericidal effects and mechanisms of visible light-responsive titanium dioxide photocatalysts on pathogenic bacteria. Arch Immunol Ther Exp 60:267–275

Liu Z, Tian J, Zeng D, Yu C, Huang W, Yang K, Liu X, Liu H (2019) Binary-phase TiO2 modified Bi2MoO6 crystal for effective removal of antibiotics under visible light illumination. Mater Res Bull 112:336–345

Liu H, Wang C, Wang G (2020) Photocatalytic advanced oxidation processes for water treatment: recent advances and perspective. Chem Asian J 15(20):3239–3253

Matsunaga T, Tomoda R, Nakajima T, Wake H (1985) Photoelectrochemical sterilization of microbial cells by semiconductor powders. FEMS Microbiol Lett 29(1–2):211–214

Mills A, Le Hunte S (1997) An overview of semiconductor photocatalysis. J Photochem Photobiol A Chem 108(1):1–35

Mutamim NSA, Noor ZZ, Hassan MAA, Olsson G (2012) Application of membrane bioreactor technology in treating high strength industrial wastewater: a performance review. Desalination 305:1–11

Nourmohammadi A, Rahighi R, Akhavan O, Moshfegh A (2014) Graphene oxide sheets involved in vertically aligned zinc oxide nanowires for visible light photoinactivation of bacteria. J Alloys Compd 612:380–385

Pérez-Álvarez DT, Brown J, Elgohary EA, Mohamed YM, El Nazer HA, Davies P, Stafford J (2022) Challenges surrounding nanosheets and their application to solar-driven photocatalytic water treatment. Mater Adv 3(10):4103–4131

Popli S, Patel UD (2015) Destruction of azo dyes by anaerobic–aerobic sequential biological treatment: a review. Int J Environ Sci Technol 12:405–420

Rajkumar D, Kim JG (2006) Oxidation of various reactive dyes with in situ electro-generated active chlorine for textile dyeing industry wastewater treatment. J Hazard Mater 136(2):203–212

Rehman S, Ullah R, Butt A, Gohar ND (2009) Strategies of making TiO2 and ZnO visible light active. J Hazard Mater 170(2–3):560–569

Ruziwa DT, Oluwalana AE, Mupa M, Meili L, Selvasembian R, Nindi MM, Sillanpaa M, Gwenzi W, Chaukura N (2023) Pharmaceuticals in wastewater and their photocatalytic degradation using nano-enabled photocatalysts. J Water Process Eng 54:103880

Samad A, Furukawa M, Katsumata H, Suzuki T, Kaneco S (2016) Photocatalytic oxidation and simultaneous removal of arsenite with CuO/ZnO photocatalyst. J Photochem Photobiol A Chem 325:97–103

Samadi S, Yousefi M, Khalilian F, Tabatabaee A (2015) Synthesis, characterization, and application of Nd, Zr–TiO 2/SiO 2 nanocomposite thin films as visible light active photocatalyst. J Nanostructure Chem 5:7–15

Sang Y, Liu H, Umar A (2015) Photocatalysis from UV/Vis to near-infrared light: towards full solar-light spectrum activity. ChemCatChem 7(4):559–573

Sharma V, Kumar A, Singh P, Verma PK, Ahamad T, Thakur S, Van Le Q, Nguyen VH, Khan AAP, Raizada P (2023) Advanced strategies for modifying the water splitting performance of MoSe2 photocatalyst: a critical review of recent progress. J Ind Eng Chem 128:55

Shukla S, Pandey H, Singh P, Tiwari AK, Baranwal V, Pandey AC (2021) Synergistic impact of photocatalyst and dopants on pharmaceutical-polluted waste water treatment: a review. Environ Pollut Bioavail 33(1):347–364

Tbessi I, Benito M, Molins E, Liorca J, Touati A, Sayadi S, Najjar W (2019) Effect of Ce and Mn co-doping on photocatalytic performance of sol-gel TiO2. Solid State Sci 88:20–28

Wenderich K, Mul G (2016) Methods, mechanism, and applications of photodeposition in photocatalysis: a review. Chem Rev 116(23):14587–14619

Wongaree M, Chiarakorn S, Chuangchote S (2015) Photocatalytic improvement under visible light in TiO2 nanoparticles by carbon nanotube incorporation. J Nanomater 16(1):295–295

Xiao Y, Tan S, Wang D, Wu J, Jia T, Liu Q, Qi Y, Qi X, He P, Zhou M (2020) CeO2/BiOIO3 heterojunction with oxygen vacancies and Ce4+/Ce3+ redox centers synergistically enhanced photocatalytic removal heavy metal. Appl Surf Sci 530:147116

Ynalvez R, Gutierrez J, Gonzalez-Cantu H (2016) Mini-review: toxicity of mercury as a consequence of enzyme alteration. Biometals 29:781–788

Chapter 7
Green Photocatalytic Air Purification

Shweta Singh, Homam Anntar, Sudhanshu Jangir, and Sasmita Chand

Abstract Biomimetics has provided solutions to many of the intricate problems of humans since time immemorial. Photocatalysis is one such application that draws inspiration from photosynthesis, the life-giving chemical reaction. Given the sedentary life choices of humans nowadays, and indoor environments being the source of a wide range of air pollutants, photocatalysis is the trending science defining the air purification process. Existing means of air purification, namely, absorption, adsorption, condensation, ionization, and filtration, have their own constraints. Photocatalytic oxidation-based air purifiers and building materials, and photocatalysis usage in heating, ventilating, and air conditioning are the emerging saviors of human health in the indoor environment. Photosensitive semiconductors absorb UV light to form reactive OH radicals in the presence of oxygen and water vapor. Generated free radicals are then involved in a series of reactions with indoor air pollutants, to mineralize them into CO_2 and water. Photocatalytic compounds are also utilized to construct concrete with air-purifying coating by integrating photocatalyst particles within a thin concrete surface layer. Such integration of photocatalysts into building materials is a good technological application for urban air pollution mitigation. After the COVID-19 pandemic, the application of photocatalytic oxidation-based air purifiers has increased even more, and extensive research is being done to improve their efficiency and reliability. For outdoor air pollution, PCO is a green technology as sunlight is a natural source that provides photonic energy as a driving force to initiate the chemical reactions. Such technological usage opens the scope of sustained solutions for air pollution in indoor and outdoor environments yet requires continuous research for enhancing the efficiency.

S. Singh · H. Anntar · S. Jangir
Indian Institute of Sustainability, Gujarat University, Ahmedabad, Gujarat, India

S. Chand (✉)
Manipal School of Architecture and Planning, Manipal Academy of Higher Education, Manipal, India

C. M. Hussain, G. Hussain (eds.), *Green Nanostructured Photocatalysts for Clean Energy*, Clean Energy Production Technologies,
https://doi.org/10.1007/978-981-95-1618-6_7

Keywords Photocatalysis · Air purifier · Air pollution · Indoor air pollution · Green technology · Biomimetics

7.1 Introduction

Industrial growth and technological intervention have eased out human life in different spheres, but at the same time, it has also increased the problem of pollution (Manisalidis et al. 2020) during the generation, usage, and end-of-life of products and services. The indoor environment, which now holds most of our time, has been stuffed with several equipment and material things, and while we are surrounded by these things, we are also exposed to the pollutants emitted by them either directly or indirectly. The quality of indoor air has deteriorated to a huge extent and is shown in several adverse health consequences. As per the findings of WHO (2022), indoor air pollution can bring more dangerous health consequences than outdoor air pollution in case of developing nations. The urban spaces in almost all countries have become more compact owing to the mushrooming urbanization in every nook and corner. This has further reduced the possibility of air exchange and thus aids in the build-up of air pollutants from various sources. In addition, the modern heating and cooling technologies have further changed people's habits of keeping the indoor completely disconnected from the outdoors (Javier Gonzalez-Martin et al. 2021), which in turn has further surmounted the indoor pollutants concentration and its exposure to the indoor dwellers. All these combined make the indoor environment conducive to a few acute and several chronic illness symptoms and in the long run impact the cognitive ability of humans (Künn et al. 2019; Wang et al. 2021), consequently reducing employee productivity (Sadick et al. 2020; Wolkoff et al. 2021) (Fig. 7.1). With the growing global concern over climate change, outdoor air pollution has fetched attention, and governments, organizations, research communities, and common people have come forward to work in tandem to limit the emissions, wherever possible. Yet, indoor air pollution, which can be a cause of both acute and chronic exposure, still requires the attention of people for its mitigation and adaptation (Sundell 2004).

The indoor air is impacted by many sources like materials (wall finishes, furniture varnishes, paints, fabrics), equipment and processes (equipment utilized for any kind of combustion, cleaning, cooking, etc.), heating/cooling and ventilation, and activities of occupants (usage of chemicals and movement of occupants) (Tham 2016). The different types of pollutants generated and found in an indoor environment are particulate matter, volatile inorganic and organic compounds, gases (like CO_2, CO, nitrogen oxide, and ozone), PAH (polycyclic aromatic hydrocarbons), and certain categories of microorganisms and pathogens. Indoor air quality becomes a serious concern when it questions the cognitive functions of children and adults, and

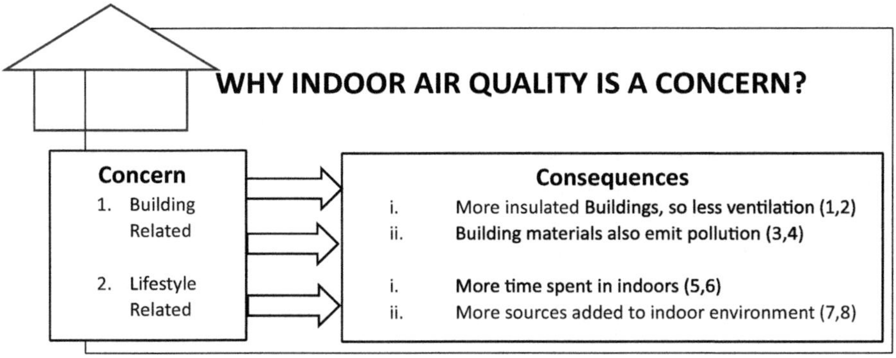

Fig. 7.1 The concerns associated with indoor air environment and their identified consequences (1) Fisk 2015; (2, 3) Weschler 2009; (4) González-Martín et al. 2021; (5) Cincinelli and Martellini 2017; (6) Marć et al. 2018; (9) Samet 1988; (10) Royal College of Physicians 2016; (11) Zewdie et al. 2023; (12) Burge 2004; (13) Ghaffarianhoseini et al. 2018; (14) Cincinelli et al. 2016; (15) Zhong et al. 2017

overall work performance of employees. Cognitive function refers to the brain-based skills that are essential for executing tasks of various difficulty levels (Angevaren et al. 2008). Understanding the concern, WHO published its first specific indoor air quality guidelines in 2010 (WHO 2010).

7.2 Current Practices for IAQ Improvement

Outdoor and indoor air pollution is a prime health concern in developing and developed countries. As many of the scientific researches have shown proportionality between improved air quality and positive impacts on human health (Sundell 2004; Coggins et al. 2013), cognitive abilities (Wang et al. 2021; Satish et al. 2012), and work performance (Sadick et al. 2020; Al Horr et al. 2016), it is important that indoor air quality in all indoor spaces where humans live, work, and spend significant amount of time requires improvement in terms of reduction of build of indoor pollutants. This reduction requires usage of air purification techniques. The air pollution management and control in outdoor as well as indoor environments rely on prevention and control methods. While the former involves relying on more renewable energy sources and energy system optimization (Ren et al. 2017), the latter involves the use of filtration and adsorption (Bolashikov and Melikov 2009; Zhang et al. 2011). The broad methods for air purification in the indoor environment involve physicochemical technologies and biological technologies (Mata et al. 2022).

7.3 Physicochemical Technologies

Technologies covered under physicochemical ones involve either a physical process of arresting the pollutants on some surface or a chemical process of combustion of pollutants. The physical process is the more commonly used method, in which pollutants are passed through a porous structure, like fabric filters, and are then arrested depending on the size of the pollutant molecules and the size of the pores present in the filtering structure. The chemical process that involves combustion is a less frequently applied method. Physicochemical technologies currently employed are filtration, adsorption, ionization, UV-photocatalytic oxidation, and UV disinfection.

7.4 Biological Technologies

Biological technologies used for air purification purposes rely on living plants or microorganisms like bacteria and fungi. Plant-based techniques are applied to arrest several kinds of pollutants. In other words, it can be considered an application of phytoremediation. This is also gaining attention nowadays owing to its lower cost and the good aesthetic value provided by plants. Another plant-based technique involves microalgae-based methods. Moreover, biofiltration techniques are also employed to filter certain gaseous pollutants, and recently, microorganisms are also used for the removal of VOCs from the air..

7.5 Biomimetics

Nature is the source of many kinds of cultural services, and they are also termed ecosystem services. Cultural services include the opportunities nature provides us for recreational activities, and inspiration for art, literature, etc. Apart from these, nowadays, humans look towards nature to solve many existing problems. In this regard, biomimetics can be considered as the technology transfer from nature to engineering (Vincent et al. 2006). Biomimetics, also known as biomimicry, or biologically inspired design, is an interdisciplinary research field that involves applying principles, strategies, and structures found in nature to solve engineering, design, and many other challenges. The various fields in which biomimetics work are art, architecture, medicine, and engineering. The biomimetics spiral is shown in Fig. 7.2.

Bioinspired photocatalysts or photocatalytic systems exploit solar energy using semiconductors and are used mainly for two purposes: one to degrade pollutants and second to generate energy (Wang et al. 2024). In this way, photocatalysis is directing sustainable solutions for two of the most pressing global concerns today, i.e., environmental pollution and energy generation.

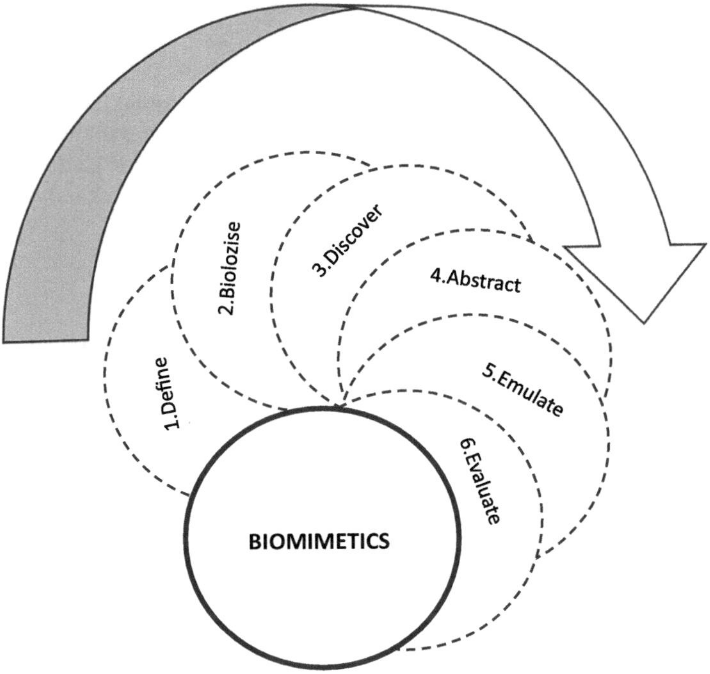

Fig. 7.2 Biomimetics spiral showcasing the steps involved in drawing inspiration from nature for problem solving. (Adopted from the Biomimicry Institute)

7.6 Photocatalysis for Air Purification

As the term photocatalysis suggests, it utilizes solar energy to oxidize or at least reduce contaminants by stimulating electron excitation-transfer in a semiconductor photocatalyst (Antonopoulou et al. 2021), thus making the photocatalyst the main efficiency driver. The best aspect of the photocatalysis process is the destruction or degradation of pollutants through the conversion of solar energy into chemical energy (Long et al. 2020). The degradation of pollutants through photocatalysis can be employed for both air and water pollution.

Research work carried out by Fujishima and Honda (1972) inspired the later studies to take up photocatalysis as the process for the destruction of liquid organic compounds. The possibility of disinfecting pathogens using photocatalysis was first demonstrated by Matsunaga et al. (1985).

Homogenous as well as heterogeneous catalysts can be employed for air purification purposes, yet heterogeneous ones are more frequently utilized. In the case of homogenous photocatalysis, the reactants and photocatalyst are in the same phase. It is used for pollutant removal in exhaust air purification, wastewater recovery, and soil decontamination resulting from industrial, agricultural, and urban activities (Mak et al. 2021). Although homogenous photocatalysis has contributed significantly towards the removal of environmental contaminants and energy generation, the associated concern is the recovery part, as recovery after reaction is difficult (Ravelli et al. 2009). The 3 Rs process, reuse, recycle, and recover, is important to make a photocatalytic process eco-friendly (Mak et al. 2021).

Heterogeneous photocatalysis involves the acceleration of a photoreaction in the presence of a catalyst, where the reactant and photocatalysts are in different phases (Ibhadon and Fitzpatrick 2013). In general, heterogeneous photocatalysis produces highly reactive species, which can oxidize various air pollutants.

7.7 Mechanism Behind Photocatalytic Air Purification

Photocatalysis can be considered as the most interesting utilization of the Sun's energy, where it gets transformed into chemical potentials that catalyze processes in usual conditions, which otherwise are very much energy- and cost-intensive. This process of photocatalysis relies on a material that can be considered a photocatalyst owing to its specific properties. These are usually semiconductors.

A semiconductor which can absorb light with energy larger than its band gap, leading to the generation of an electron–hole pair, is called a photocatalyst. After the generation of the electron–hole pair, production of reactive oxygen species (ROS) takes place, which is utilized to convert air pollutants into less harmful compounds (Mamaghani et al. 2017; Ren et al. 2017).

The most commonly used photocatalyst is TiO_2. Reasons for its persistent performance are its very high oxidation power, easy availability for mass production at an industry level, easy synthesis in the laboratories, and its stability, which keeps it chemically and photochemically inert (Lee and Park 2013; Horikoshi and Serpone 2020; Lee and Park 2013; Antonopoulou et al. 2021). The mechanism is graphically depicted in Fig. 7.3.

7.8 Photocatalytic Air Purifiers

The main features of a photocatalytic air purifier system involve the photocatalyst, UV source, reactor design and operation, and contaminants. Although all photocatalytic materials can be excited by UV radiation, those excited by visible light showcase absorption in a limited wavelength range of visible light. As the solar radiation has less than 5% UV and around 50% visible light wavelengths, currently ongoing research are trying to expand this absorption band under visible light wavelengths so that solar radiation can activate the photocatalyst. Most of the currently employed photocatalytic air purifiers use a UV light source, thus increasing cost and safety concerns. Some of the important recent research depicting functioning and efficiency of air purifiers based on photocatalysis are shown in Table 7.1. The major types of photocatalysts utilized in air purification are binary metal oxides, complex

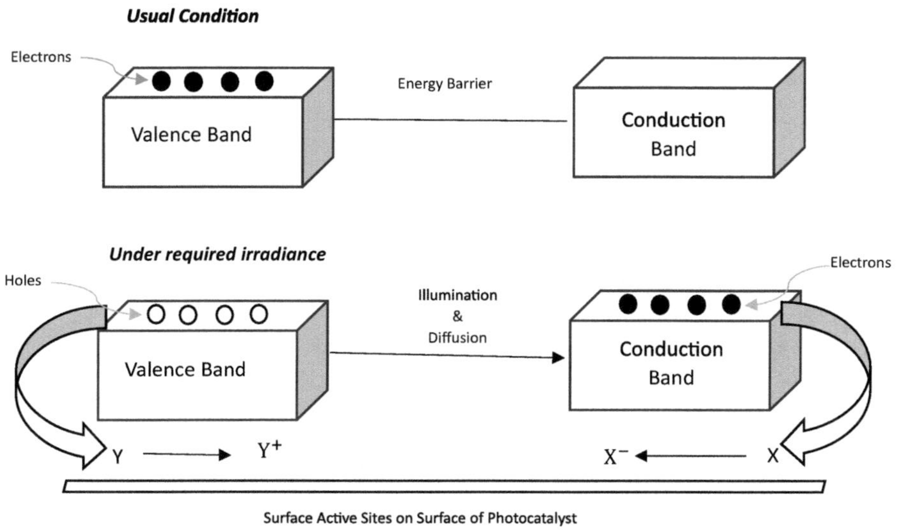

Fig. 7.3 Mechanism for photocatalysis

metal oxides, metal sulfides, and metal-free materials (Zhao and Yang 2003; Mo et al. 2009; Cao et al. 2015; Wang et al. 2015). UV-based photocatalytic air purifiers help in reducing the indoor pollutant load significantly, but in order to make it widely acceptable, the safety concerns of UV radiation emitted should be researched. Although UV light is utilized in the healthcare sector for disinfecting the air in buildings, many researchers have supported its usage in other indoor environments as well. Based on such scientific back-up, ASHRAE and EPA have also recognized the importance of methods that include filtration and UV-C-based systems in indoor environment.

7.9 Market Research for Air Purifiers

Regarding the market research that was conducted on the international e-commerce platforms (like eBay, Amazon, OTTO, and Flipkart), the first finding is that most of the air purifiers are filter-based units using HEPA-type (High Efficiency Particulate Air) filters that filter allergens, dust, bacteria, and smoke using several layers of filters as follows:

1. *Pre-filter layer*: This layer captures larger particles to prevent them from clogging the HEPA filter and extends its life.
2. *Main HEPA layer*: Made from fine fiberglass or synthetic fibers, this layer is responsible for filtering out particles as small as 0.3 microns with 99.97% efficiency.

Table 7.1 Compilation of some of the latest important research in the field of photocatalytic air purification

S. no.	Title	Year of publication	Study area	Type of air purifier	Useful for indoor/ outdoor/ both	Working principle	Whether embedded in building material	Efficiency	Working life
1.	A green and direct synthesis of photosensitized CoS2–graphene/ TiO2 hybrid with high photocatalytic performance	2014	Korea and China	Titania (TiO2) as a photocatalyst in the degradation of organic and inorganic pollutants in air and water	N/A	Photocatalysts using CoS2, CoS2–graphene, and CoS2–graphene/TiO2 composites were tested for the effects in visible light region. The photocatalytic activity of the sample was evaluated by measuring the degradation of organic pollutants such as methylene blue (MB) and industrial dyes such as Texbrite BA-L (TBA) under visible light	NA	As a result, patterns indicated that CoS2–graphene/ TiO_2 has good photoinduction effects in the visible light region	N/A

2.	TiO_2-impregnated porous silica tube and its application for compact air and water purification units	2015	Japan	Novel one-end sealed porous TiO2-coated amorphous-silica (a-silica) tubes with large porosity using the outside vapor deposition (OVD) method	N/A	Simple fabrication of novel one-end sealed porous TiO_2-coated amorphous-silica (a-silica) tubes with large porosity using the outside vapor deposition (OVD) method. The impregnation method was used to fill TiO_2 precursor deep into the pores of one-end sealed porous a-silica tubes. The porous a-silica tubes were assayed for their VOC removal ability through an acetaldehyde decomposition test. In addition, the water purification ability of these tubes was preliminarily evaluated through the methylene blue decolorization test	Not embedded in building materials; its tubes have a great potential for compact and in-line VOC removal and water purification units	The tube (8.5 mm e.d. × 150 mm) demonstrated a 93% removal rate for high concentrations (ca. 300 ppm) of acetaldehyde gas at a single-pass condition with a 250 mL/min flow rate under UV irradiation. The tube also demonstrated water purification capacity at a rate 2.0 times higher than a-silica tube without TiO_2 impregnation	N/A

(continued)

Table 7.1 (continued)

S. no.	Title	Year of publication	Study area	Type of air purifier	Useful for indoor/ outdoor/ both	Working principle	Whether embedded in building material	Efficiency	Working life
3.	Photocatalytic air cleaners and materials technologies—Abilities and limitations, building and environment	2015	Canada	Heterogeneous photocatalytic oxidation	Indoor	One of the most highly anticipated uses is photocatalytic oxidation (PCO) technology, which utilizes a nano-catalyst that is stimulated when illuminated by UV light to provide an alternative and energy-efficient solution to air quality challenges. Photosensitive semiconductors, such as titanium dioxide (TiO_2), absorb ultraviolet (UV)light to form reactive hydroxyl radicals (•OH) in the presence of oxygen and water vapor. These free radicals undertake a series of reactions with pollutants, such as containing bond cleavage, substitution, and electron transfer, to mineralize them into CO_2 and water	Photocatalysts integrated into construction materials	The paper itself is a comparison between test methods and conditions for PCO air cleaners in degradation of gaseous pollutants of a number of papers	

4.	An advanced semimetal–organic bi spheres–g-C3N4 nanohybrid with SPR-enhanced visible-light photocatalytic performance for NO purification	2015	China	Elemental bismuth nanospheres (BiNSs) were introduced onto the surface of g-C3N4 nanosheets through an in situ deposition method. The thin g-C3N4 nanosheets were used as a substrate for the green synthesis of BiNSs-loaded g C3N4 (bi–g-C3N4)	Indoor and outdoor	All the reagents employed in this study were of analytical grade (sigma-Aldrich) and used without further purification. The g-C3N4 was synthesized by the polymerization of urea under high temperature	N/A	42.0% NO removal in a continuous reactor to demonstrate the capability in air purification.	N/A
5.	The viability of photocatalysis for air purification	2015	USA	A photocatalytic air purifier uses a catalyst, a substrate to support the catalyst in the air flow, and a light source	Indoor	United Technologies research center (UTRC) uses a modular approach to the purifier design utilizing aluminum honeycomb as a catalyst substrate. The basic module is a light source and a titania catalyst supported on the honeycomb	No, integrated with heating, ventilation, and air conditioning or HVAC systems	N/A	Sun: Exists with daylight Black light fluorescent: 5000–12,000 h, usually limited by phosphor degradation Germicidal fluorescent: 10,000–20,000 h LEDs:Wavelength-dependent, a few thousand hours at short wavelengths

(continued)

Table 7.1 (continued)

S. no.	Title	Year of publication	Study area	Type of air purifier	Useful for indoor/outdoor/both	Working principle	Whether embedded in building material	Efficiency	Working life
6.	Photocatalytic degradation of malachite green by pyrite and its synergism with Cr(VI) reduction: Performance and reaction mechanism	2015	China	A photocatalyst for the simultaneous removal of organic compounds and heavy metals from water	N/A	Pyrite (FeS_2) is a tailing waste from the mineral separation process that is becoming a promising candidate for adsorption and photocatalysis of the pollutants malachite green (MG) and Cr(VI) under UV light irradiation and oxygen supply	N/A	The removal efficiencies of malachite green (MG) and Cr(VI) were 96.7% and 53.1%, respectively, after 120 min. The reduction ratio of Cr(VI) was significantly accelerated in the presence of MG	It could be used 5 times for 120 min
7.	Photocatalytic activities of titanium dioxide incorporated architectural mortars: Effects of weathering and activation light	2015	China and Colombia	Self-cleaning and air-purifying through photocatalytic reactions on building materials containing TiO_2	Indoor	Two methods for introducing TiO_2 particles onto building materials were carried out, and their photocatalytic performances, in particular the resistance to weathering, were compared: 1. A novel transparent photocatalytic coating applied on the architectural mortar. 2. TiO_2 intermixed samples	Embedded in building material	RhB removal efficiencies (R4 = 40% and R26 = 51%) for the PC-S7-coated samples	20 years under Hong Kong conditions

8.	Synthesis, characterization and photocatalytic performance of Yb-doped CdTe nanoparticles	2015	South Korea and USA	Doped CdTe nanoparticles as a photocatalyst with different Yb contents	N/A	The absorption edges of Yb-doped CdTe were shifted to longer wavelengths (red-shift) in comparison to pure CdTe. The photocatalytic performance of Yb-doped CdTe compounds was also determined by malachite green (MG) degradation Under visible light irradiation	N/A	The color removal efficiencies of $Yb_{0.1}Cd_{0.9}Te$ and pure CdTe Were 84.65% and 15% after 120 min of treatment, respectively	N/A
9.	Photocatalytic degradation of brilliant green using undoped and Zn-doped SnO2 nanoparticles under sunlight irradiation	2015	India	Treatment of wastewater polluted with organic compounds	N/A	Photocatalytic activity of undoped and Zn-doped SnO_2 was evaluated by studying photodegradation of brilliant green dye solution under sunlight radiation to measure the percentage of degradation	N/A	The study compared different densities of SnO_2: Zn to determine which gave the highest degradation, and the best was SnO_2: Zn (0.075), which achieved 98% after 105 min	N/A
10.	Photocatalytic air cleaners and materials technologies—Abilities and limitations	2015	Canada	A study in general photocatalyst and in the end a similar table to this one					

(continued)

Table 7.1 (continued)

S. no.	Title	Year of publication	Study area	Type of air purifier	Useful for indoor/ outdoor/ both	Working principle	Whether embedded in building material	Efficiency	Working life
11.	Long-lifetime water-washable ceramic catalyst filter for air purification	2023	Korea, China, Russia, and the USA	Water-washing regenerable ceramic catalyst filters (CCFs)	Indoor and outdoor air purification	Coating the inner and outer filter channels with an inorganic membrane and a Cu2O/TiO2 photocatalyst respectively	N/A	Removes particulate matter (PM) (>95%) and volatile organic compounds (VOCs) (>82%) in a single pass	20 years
12.	Recent developments in photocatalytic nanotechnology for purifying air polluted with volatile organic compounds: Effect of operating parameters and catalyst deactivation	2023	Switzerland	A study of general photocatalysts and a comparison of the different types of TiO_2-based photocatalysts					

13.	Photocatalytic degradation of organic and inorganic pollutants to harmless end products: Assessment of practical application potential for water and air cleaning	2023	Switzerland	Comparison study between different kinds of photocatalysts like (ZnO, CuO, TiO_2, Bi_2O_3, etc.) and their ability to clean volatile organic compounds (VOCs) from air and wastewater
14.	Photocatalytic fibers for environmental purification: Challenges and opportunities in the post-pandemic era	2023	Taiwan and Japan	Studying the necessity of photocatalysts during the COVID-19 pandemic and an overview on how they work, what challenges they face, and what further research is required to improve them

3. *Activated carbon layer (optional)*: Some HEPA filters have an additional layer of activated carbon that absorbs odors and harmful gases.

On average, companies suggested a 6-month replacement interval of the filters while some of them gave a 6–8 month replacement period. As per the new filter measured in dollars, the average price found was 226.2 USD. One company, called Airzone, based in the USA mentioned the possibility of having a bad impact of using photocatalyst-based air purifiers. "Most air purifiers or depolluting materials using photocatalysis employ nanoscale titanium dioxide (TiO_2), which the International Agency for Research on Cancer (IARC) classified as a "possible human carcinogen" in 2007. The potential release of TiO_2 particles from these products and their effects on human health are not fully understood."

HIGMARK, another USA-based company, mentioned that there are many more effective technologies to accomplish this task which has pushed UV alone to be a more outdated technology since the way needs proper exposure to the ventilation line. The one company that was found to sell photocatalyst air purifiers, "the airpurifiers," at the revealed price of 1199.98 USD.

To sum up, the main international e-commerce platforms sell mostly HEPA-type filters since it is lower in cost, as well as one report mentioned the hazards of using photocatalytic air purifiers which can be a reason for the low percentage of selling them online (Table 7.2).

7.10 Conclusion

Indoor air quality has gained little attention till date when compared to the outdoor air quality. While governments across countries have provided regulations for outdoor air quality and emission limits, regulations for indoor air quality are yet to be worked upon largely. The emergence of COVID-19 highlighted the importance of indoor air quality, and the role of ventilation was understood. Among the various technologies available for indoor air quality management, filters are the most widely utilized method. They are efficient in removing particles but are not effective for the organic ones. Also, timely removal and replacement are essential, else it can also become a source of indoor pollution. Photocatalytic oxidation is an emerging technology, which is relatively clean as well, and is considered to be efficient in removing organic particles, inorganic particles, and microorganisms. However, currently available photocatalytic air purifier essentially comprises a UV source, which makes the purifier an expensive one, and can also question the safety issues of UV radiation. Ongoing research to implement semiconductors that can absorb in the widely available visible wavelength range of the sun, to initiate the photocatalysis, will be a more sustainable solution.

Table 7.2 Market research about the current air purifiers available globally

Platform	Company	Company location	Model	Price ($)	Does it use a photocatalyst?	Cleaning percentage	Environmental impact
Ebay	Honeywell	USA	Honeywell HPA3300B Plug-In Air Purifier	239.99	No (only filters)		
			Honeywell HPA5100B Air Purifier	81.22	No (only filters)		
			Honeywell HPA5300B HEPA Air Purifier	169.99	No (only filters)		
	Dyson	England	Dyson BP03 Purifier	900	No (only filters)		
			Dyson TP07 Smart Cool Air Purifier and Fan	500	No (only filters)		
			Dyson Purifier Cool Gen1 TP10 Tower Air Purifier—664609-01	340	No (only filters)		
	LEVOIT	China	Levoit LAP-V201S-WUS Vital 200S-P 100s Smart True HEPA Air Purifier	273.62	No (only filters)	99.97%	
			Levoit PlasmaPro 400s Smart 403 ft^2 True HEPA Air Purifier	157.95	No (only filters)		
	Blueair	Sweden	Blueair Blue Pure 211i Max Air Purifier	224.95	No (only filters)		
			Blueair Blue Pure 211+ Console Air Purifier	275	No (only filters)		
			Blueair Pro M Air Purifier—White (PROMK120PAW)	220	No (only filters)		
	Winix	South Korea	Winix AM90 360 ft^2 Capacity Wi-fi Air Purifier work with Amazon Alexa	204.66	No (only filters)	99. 97%	

(continued)

Table 7.2 (continued)

Platform	Company	Company location	Model	Price ($)	Does it use a photocatalyst?	Cleaning percentage	Environmental impact
			Winix C545 Air Purifier with Replacement Filters	170	No (only filters)		
			Winix Dark Grey 360 ft^2. 4-stage True HEPA Air Purifier Am80	222.59	No (only filters)		
	Shark	USA	Just UVC lamps for other purifiers				
	Coway	South Korea	Coway AP-1518R Air Purifier	169.99	No (only filters)		
			Coway Airmega 400 Smart Air Purifier with 1560 ft^2 Coverage—Graphite (AP-2015F(G))	349.99	No (only filters)		
			Airmega Ap-2015f 400—The Smarter Air Purifier with Max2 Filter	450	No (only filters)		
	GermGuardian	USA	GermGuardian AC5350 28″ 4-in-1 Tower Air Purifier	149	No (only filters)	99.97%	
			GermGuardian AC4825 4-In-One Air Purifier	69.99	No (only filters)		
	Rainbow	USA	2024 Rainbow Rainmate IL LED Air Purifier—Allergies—Asthma—Sealed!	94	Led ionizer		

Flibkart	Coway	South Korea	Coway AirMega Aim Portable Room Air Purifier	119.42	No (only filters)		
	Honeywell	USA	Honeywell Air Touch V3 Air Purifier with H13 HEPA	106.17	No (only filters)	99.99%	
	ECO GUARD	USA	ECO GUARD UV-C Photocatalyst Portable Room Air Purifier	179.16	Photocatalyst		
	Qubo	India	Qubo Q500 Smart Room Air Purifier	119.32	No (only filters)	99%	
OTTO	DYSON	England	DYSON Luftreiniger Dyson Purifier Hot+Cool™ Gen1	499	No (only filters)		
			DYSON Luftreiniger Purifier Humidify+Cool PH03	749,00	No (only filters)		
	Philips	Netherland	Philips air purifier AC0820/10800 series, for 48 m^2 rooms, suitable for allergy sufferers	99,90	No (only filters)		
			Philips Air Purifier AC0850/11 800 Series, for 49 m^2 rooms, pre-HEPA and activated carbon filters, with app connection	149,90	No (only filters)		
			Philips air purifier AC1715/111000i series, for 78 m^2 rooms, with app connection and suitable for allergy sufferers	219,90	No (only filters)		

(continued)

Table 7.2 (continued)

Platform	Company	Company location	Model	Price ($)	Does it use a photocatalyst?	Cleaning percentage	Environmental impact
Amazon	WINIX	South Korea	WINIX 5500–2 Air Purifier for Home Large Room Up to 1740 ft^2 in 1 Hr With Air Quality Monitor	132.99	No (only filters)	99.97%	
			WINIX T810 Air Purifier for Home Extra Large Room	171.45	No (only filters)		
	Jafanda	USA	Jafanda Air Purifiers for Home Bedroom, H13 True HEPA	37.99	No (only filters)		
			Jafända Air Purifiers for Home Large Room Up To 1190ft^2, True HEPA 13 Filter	79.99	No (only filters)		
			Jafända® JF500 Air Purifiers for Home	269.99	No (only filters)		
			Jafända® JF260S Wifi Air Purifier	139.99	No (only filters)		
			Jafända® JF239 Air Purifier	119.99	No (only filters)	99.97%	

	Airzone	USA	Airzone does not manufacture or distribute air purification or disinfection products		Photocatalytic oxidation (PCO) uses UV light and a titanium dioxide (TiO_2) catalyst to convert harmful contaminants into water, carbon dioxide, and harmless residues. This air purification method typically integrates into an existing air system, where air passes through UV light/TiO_2 chambers. Here, free hydroxy radicals and superoxide ions generated by the UV light and TiO_2 attack and destroy the contaminants' cellular structure, including DNA/RNA		Most air purifiers or depolluting materials using photocatalysis employ nanoscale titanium dioxide (TiO_2), which the International Agency for Research on Cancer (IARC) classified as a "possible human carcinogen" in 2007. The potential release of TiO_2 particles from these products and their effects on human health are not fully understood. Additionally, incomplete degradation of pollutants through photocatalysis can produce harmful by-products like ketones and aldehydes. Even complete degradation can lead to the formation of other pollutants, such as nitrates

(continued)

Table 7.2 (continued)

Platform	Company	Company location	Model	Price ($)	Does it use a photocatalyst?	Cleaning percentage	Environmental impact
	HIGHMARK	USA	Genesis Air 2008 CU unit	UV lights need to be replaced every 12,000 h. Lights are more expensive to replace than M8/M13 filters, but less expensive than activated carbon/charcoal or HEPA filters	Using UV light to energize TiO_2 electrons as a TiO_2-coated surface to generate hydroxyl radicals that chemically react in the air cleaner only to destroy bacteria, viruses, and pollutants and produce harmless byproducts (H_2O, CO_2). This can be used to remove viruses and other harmful contaminants from the airstream if proper exposure is allowed; however, there are much more effective technologies to accomplish this task which has pushed UV alone to be a more outdated technology	The effectiveness of contaminant removal in a single pass without a filter was tested by RTI International, an independent nonprofit research institute. RTI found that the PCO air purifier offers tremendous efficacy in removing contaminants, including: MRSA (se): 99.9%, tuberculosis (Mp): 88.7%, common virus/influenza (MS2): 43.6%, bacteria (MS2): 34.6%, Mold (Av): 11.4%. The full RMI international report can be accessed at the website	

	Olansi	China	Olansi A12B 2021 Desktop UVC Air Purifier HEPA Filter Purifying CE Air Purifier Clean Air Quality PM2.5	N/A	No (only filters)	N/A	HEPA air purifiers are the best in the market, especially when they have true HEPA filters within. These kinds of purifiers catch mold, thereby restricting its movement and reducing pollution within the air
			Olansi A17A High Efficiency Washable Non-consumable Air Purifier	N/A	No (only filters)	N/A	
			Olansi A17 Portable Home Remove Smog PM2.5 UV Air Cleaner H13 Office HEPA Filter Air Purifier	N/A	No (only filters)	N/A	
			Olansi K08D 220 V/50HZ True Hepa Air Purifier H13 Humidifier Air	N/A	No, filters with UV light to break down the harmful formaldehyde, ammonia, benzene, TVOC, hydrogen sulfide, and other harmful gases	99%	
	Theairpurifiers	Canada	AirPura P600	1199.98	The photocatalytic oxidation process breaks down the complex, hard-to-remove chemical compounds into shorter, simpler chemicals. This process repeats itself multiple times until only basic chemicals remain, which can be easily adsorbed by regular carbon	N/A	

References

Al Horr Y, Arif M, Kaushik A, Mazroei A, Katafygiotou M, Elsarrag E (2016) Occupant productivity and office indoor environment quality: a review of the literature. Build Environ 105:369–389

Angevaren M, Aufdemkampe G, Verhaar HJ, Aleman A, Vanhees L (2008) Physical activity and enhanced fitness to improve cognitive function in older people without known cognitive impairment. Cochrane Database Syst Rev 2

Antonopoulou M, Kosma C, Albanis T, Konstantinou I (2021) An overview of homogeneous and heterogeneous photocatalysis applications for the removal of pharmaceutical compounds from real or synthetic hospital wastewaters under lab or pilot scale. Sci Total Environ 765:144163

Bolashikov ZD, Melikov AK (2009) Methods for air cleaning and protection of building occupants from airborne pathogens. Build Environ 44(7):1378–1385

Burge PS (2004) Sick building syndrome. Occup Environ Med 61(2):185–190

Cao S, Low J, Yu J, Jaroniec M (2015) Polymeric photocatalysts based on graphitic carbon nitride. Adv Mater 27(13):2150–2176

Cincinelli A, Martellini T, Amore A, Dei L, Marrazza G, Carretti E, Leva P (2016) Measurement of volatile organic compounds (VOCs) in libraries and archives in Florence (Italy). Sci Total Environ 572:333–339

Coggins MA, Semple S, Hurley F, Shafrir A, Galea KS, Cowie H, Sanchez-Jimenez A, Garden C, Whelan P, Ayres JG (2013) Indoor Air Pollution and Health (IAPAH) (2008-EH-MS-8-S3); STRIVE report series no.104. EPA STRIVE Programme 2007–2013. EPA—Environmental Protection Agency, Wexford. ISBN:9781840954852

Fisk WJ (2015) Review of some effects of climate change on indoor environmental quality and health and associated no-regrets mitigation measures. Build Environ 86:70–80

Fujishima A, Honda K (1972) Electrochemical photolysis of water at a semiconductor electrode. Nature 238(5358):37–38

Ghaffarianhoseini A, AlWaer H, Omrany H, Ghaffarianhoseini A, Alalouch C, Clements-Croome D, Tookey J (2018) Sick building syndrome: are we doing enough? Archit Sci Rev 61(3):99–121

González-Martín J, Kraakman NJR, Pérez C, Lebrero R, Muñoz R (2021) A state–of–the-art review on indoor air pollution and strategies for indoor air pollution control. Chemosphere 262:128376

Ibhadon AO, Fitzpatrick P (2013) Heterogeneous photocatalysis: recent advances and applications. Catalysts 3(1):189–218

Künn S, Palacios J, Pestel N (2019) Indoor air quality and cognitive performance

Long Z, Li Q, Wei T, Zhang G, Ren Z (2020) Historical development and prospects of photocatalysts for pollutant removal in water. J Hazard Mater 395:122599

Mak CH, Han X, Du M, Kai JJ, Tsang KF, Jia G, Hsu HY (2021) Heterogenization of homogeneous photocatalysts utilizing synthetic and natural support materials. J Mater Chem A 9(8):4454–4504

Mamaghani AH, Haghighat F, Lee CS (2017) Photocatalytic oxidation technology for indoor environment air purification: the state-of-the-art. Appl Catal B Environ 203:247–269

Marć M, Śmiełowska M, Namieśnik J, Zabiegała B (2018) Indoor air quality of everyday use spaces dedicated to specific purposes – a review. Environ Sci Pollut Res 25:2065–2082

Mata TM, Martins AA, Calheiros CS, Villanueva F, Alonso-Cuevilla NP, Gabriel MF, Silva GV (2022) Indoor air quality: a review of cleaning technologies. Environments 9(9):118

Matsunaga T, Tomoda R, Nakajima T, Wake H (1985) Photoelectrochemical sterilization of microbial cells by semiconductor powders. FEMS Microbiol Lett 29:211–214

Mo J, Zhang Y, Xu Q, Lamson JJ, Zhao R (2009) Photocatalytic purification of volatile organic compounds in indoor air: a literature review. Atmos Environ 43(14):2229–2246

Ravelli D, Dondi D, Fagnoni M, Albini A (2009) Photocatalysis. A multi-faceted concept for green chemistry. Chem Soc Rev 38(7):1999–2011

Ren H, Koshy P, Chen WF, Qi S, Sorrell CC (2017) Photocatalytic materials and technologies for air purification. J Hazard Mater 325:340–366

Royal College of Physicians (2016) Every breath we take: the lifelong impact of air pollution. Report of a Working Party, London

Sadick AM, Kpamma ZE, Agyefi-Mensah S (2020) Impact of indoor environmental quality on job satisfaction and self-reported productivity of university employees in a tropical African climate. Build Environ 181:107102

Samet JM (1988) Health effects and sources of indoor air pollution. Am Rev Respir Dis 137:221–242

Satish U, Mendell MJ, Shekhar K, Hotchi T, Sullivan D, Streufert S, Fisk WJ (2012) Is CO2 an indoor pollutant? Direct effects of low-to-moderate CO2 concentrations on human decision-making performance. Environ Health Perspect 120(12):1671–1677

Sundell J (2004) On the history of indoor air quality and health. Indoor Air 14:51

Tham KW (2016) Indoor air quality and its effects on humans – a review of challenges and developments in the last 30 years. Energ Buildings 130:637–650

Wang W, Tadé MO, Shao Z (2015) Research progress of perovskite materials in photocatalysis- and photovoltaics-related energy conversion and environmental treatment. Chem Soc Rev 44(15):5371–5408

Wang C, Zhang F, Wang J, Doyle JK, Hancock PA, Mak CM, Liu S (2021) How indoor environmental quality affects occupants' cognitive functions: a systematic review. Build Environ 193:107647

Wang X, Shi X, Yin S, Zheng J, Song Y, Sun H (2024) Bioinspired material for high-performance photocatalytic systems. ACS Mater Lett 6

Weschler CJ (2009) Changes in indoor pollutants since the 1950s. Atmos Environ 43(1):153–169

Wolkoff P, Azuma K, Carrer P (2021) Health, work performance, and risk of infection in office-like environments: the role of indoor temperature, air humidity, and ventilation. Int J Hyg Environ Health 233:113709

World Health Organization (2010) WHO guidelines for indoor air quality: selected pollutants

Zewdie A, Degefa GH, Donacho DO (2023) Health risk assessment of indoor air quality, sociodemographic and kitchen characteristics on respiratory health among women responsible for cooking in urban settings of Oromia region, Ethiopia: a community-based cross-sectional study. BMJ Open 13(6):e067678

Zhang Y, Mo J, Li Y, Sundell J, Wargocki P, Zhang J, Sun Y (2011) Can commonly-used fan-driven air cleaning technologies improve indoor air quality? A literature review. Atmos Environ 45(26):4329–4343

Zhao J, Yang X (2003) Photocatalytic oxidation for indoor air purification: a literature review. Build Environ 38(5):645–654

Zhong L, Su FC, Batterman S (2017) Volatile organic compounds (VOCs) in conventional and high performance school buildings in the US. Int J Environ Res Public Health 14(1):100

Chapter 8
Metal-Based Nanostructures for Sustainable Photocatalysis

Kutloano Edward Sekhosana and Xolile Fuku

Abstract The increasing demands for clean, green energy driven by urbanization and modernization while reducing atmospheric pollution are among the current biggest global challenges. Current technologies employed in the production of energy involve the burning of fossil fuels and directly impact our ecosystem, with 92% of carbon emissions emanating from such exercises. Thus, a drastic reduction in greenhouse gas (GHG) emissions around the globe is highly recommended to produce clean and green photocatalytic energy. The environment may also be impacted by water pollution from industrial effluent and volatile airborne compounds (VOCs). The implementation of these advanced technologies is a foreseeable revolutionization for energy production and purification of the environment. In addition, standard procedures and enactment for monitoring, immediate rectification, and other propitious solutions to these environmental and energy-related concerns must be the forefront of research at both institutional and industrial levels. One of such solutions is the metal-based semiconductor nanostructures for enhanced photocatalysis. These nanostructures could be exploited for their solar energy-driven redox reactions to produce green and renewable energy. Despite the reported rapid recombination of photo-excited charge carriers and slow interfacial transfer of electron-hole pairs in semiconductors, expediting the segregation of charge carriers, incorporation of noble metals, engineering of nanocomposites, and morphology modulation suppress these shortfalls. In addition, the formation of junctions inserted amidst two semiconductors can enhance the segregation of photo-excited charge carriers, resulting in homojunction- and heterojunction-based semiconductor photocatalysts. This chapter provides a comprehensive review and discussion of photocatalysis driven by metal-based nanostructures for addressing global environmental and sustainable green energy production-related challenges. Moreover, this chapter provides an overview of the mechanisms behind the photocatalytic activity of

K. E. Sekhosana (✉) · X. Fuku
Institute for Nanotechnology and Water Sustainability (iNanoWS), College of Science, Engineering and Technology, University of South Africa, Roodepoort, Florida, South Africa
e-mail: sekhoke@unisa.ac.za

C. M. Hussain, G. Hussain (eds.), *Green Nanostructured Photocatalysts for Clean Energy*, Clean Energy Production Technologies,
https://doi.org/10.1007/978-981-95-1618-6_8

metal-based nanostructures, their potential applications, challenges, and future prospects in renewable energy, energy production, and CO_2 utilization.

Keywords Sustainable · Implementation · Energy production · Global challenges · Solutions · Photocatalysis

8.1 Introduction

It is well-documented that coal, oil, and natural gas are a class of deposits accumulated through the living tissue produced from the energy of sunlight and buried in the depths of the earth (Omer 2008). These three chemical substances find application in the energy sector, with coal having been used for energy generation globally for many decades, while oil and natural gas are applied as liquid fuel and electricity generation, respectively. These complex substances are close competitors. For example, natural gas has the potential to substitute coal particularly in the US power sector as a result of increasing natural gas production and a decline in production costs. In addition, policies advocating for lower global greenhouse gas and pollutant emissions support a drift from coal-fired power generation (Gagarin et al. 2020). The combustion of both coal and natural gas contributes to CO_2 emissions (Ren and Patel 2009). Indeed, the landscape of energy production and use is changing, as many alternative technologies are being developed. For example, there is a notable application growth of wind, solar, and hydropower, largely due to price decreases, and many governments have policy mechanisms that support fuel switching to these renewable energy sources for power generation (Gagarin et al. 2020). There is also a realization of the human development index which relies on per capita energy consumption (Samokhvalov 2017). Thus, the use of clean solar energy for securing future energy requirements has been recommended (Aziz Aljar et al. 2020). Thus, dependency on the conventional energy sources including coal and natural gas can be reduced. Hydrogen is considered an ideal clean fuel that can be obtained from water, which is an abundantly available renewable resource. Researchers are already studying the hydrogen energy economy, but have found that production methods such as steam and plasma reforming, oxidation, and electrolysis to be more costly, thereby contributing to the high prices (Zheng et al. 2019a). Hydrogen generated through water electrolysis presents a high purity of 99.999 vol.% when dried up and oxygen impurities have been removed (Ursúa et al. 2012). Thus, hydrogen, as an energy fuel, is worth being focused on. Researchers have developed approaches of hydrogen generation via photocatalytic water splitting using a variety of semiconductors such as PANI/MXene/MnO_2 (Sharma et al. 2025), MnO_2@SnO_2 (Ali et al. 2024), TiO_2 (Pahlevanpour and Bashiri 2022), and graphitic-carbon nitride (Dankawu et al. 2024). Photocatalytic properties of certain materials have been taken advantage of converting solar energy into chemical energy, leading to the oxidation or reduction of materials for the production of hydrogen (Nakata and Fujishima 2012; Kudo and Miseki 2009; Maeda 2011; Jafari et al. 2016) and hydrocarbons (Tu et al. 2014; Gong et al.

2021) and even for the removal of pollutants and bacteria (Zuo et al. 2021; Fagan et al. 2016; Wang et al. 2015) on wall surfaces and in air and water. In this chapter, our focus will be on the fundamentals of photocatalysis in photodegradation, energy conversion, and storage using metal-based photocatalytic nanostructures.

In this context, the integration of metal-based nanostructures in photocatalysis holds considerable promise for the advancement of renewable energy technologies, such as hydrogen production via water splitting, the enhancement of fuel cell performance, and the effective conversion of CO_2 into valuable chemicals as well as energy storage. In this chapter, we explore the role of metal-based nanostructures in synergistic photocatalysis, focusing on their applications for sustainable energy solutions and environmental remediation.

8.2 Photocatalysis, Principle, and Mechanism

Photocatalysis refers to the acceleration of a chemical reaction by a catalyst under the influence of light. In this process, light energy is absorbed by a photocatalyst, generating electron–hole pairs that participate in redox reactions. This principle is central to various applications, including solar-driven energy production, water splitting, and CO_2 reduction. The process has gained immense attention for its potential to harness solar energy, reduce environmental pollution, and contribute to sustainable energy solutions.

8.2.1 Photocatalysis Mechanism

- *Photon absorption*: The photocatalytic process begins when the catalyst absorbs light, typically ultraviolet (UV) or visible light, depending on the properties of the material. When a photon with sufficient energy hits the surface of the catalyst, an electron in the valence band of the material is excited to the conduction band, leaving behind a hole in the valence band. This results in the generation of an electron–hole pair, which is essential for initiating the catalytic reaction.
- *Charge separation*: For photocatalysis to be effective, the electron–hole pairs generated upon light absorption must be efficiently separated and remain stable long enough to participate in chemical reactions. If these charge carriers recombine rapidly, the photocatalytic efficiency decreases. Therefore, photocatalysts with structures that facilitate the separation and movement of these charge carriers are crucial for improving performance.
- *Redox reactions*: The separated electrons and holes participate in redox reactions on the surface of the catalyst. The electrons in the conduction band are often involved in reduction reactions, such as the reduction of oxygen to generate reactive oxygen species (ROS), while the holes in the valence band are involved in

oxidation reactions, such as the oxidation of water to form hydroxyl radicals or other oxidizing species.

(i) *Reduction reactions*: Electrons reduce substances like metal ions, oxygen molecules (O_2) to superoxide ($O_2^{\bullet -}$), or other substrates, depending on the application.
(ii) *Oxidation reactions*: Holes oxidize substances such as water molecules (H_2O) to produce hydroxyl radicals (•OH) or degrade organic pollutants, depending on the photocatalytic application.

- *Surface interaction and reaction pathways*: The photocatalytic reactions typically occur on the surface of the catalyst, where the reactants interact with the excited electron–hole pairs. In the case of water splitting, the photocatalyst facilitates the oxidation of water molecules at the surface to generate oxygen, while the reduction of protons forms hydrogen. Similarly, in the photocatalytic reduction of CO_2, the electrons participate in the reduction of CO_2 to value-added chemicals like methane or methanol.

8.3 Metal-Based Heterogeneous Photocatalysts

As already alluded to, photocatalysts would refer to materials such as semiconductors which are capable of absorbing and interacting with entities such as H^+ or CO_2 to produce fuels via photooxidation or photoreduction in the valence band and conduction band, respectively. Guo et al. (2023) demonstrated the photoreaction processes using the equations below:

$$\mathrm{ZnO} + h\nu \rightarrow \mathrm{ZnO}\left(e^{-}_{\mathrm{CB}} + \mathrm{h}^{+}_{\mathrm{VB}}\right) \tag{8.1}$$

$$e^{-}_{\mathrm{CB}} + \mathrm{M}^{n+} \rightarrow \mathrm{M}^{(n-1)+} \tag{8.2}$$

$$\mathrm{O}_2 + \mathrm{M}^{(n-1)+} \rightarrow \mathrm{O}_2 + \mathrm{M}^{n+} \tag{8.3}$$

$$\mathrm{ZnO}\left(\mathrm{h}+_{\mathrm{VB}}\right) + \mathrm{OH}^{-} \rightarrow \mathrm{ZnO} + \cdot\mathrm{OH} \tag{8.4}$$

$$\cdot\mathrm{OH} + \text{organic contaminant} \rightarrow \text{degradation products} + \mathrm{CO}_2 + \mathrm{H}_2\mathrm{O} \tag{8.5}$$

Hydrogen as a fuel is produced when the reactants or pollutants such as feedstocks or biomass are photodegraded or decomposed, with the addition of a photocatalyst only improving the production efficiency.

Industries such as agricultural sectors and coal-processing entities should be beneficial in clean energy production via the decomposition of biomass or coal combustion and, consequently, carbon capture using the photocatalysis process. For example, CO_2 that pollutes the environment should be captured and photoreduced into carbonaceous fuel products. For this reason, it has been suggested that biomass, especially from agricultural residues or dedicated energy crops, should be converted through various processes yielding energy-rich gases or liquid products (Guo et al. 2023). The major benefit of combining these conversion technologies should be the achievement of simultaneous production of both hydrogen and other valuable energy products. Another synergistic effect in photocatalysis is observed during photocatalytic oxidation and reduction which should overcome the efficiency defects of photocatalytic reactions, thus improving the degradation production capacity. This process is pollution-free and has no additional energy consumption. Additionally, researchers should take advantage of the abundance of organic pollutants in wastewater as compared to the inorganic counterparts, thus not only decomposing these organics but also producing liquid and gaseous fuel products.

The nature of semiconductors in these processes is very important. For example, heterogeneous photocatalysts based on semiconductors absorb a flux of energy that exceeds the bandgap energy (E_g). This type of absorption propels the electrons into the conduction band (CB), while, at the same time, creating holes in the valence band (VB), with separated charges participating in various reduction and oxidation reactions (Yaemsunthorn et al. 2024). Stability and toxicity of semiconductors should also be taken into account. The former directly impacts the long-term performance and durability of semiconductor materials under operating conditions, while the latter can affect the environmental and human health. For example, metal chalcogenides are susceptible to photodegradation, also known as photocorrosion, despite their remarkable bandgap energies and efficient solar light absorption, while metal oxides are more stable and less prone to photocorrosion (Yaemsunthorn et al. 2024). These photocatalysts should, by no means, leach into the environment. The performance of semiconductors may also be affected by the extreme bandgap width.

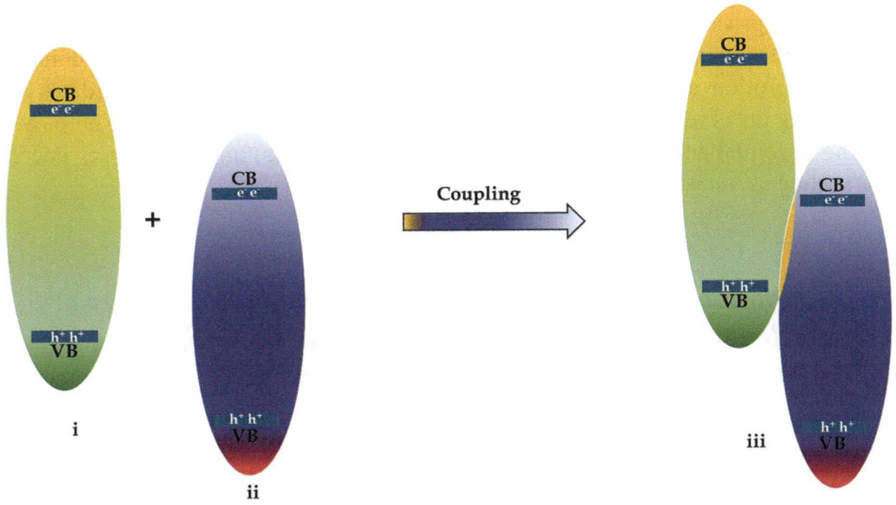

Scheme 8.1 Coupling two photocatalysis (i and ii) of different bandgaps and levels of VB and CB

Such semiconductors can be coupled with other materials (Scheme 8.1) of varying bandgaps, resulting in heterojunctions of different types which influence the redox reaction strength and rely on the CB and VB of both materials.

Other studies combine photocatalysis and electrocatalysis of heterojunctions when photodegrading organic pollutants (An et al. 2025). The design of these heterojunctions involves the integration of Bi_2MoO_6 and Cu_2O as photoanodes in a self-powered photoelectrocatalytic system to remove tetracycline hydrochloride (TCH) (Fig. 8.1). The researchers of this work designed a schematic of the photoelectrocatalytic system (Fig. 8.1a), assessed the performance of the self-powered photoelectrocatalytic system (Fig. 8.1b–e), and analyzed the removal efficiencies of different heterojunction photoanodes (Fig. 8.1b–e). The data present different rate constants which were larger for triboelectric-electromagnetic nanogenerator (TENG-EMG) photoelectrocatalysis compared to photocatalysis. Indeed, there are many approaches to synergistic photocatalysis known to date and must be further explored to mitigate the environmental issues such as pollution, while producing clean, rich-energy fuels.

Several photocatalytic heterojunctions including Bi_2WO_6 nanosheet/TiO_2 nanobelt (Ahmadi et al. 2023), $CeCuO_3$/g-C_3N_4 (Zhu et al. 2025), Fe_3O_4 bridged Pt/g-C_3N_4 (Asiri et al. 2022), $CuBi_2O_4$/$AgIO_3$ (Yan et al. 2025), CdS@N-NiCoO (Yang et al. 2025), In_2O_3/TiO_2 (Zhang et al. 2025), and $Bi_4O_5I_2$/CuO (Dhiman et al. 2025) have demonstrated their superior photocatalytic performance in both CO_2 conversion and photodegradation.

8.4 Applications in Renewable Energy

Photocatalysis offers a sustainable route for hydrogen production through water splitting, a process where water molecules are dissociated into hydrogen and oxygen using solar energy. Metal-based nanostructures, particularly those that enhance light absorption and provide efficient charge transfer, are ideal candidates for improving the efficiency of this process. Furthermore, these materials are employed in the development of dye-sensitized solar cells (DSSCs) and other solar energy harvesting systems.

8.4.1 Photocatalysis in Fuel Production and Clean Energy or Renewable Energy Production: Solar Water Splitting and Photocatalytic Hydrogen Production

One of the most promising applications of metal-based nanostructures in photocatalysis is in renewable energy production, particularly in the splitting of water into hydrogen and oxygen (solar-driven water splitting). Metal-based photocatalysts can

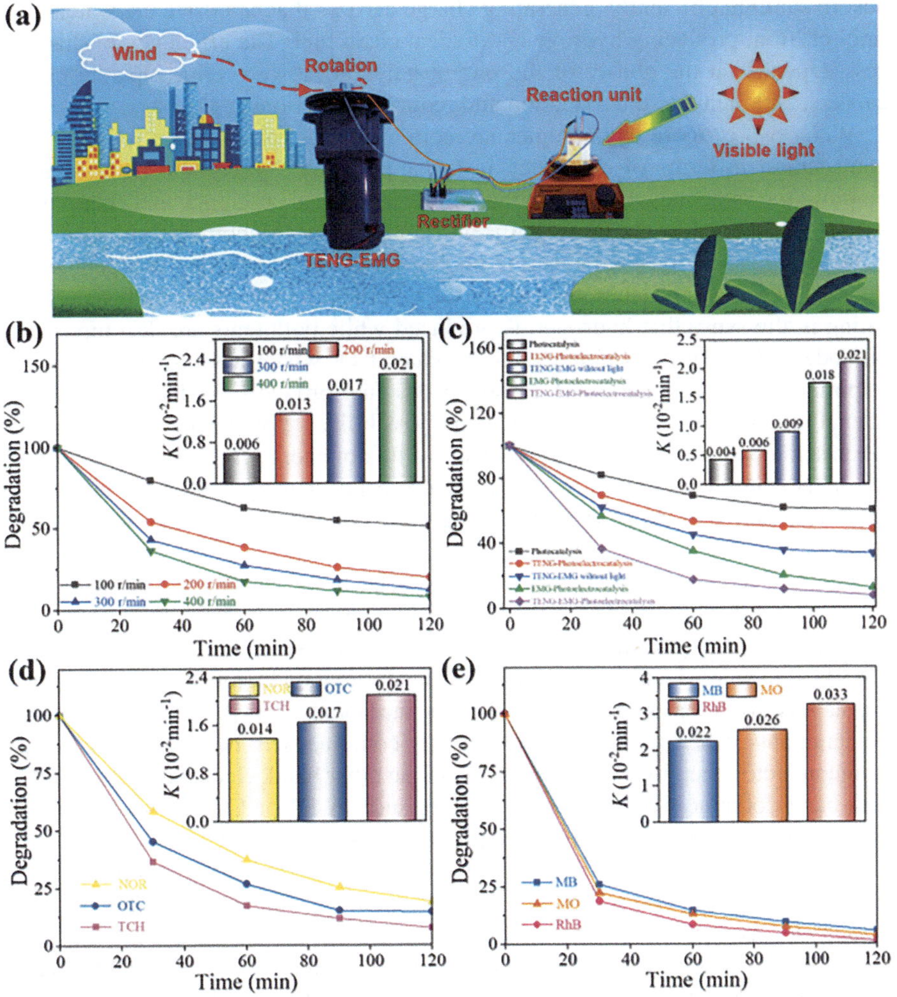

Fig. 8.1 Self-powered wastewater photoelectrocatalytic treatment system. (**a**) Schematic diagram of a self-powered photoelectrocatalytic system. (**b**) Degradation of TCH through photoelectrocatalysis with TENG-EMG at different rotation rates. (**c**) TCH removal through photocatalysis, TENG-photoelectrocatalysis, TENG-EMG without light, EMG-photoelectrocatalysis, and TENG-EMG-photoelectrocatalysis (insert is the reaction rate constant k). (**d**) Degradation of norfloxacin (NOR), oxytetracycline (OTC), and TCH. (**e**) Methylene blue (MB), methyl orange (MO), and rhodamine (RhB) through photoelectrocatalysis with TENG-EMG at a speed of 400 r/min, insert is the reaction rate constant k. Reproduced with permission from ref (An et al. 2025). Copyright 2024 Elsevier Ltd.

absorb sunlight efficiently, generating charge carriers that can drive the water splitting reaction, producing hydrogen—an ideal clean fuel. The efficiency of this process depends on the ability of the photocatalyst to absorb light, separate charge carriers, and facilitate the oxidation and reduction reactions.

When two or more technologies are combined, the purpose is to improve their contribution in terms of efficiency in the application. Synergistic photocatalysis employs the same idea to enhance the pollutant removal efficiency and energy utilization efficiency, leading to a significant improvement in photocatalytic efficiency (Guo et al. 2023). An ideal synergic photocatalysis would be one that enables the removal of pollutants using low energies at an efficient rate. Photocatalysis is interesting in the sense that hydrogen is produced while pollutants are decomposed or photodegraded. These pollutants also produce CO_2 which has proved to be more important in the production of carbonaceous fuel products as shall be discussed later. It should be noted though that the possibility of photocatalysis is realized when semiconductor photocatalysts absorb a flux of energy from the source of light to photoexcite the electrons from the valence band (leaving behind the holes (h^+)) into the conduction band, thereby pairing these two entities to create the electron-hole pairs.

Photoreduction happens when photodegraded by-products such as CO_2 accept electrons, while photooxidation takes place when holes come into contact with H_2O to produce hydroxyl radicals and protons. These photocatalytic reactions will be discussed in some detail later. On the one hand, semiconductors are capable of acting as electron donors (also referred to as sacrificial agents) to the pre-absorbed contaminants which rapidly and irreversibly combine with holes generated in the semiconductor, thereby oxidizing them. On the other hand, electrons in the conduction band combine with electron acceptors such as H^+ or CO_2 in the solution to produce hydrogen or hydrocarbons to produce clean energy or electricity (Guo et al. 2023). It is, therefore, important to understand by photooxidation and photoreduction whether in the photodegradation of pollutants or combustion mitigation processes such as carbon capture.

Clean energy is very important in human lives for the reduction of health hazards caused by technologies such as those that produce fuels that litter the environment. In addition, pollutants affect the lives of not only humans but also aquatic life. In light of these challenges, researchers take advantage of photocatalysis to convert pollutants into clean energy. Photocatalysis involves several reaction steps including the following: (Guo et al. 2023)

1. Light absorption.
2. Excitation and separation of the carriers.
3. Transfer of electrons and holes to the photocatalytic material surface.
4. Recombination of electron-hole pairs inside and on the surface of the photocatalytic material.
5. Oxidation of holes on the material surface and reduction of electrons.

Guo et al. (2023) demonstrated the phenomenon of electron acceptors and donors, the actual reaction mechanisms, the number of electrons involved in the process, and the reaction products, thus dividing photocatalysis into two main categories:

1. Direct reduction of pollutants to produce energy.
2. Synergistic photocatalytic degradation of pollutants to produce energy.

The direct photocatalytic reduction of air pollutants such as CO_2 and, while other pollutants can be such as halogenated alkanes to hydrocarbon fuels and valuable chemicals is possible. The second category of photocatalysis can be achieved by coupling two or more technologies (Guo et al. 2023). metal-based photocatalysts find application in all these categories of photocatalysis.

8.4.2 *Direct Reduction of CO_2/CO_2 Reduction and Climate Change Mitigation*

While fossil fuels may have been produced by photosynthesis and stored underground as energy sources, it has become apparent that their burning results in the production of CO_2 that litters the atmosphere. Progress has been made in the artificial photosynthesis systems converting CO_2 to useful and energy-rich compounds using solar light as the energy source and water as the reductant as ideal solutions (Berardi et al. 2014; Knör 2015; Zhang and Lin 2014; Singh et al. 2014). These artificial systems can convert CO_2 into glucose and oxygen, or hydrogen, alkanes, alcohols, and aldehydes (Fig. 8.2) (Singh et al. 2014). Despite this, more research to find other alternative technologies is being conducted globally. For example, the research group of Ishitani (Sahara and Ishitani 2015) has developed the photocatalysts to support artificial photosynthesis systems for CO_2 reduction. Interestingly, this research found that the systems can be divided into three parts:

1. Two-component systems.
2. Supramolecular photocatalysts.

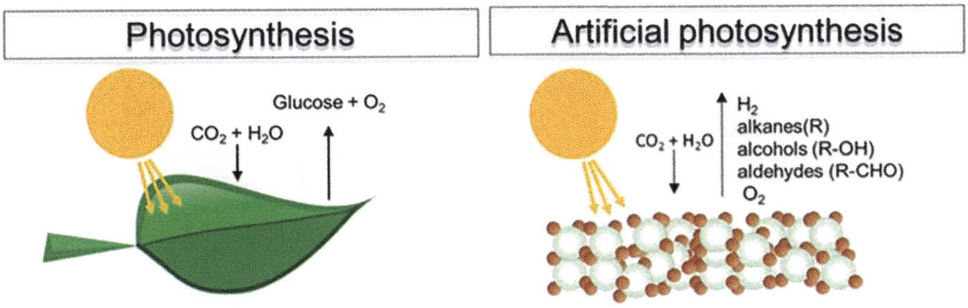

Fig. 8.2 Schematic of the proposed artificial photosynthesis in TiO_2-QD-based catalysts for selectively producing alkanes, alcohols, and aldehydes. Reproduced with permission from ref (Singh et al. 2014). Copyright 2014 American Chemical Society

3. Hybrids of the supramolecular photocatalyst with a semiconductor photocatalyst.

Sahara et al. (Sahara and Ishitani 2015) observed that molecular photocatalytic systems for CO_2 reduction pose a conundrum: when the photochemical electron-transfer process is applied, excitation by one photon usually effects only one-electron transfer but not multielectron transfer; reduction of CO_2 through a one-electron transfer process requires a highly negative potential ($E°$ = −1.9 V vs NHE at pH = 7) and produces an active $CO_2^{\bullet-}$ that impedes the administration of the selective product (Eq. 8.6) (Tanaka and Ooyama 2002).

$$CO_2 + e^- \rightarrow CO_2^{\bullet-} \quad (E° = -1.9\,V) \tag{8.6}$$

This challenge is tackled using two-component photocatalytic systems constructed with a redox photosensitizer that photochemically mediates one-electron transfer and a catalyst that converts one-electron transfer to multielectron reduction of CO_2. Rigorous reduction potential of CO_2 is achieved through multielectron reduction of CO_2 while producing stable and useful products such as HCOOH and CO (Eqs. 8.7 and 8.8) (Sahara and Ishitani 2015).

$$CO_2 + 2H^+ + 2e^- \rightarrow HCOOH \quad (E° = -0.61\,V) \tag{8.7}$$

$$CO_2 + 2H^+ + 2e^- \rightarrow CO + H_2O \quad (E° = -0.53\,V) \tag{8.8}$$

The general mechanism of photocatalysis is summarized in Fig. 8.3. The central is a photocatalyst characterized by valence band (VB) and conduction band (CB)

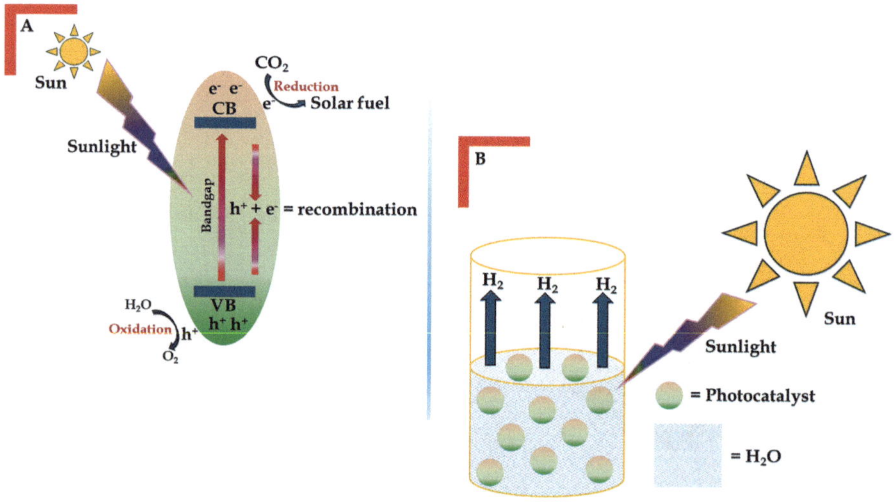

Fig. 8.3 (**a**) Photoreduction of CO_2 into fuels and (**b**) hydrogen evolution via the absorption of sunlight by the photocatalysts

(Fig. 8.3a). Upon pumping sufficient energy flux (usually sunlight from the sun), an electron (e^-) moves from VB to CB, leaving the hole (h^+). The e^- can relax and recombine with the hole. For the purpose of energy generation, water is reduced by holes to release hydrogen while also producing oxygen. This oxidation process is further demonstrated in Fig. 8.3b where water is split into its fundamental elements, with hydrogen being the fuel. On the other hand, CO_2 can be converted to protect the environment while producing energy through reduction. Both oxidation and reduction are efficient when the bandgap of an electrocatalyst is small.

Although thermocatalytic and electrocatalytic reduction of CO_2 has been reported (Zhang et al. 2024; Gao et al. 2023; Hu et al. 2024; Etim and Zhong 2023), especially for the production of hydrocarbons and alcohols, photoreduction is considered the most efficient approach owing to its merits of nonpollution, inexhaustibility, and non-energy inputs (Bian et al. 2022; Li et al. 2019a). Low conversion efficiency in previously reported photocatalysts restricts their widespread applications in CO_2 photoreduction, despite these merits (Jiao et al. 2020). As demonstrated in Fig. 8.3a, b, carbonaceous fuels can be obtained through photoreduction and conversion of CO_2 in the CB, while H_2O would be photo-oxidized into O_2 in the VB by the photo-excited electrons and photogenerated holes, respectively. Jiao et al. (2020) suggested that the CO_2 photoreduction can be realized if the position of a photocatalyst is negative as compared to the standard potentials of CO_2 reduction (Table 8.1), while its VB position is positive as compared to the standard potential of H_2O oxidation.

Photocatalysis of CO_2 into fuels may also be aided by the incorporation of sacrificial agents especially when the band edge positions of photocatalysts are unable to simultaneously match the standard potentials of CO_2 reduction and H_2O oxidation. This challenge in photocatalysis has been observed in semiconductors with extremely narrow band gaps that can absorb the IR light that occupies ca. 52%of full sunlight (Jiao et al. 2020). It has been reported that the CO_2 photoreduction can

Table 8.1 Conversion of CO_2 and H_2O into various products via reduction and the corresponding reduction potentials with reference to NHE at pH 7 in aqueous solution

Reaction	Product	E° (V vs. NHE)
$H_2O \rightarrow ½O_2 + 2H^+ + 2e^-$	Oxygen	+0.82
$H_2O + 2e^- \rightarrow 2OH^- + H_2$	Hydrogen	−0.41
$CO_2 + 8H^+ + 8e^- \rightarrow CH_4 + 2H_2O$	Methane	−0.24
$CO_2 + 2H^+ + 2e^- \rightarrow CO + H_2O$	Carbon monoxide	−0.51
$CO_2 + 4H^+ + 4e^- \rightarrow HCHO + H_2O$	Formaldehyde	−0.55
$CO_2 + 6H^+ + 6e^- \rightarrow CH_3OH + H_2O$	Methanol	−0.39
$CO_2 + 2H^+ + 2e^- \rightarrow HCOOH$	Formic acid	−0.58
$2CO_2 + 14H^+ + 14e^- \rightarrow C_2H_6 + 4H_2O$	Ethane	−0.27
$2CO_2 + 12H^+ + 12e^- \rightarrow C_2H_5OH + 3H_2O$	Ethanol	−0.33
$2CO_2 + 2H^+ + 2e^- \rightarrow H_2C_2O_4$	Oxalate	−0.87
$3CO_2 + 16H^+ + 16e^- \rightarrow CH_3CH_2CHO + 5H_2O$	Propionaldehyde	−0.32
$3CO_2 + 18H^+ + 18e^- \rightarrow CH_3CH(OH)CH_3 + 5H_2O$	2-propanol	−0.30

occur as a result of sacrificial agents provided the VB position of photocatalysts is negative when compared to the standard potential of H_2O oxidation, while the separation rates of electron–hole pairs are also sped up to improve the CO_2 photoreduction property (Yin et al. 2021, 2023; Alhebshi et al. 2022; Zhang et al. 2022). Economic problems and association to pollution, and the complexity of photocatalytic systems of the sacrificial agents has been reported to inhibit the exploration of the internal mechanism of CO_2 reduction (Jiao et al. 2020). It is for these reasons that researchers should come up with strategies to obtain the photocatalysts that can exhibit broad-spectral-response to achieve efficient CO_2 photoreduction in aqueous solutions or H_2O if the use of the sacrificial agent is to be avoided. In addition, the multielectron-transfer process is preferred over the one-electron transfer process. To further understand the multielectron-transfer processes for the reduction of processes, we delve into the possible mechanisms.

In direct photocatalysis, a semiconductor material would normally absorb a flux of sunlight energy with a wavelength less than the bandgap width, causing photoexcitation and the transitioning of electrons in the VB into the CB, forming electron-hole pairs at their original positions. This process results in the photoreduction of CO_2 absorbed on the semiconductor surface as an electron acceptor. The number of photogenerated electrons required in CO_2 reduction varies and, thus, results in different products such as CH_3OH, HCHO, HCOOH, CH_4, C_2H_4, and CO as presented in Table 8.1 (Hong et al. 2013; Chang et al. 2016).

It is interesting to note that CO_2 reduction reactions taking place by multielectron transfer result in different carbonaceous fuel products, albeit with the assistance of many protons. Another path may be more attractive and environmentally significant.

8.4.3 Mechanisms of Photoreduction of CO_2 into Useful Fuels

Extensive demonstration of possible mechanistic pathways to photo-covert CO_2 into useful fuels via intermediates including *CO, *CHO, and *OCCO(H) has been provided in the previous studies (Yang et al. 2023; Meng et al. 2022; Wang et al. 2022; Li et al. 2019b). CO_2 reduction reactions (CO_2RR) are normally classified into photothermal, photocatalytic (PC), and photoelectrochemical (PEC) approaches (Yang et al. 2023). Among these three main approaches, PC reduction of CO_2 into hydrocarbons has shown a remarkable feat for pursuing direct sunlight energy input without any external bias (Yang et al. 2023). The importance of photocatalytic reduction is also observed in a PEC approach where the reaction is controlled by aiding the external bias potential through the input of solar light energy (Jiao et al. 2020). The same cannot be observed about photothermal reduction of CO_2 which requires high-temperature durable reactors as it is not a mild reaction comparable to natural photosynthesis (Liu et al. 2021; Creissen and Fontecave 2021; He and Janáky 2020).

The initial electron transfer to form $CO_2^{\bullet-}$ is rate-determined by the redox potential (between −1.90 V and − 1.85 V) (vs. NHE at pH = 7) in the CO_2RR. Injecting an electron into the orbitals of CO_2 in such a process I unfavorable due to the thermodynamic stability of CO_2 which possesses strong double bonds between C and O (Morris et al. 2011; Deng et al. 2022). In other instances, CO_2 may be bonded and activated in a bent configuration with lower stability, thus exhibiting a lower LUMO level. This configuration improves electron transfer from photogenerated or electricity-injected electrons on the surface of photocatalysts in CO_2RR (Yang et al. 2023). Thus, CO_2 activation is always the most important first reaction step with which to convert the lower energy state of linear CO_2 molecules into an activated higher energy state with the bent structure of CO_2 molecules (Chang et al. 2016; Álvarez et al. 2017). Two modes of bonded CO_2 molecules are known: the carbon-bond (C bond) and oxygen-bond (O-bond) intermediates (Fig. 8.4) (Chang et al. 2016; Chernyshova et al. 2018). The activated CO_2 molecules are capable of accepting one electron, which converts them into $CO_2^{\bullet-}$ intermediates. This conversion step is accompanied by a multielectron transfer (PCET) process coupled with the sequential multiproton step to form a series of reduction products via intermediates (Table 8.1). In a photo/electrochemical technology, the catalyst enables high-energy reactions by lowering the activation energy, thus enabling high-energy reactions. Meanwhile, the photo/electrical and chemical energy is interconverted in the process (Yang et al. 2023). The adsorption, activation, and transfer of photo/electrogenerated active electrons to achieve CO_2 reduction take place at the catalyst/electrode surface. Therefore, such photo−/electrocatalytic CO_2 reduction can be considered heterogeneous or interfacial catalysis that involves a redox reaction between the electrode and reactants (Zhang et al. 2020) (Table 8.2).

It is very important to study the mechanistic pathways for the CO_2 reduction reactions involved in the production of fuels as encouraged by multielectron transfer processes, thanks to the previous demonstrations in other studies (Yang et al. 2023; Todorova et al. 2020; Lu and Jiao 2016; Gong et al. 2022; Karamian and Sharifnia 2016; Mehla et al. 2021). Such CO_2RR mechanistic pathways can be summarized in Fig. 8.5 (Chernyshova et al. 2018; Zheng et al. 2019b; Kortlever

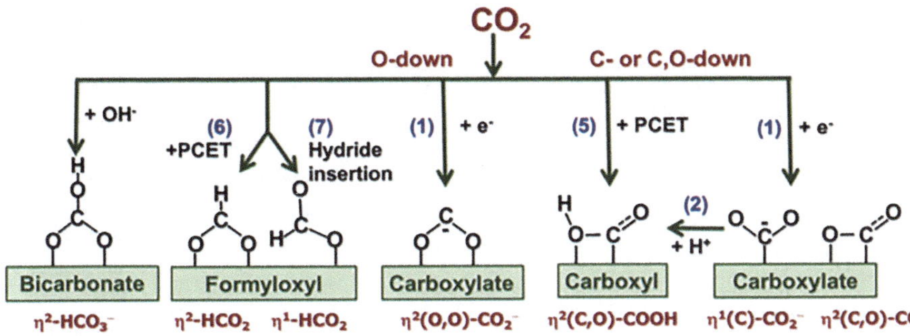

Fig. 8.4 Possible electroreduction of first CO_2 intermediates into carbonaceous fuels in aqueous media. Reproduced with permission from reference (Chernyshova et al. 2018). Copyright 2018 the National Academy of Sciences

Table 8.2 Some important intermediates generated during CO_2 reduction reactions

Intermediate	C/O	From	To	Product	Technique	Reference
$*CO_2^{\bullet-}$	C bound	CO_2	*COOH, $*CO_2–CO_2^-$, HCO_3^-	$CH3COO^-$, $HCOO^-$	UV-Vis, Raman, ESR, IR	Grills and Lymar (2018) and Flyunt et al. (2001)
*CO	C bound	*COOH	*OC–CO, *CHO	C_{2+}, CO, CH_3OH	VSFG, DFT	Neri et al. (2018)
*COOH	C bound	$*CO_2^{\bullet-}$	*CO	CO	Raman	An et al. (2021)
*CHO	C bound	*CO	*CHOH, $*CH_2O$	CH_4, C_2H_4	EPR, DFT	Luo et al. (2016) and Shkrob et al. (2012)
*OCHO	O bound	*CO	$HCOO^-$	$HCOO^-$	Raman, DFT	Vasileff et al. (2019)
*CHOH	C bound	*CHO	$*CH_2OH$, *CH	CH_4	IR, DFT	Xue et al. (2022)
$*CH_2OH$	C bound	*CHO, *CHOH	$*CH_2$	CH_3OH, CH_4	DFT	Zhang et al. (2018)
*CH	C bound	*COH, *CO	$*CH_2$	CH_4	DFT	Nie et al. (2013)
$*CH_2$	C bound	$*CH_2OH$, *COH	$*CH_2CO$, $*CH_3$	CH_4	XPS	Yang et al. (2021)
$*CH_2CO$	C bound	$*CH_2$, *CO	CH_3COO^-	C_{3+}, CH_3COOH	DFT	Ting et al. (2020)
*OC–CO	O bound	*OC	$*C_{2+}$	C_{2+}	EPR	Shkrob et al. (2012)
$*OC–CH_2$	O bound	*CH, *COOH	$*CH_xCO$, $*CH_3CHO$, $*CHCH_2$	C_{2+}	DFT	Zou et al. (2017)
$*CH_2–CH_2$	C bound	$*CH_2$, *COOH	CH_4, HCOOH, CO	C_2H_4, propanol	EIS	Shah et al. (2018)
$*OCH_2CH_3$	O bound	$*CH_2$, *CO, $*CH_2CHO$	C_2H_5OH, C_2H_4	C_2H_5OH, C_2H_4	Raman, DFT	Ren et al. (2018)
$*COH_2$	C bound	*CO	$*CH_3O$	C_2H_4	–	Li et al. (2016)
HO*C–*CO	C bound	*CO, *COH	*CH–CO, $*CH_2CHO$	C_{2+}	DFT	Pérez-Gallent et al. (2017)
H*C–CO	C bound	HO*C–CO	CH_2*CHO	C_2H_4	EC-MS, IR, DFT	Pérez-Gallent et al. (2017) and Schouten et al. (2012)
$*CO_2^{\bullet-}–OH$	C bound	$*CO_2^{\bullet-}/H_2O$	–	HCOOH	–	Haubrock et al. (2007)

(continued)

Table 8.2 (continued)

Intermediate	C/O	From	To	Product	Technique	Reference
*OC–CHO	O bound	*CO, *CHO	CH_3CHO	C_{2+}	DFT	Zhi et al. (2020) and Choi et al. (2019)
$C(=O)OCO_2^-$	C bound	$*CO_2^{\bullet-}$	CH_3COO^-	CH_3COO^-, CO, HCO_3^-	IR	Liu et al. (2015)
*CO–CO	C bound	*CO	$*CH_2CHO$	C_2H_4, C_2H_5OH, C_2H_6	DFT, Raman	Yu et al. (2018) and Yang et al. (2020)
$*CH_3–CH_3$	C bound	$*CH_3$	C_2H_6	C_2H_6	IR	Sorcar et al. (2019)

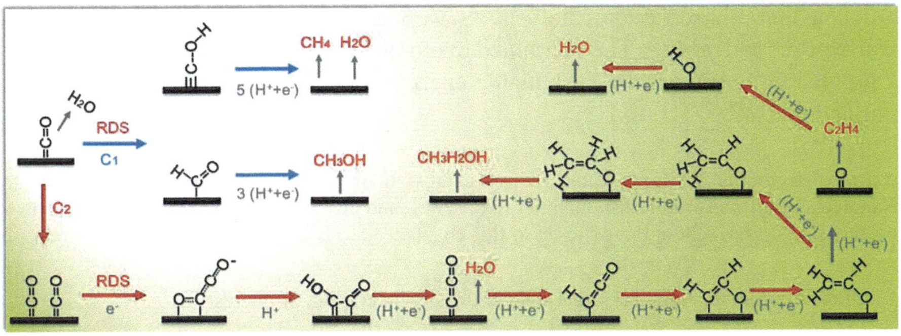

Fig. 8.5 The reaction mechanisms for CO_2RR to multi-electron reduction products. Reproduced with permission from ref (Zhang et al. 2021). Copyright 2021 Wiley-VCH GmbH

et al. 2015; Lum et al. 2018; Zhang et al. 2021). Such reduction reaction of CO_2 takes place via the key intermediates including *CO, *CHO, and *OCCO(H), with*CO being considered the initial key intermediate in multi-electron reduction pathways. The protonation of *CO is the rate-determining step in the formation of *CHO or *COH intermediates to produce fuels such as CH_4 and CH_3OH via the C1 pathway (Lum et al. 2018; Zhou et al. 2020). The steady-state adsorbed *CO intermediate on catalyst surface in close proximity at a high coverage, For C_{2+} pathway, the steady-state adsorbed *CO intermediate on catalyst surface should be in close proximity at a high coverage, dimerization two adsorbed *CO intermediates can be the rate-determining step for the generation of C_2H_4 and alcohols (Yi et al. 2020; Deng et al. 2024; Zhou et al. 2024). After a series of concerted proton–electron transfer steps, the formed $*CH_2CHO$ intermediate determines the selectivity pathway for C_2H_4 or alcohols. Low-energy barrier is observed for $*CH_2CHO$ to C_2H_4 when compared to that for the reaction of $*CH_3CHO$ to C_2H_5OH (Zhang et al. 2021). It can be deduced that key reaction intermediates, such as *CO, *CHO, and *OCCO, are very important for CO_2RR processes. Many studies on the photocatalytic reduction of CO_2 exhibit the importance and role of metal catalysts.

8.5 Conclusions

The discussion in this chapter was focused on heterogeneous catalysis either in the form of combining different catalysts or processes. This type of catalysis was found interesting as it alleviates failures of homogeneous catalysts. Photoreduction of water to produce hydrogen or carbon dioxide to form CO and other carbonaceous products was also demonstrated. Single-electron transfer has been found to be the major problem with photocatalysis, despite its importance in the rate-determining step. Thus, multielectron transfer processes can be preferred, and these reactions result in various fuels, depending on the stoichiometric ratio of CO_2 and the assisting proton when photocatalysts absorb the energy flux of the sunlight. Another observation is the photodegradation of organic pollutants while producing clean energy.

Future research should focus on developing more durable, cost-effective, and efficient metal-based photocatalysts, as well as optimizing their integration into real-world applications. The continued exploration and innovation in this area offer a promising path toward a sustainable energy future and effective climate change mitigation.

Acknowledgments Authors would like to express their sincere gratitude to the Institute for Nanotechnology and Water Sustainability (iNanoWS) and the University of South Africa (UNISA) for the support throughout the write-up of this chapter.

Conflict of Interest Authors of this chapter declare no conflict of interest.

References

Ahmadi M, Alavi SM, Larimi A (2023) UV–vis light responsive Bi2WO6 nanosheet/TiO2 nanobelt heterojunction photo-catalyst for CO2 reduction. Catal Commun 179:106681

Alhebshi A, Sharaf Aldeen E, Mim RS, Tahir B, Tahir M (2022) Recent advances in constructing heterojunctions of binary semiconductor photocatalysts for visible light responsive CO2 reduction to energy efficient fuels: a review. Int J Energy Res 46:5523–5584

Ali S et al (2024) Study of an enhanced photocatalytic hybrid MnO2@SnO2 core-shell Z-scheme nano heterojunction for efficient H2 production. Int J Hydrog Energy 89:424–433

Álvarez A et al (2017) CO2 activation over catalytic surfaces. ChemPhysChem 18:3135–3141

An H et al (2021) Sub-second time-resolved surface-enhanced Raman spectroscopy reveals dynamic CO intermediates during electrochemical CO2 reduction on copper. Angew Chemie Int Ed. 60:16576–16584

An P et al (2025) Triboelectric-electromagnetic nanogenerator coupled type-II heterojunction enhancing photoelectrocatalysis for wastewater degradation. Nano Energy 134:110589

Asiri AM et al (2022) Construction of Fe3O4 bridged Pt/g-C3N4 heterostructure with enhanced solar to fuel conversion. Appl Surf Sci 592:153159

Aziz Aljar MA, Zulqarnain M, Shah A, Akhter MS, Iftikhar FJ (2020) A review of renewable energy generation using modified titania for photocatalytic water splitting. AIP Adv 10

Berardi S et al (2014) Molecular artificial photosynthesis. Chem Soc Rev 43:7501–7519

Bian J et al (2022) Strategies and reaction systems for solar-driven CO2 reduction by water. Carbon Neutrality 1:1–20
Chang X, Wang T, Gong J (2016) CO2 photo-reduction: insights into CO2 activation and reaction on surfaces of photocatalysts. Energy Environ Sci 9:2177–2196
Chernyshova IV, Somasundaran P, Ponnurangam S (2018) On the origin of the elusive first intermediate of CO2 electroreduction. Proc Natl Acad Sci USA 115:E9261–E9270
Choi C et al (2019) A highly active star decahedron Cu nanocatalyst for hydrocarbon production at low overpotentials. Adv Mater 31:1–7
Creissen CE, Fontecave M (2021) Solar-driven electrochemical CO2 reduction with heterogeneous catalysts. Adv Energy Mater 11:1–12
Dankawu UM et al (2024) Recent advances and perspective of g–C3N4– based materials for efficient solar fuel (hydrogen) generation via photocatalytic water-splitting. Int J Hydrog Energy 67:1218–1242
Deng W, Zhang P, Seger B, Gong J (2022) Unraveling the rate-limiting step of two-electron transfer electrochemical reduction of carbon dioxide. Nat Commun 13:1–9
Deng W et al (2024) Unraveling the rate-determining step of C2+ products during electrochemical CO reduction. Nat Commun 15
Dhiman P et al (2025) Fabrication of Bi4O5I2/CuO S-scheme heterojunction photocatalyst with enhanced visible light degradation of Rhodamine B. J Alloys Compd 1014:178793
Etim UJ, Zhong Z (2023) Supported nanosized metal catalysts for thermocatalytic CO2 reduction to hydrocarbons and alcohols. Encycl Nanomater:678–698. https://doi.org/10.1016/B978-0-12-822425-0.00095-6
Fagan R, McCormack DE, Dionysiou DD, Pillai SC (2016) A review of solar and visible light active TiO2 photocatalysis for treating bacteria, cyanotoxins and contaminants of emerging concern. Mater Sci Semicond Process 42:2–14
Flyunt R, Schuchmann MN, Von Sonntag C (2001) A common carbanion intermediate in the recombination and proton-catalysed disproportionation of the carboxyl radical anion, CO2.-, in aqueous solution. Chem A Eur J 7:796–799
Gagarin H, Sridhar S, Lange I, Bazilian MD (2020) Considering non-power generation uses of coal in the United States. Renew Sust Energ Rev 124:109790
Gao J, Choo Sze Shiong S, Liu Y (2023) Reduction of CO2 to chemicals and fuels: thermocatalysis versus electrocatalysis. Chem Eng J 472:145033
Gong E et al (2021) Solar fuels: research and development strategies to accelerate photocatalytic CO2 conversion into hydrocarbon fuels. Energy Environ Sci 15:880–937
Gong S et al (2022) Fabrication of hetero-metal oxide NiCo2V2O8 hollow nanospheres for efficient visible light-driven CO2 photoreduction. Appl Catal B Environ 316:121663
Grills DC, Lymar SV (2018) Radiolytic formation of the carbon dioxide radical anion in acetonitrile revealed by transient IR spectroscopy. Phys Chem Chem Phys 20:10011–10017
Guo W, Guo T, Zhang Y, Yin L, Dai Y (2023) Progress on simultaneous photocatalytic degradation of pollutants and production of clean energy: a review. Chemosphere 339:139486
Haubrock J, Hogendoorn JA, Versteeg GF (2007) The applicability of activities in kinetic expressions: a more fundamental approach to represent the kinetics of the system CO2–OH-–salt in terms of activities. Chem Eng Sci 62:5753–5769
He J, Janáky C (2020) Recent advances in solar-driven carbon dioxide conversion: expectations versus reality. ACS Energy Lett 5:1996–2014
Hong J, Zhang W, Ren J, Xu R (2013) Photocatalytic reduction of CO2: a brief review on product analysis and systematic methods. Anal Methods 5:1086–1097
Hu Q et al (2024) Light-driven thermocatalytic CO2 reduction by CH4 on alumina-cluster-modified Ni nanoparticles with excellent durability and high light-to-fuel efficiency promoted by the photoactivation effect. J Colloid Interface Sci 657:942–952
Jafari T et al (2016) Photocatalytic water splitting – the untamed dream: a review of recent advances. Molecules 21

Jiao X, Zheng K, Hu Z, Sun Y, Xie Y (2020) Broad-spectral-response photocatalysts for CO2 reduction. ACS Cent Sci 6:653–660

Karamian E, Sharifnia S (2016) On the general mechanism of photocatalytic reduction of CO2. J CO2 Util 16:194–203

Knör G (2015) Recent progress in homogeneous multielectron transfer photocatalysis and artificial photosynthetic solar energy conversion. Coord Chem Rev 304–305:102–108

Kortlever R, Shen J, Schouten KJP, Calle-Vallejo F, Koper MTM (2015) Catalysts and reaction pathways for the electrochemical reduction of carbon dioxide. J Phys Chem Lett 6:4073–4082

Kudo A, Miseki Y (2009) Heterogeneous photocatalyst materials for water splitting. Chem Soc Rev 38:253–278

Li Q et al (2016) Controlled assembly of Cu nanoparticles on pyridinic-N rich graphene for electrochemical reduction of CO2 to ethylene. Nano Energy 24:1–9

Li X, Yu J, Jaroniec M, Chen X (2019a) Cocatalysts for selective photoreduction of CO 2 into solar fuels. Chem Rev 119:3962–4179

Li C et al (2019b) Carbon dioxide photo/electroreduction with cobalt. J Mater Chem A 7:16622–16642

Liu Y, Chen S, Quan X, Yu H (2015) Efficient electrochemical reduction of carbon dioxide to acetate on nitrogen-doped nanodiamond. J Am Chem Soc 137:11631–11636

Liu L, Zhang Y, Huang H (2021) Junction engineering for photocatalytic and photoelectrocatalytic CO2 reduction. Sol RRL 5:1–19

Lu Q, Jiao F (2016) Electrochemical CO2 reduction: electrocatalyst, reaction mechanism, and process engineering. Nano Energy 29:439–456

Lum Y, Cheng T, Goddard WA, Ager JW (2018) Electrochemical CO reduction builds solvent water into oxygenate products. J Am Chem Soc 140:9337–9340

Luo W, Nie X, Janik MJ, Asthagiri A (2016) Facet dependence of CO2 reduction paths on cu electrodes. ACS Catal 6:219–229

Maeda K (2011) Photocatalytic water splitting using semiconductor particles: history and recent developments. J Photochem Photobiol C: Photochem Rev 12:237–268

Mehla S et al (2021) Porous crystalline frameworks for thermocatalytic CO2 reduction: an emerging paradigm. Energy Environ Sci 14:320–352

Meng SL, Ye C, Li XB, Tung CH, Wu LZ (2022) Photochemistry journey to multielectron and multiproton chemical transformation. J Am Chem Soc 144:16219–16231

Morris AJ, McGibbon RT, Bocarsly AB (2011) Electrocatalytic carbon dioxide activation: the rate-determining step of pyridinium-catalyzed CO2 reduction. ChemSusChem 4:191–196

Nakata K, Fujishima A (2012) TiO2 photocatalysis: design and applications. J Photochem Photobiol C: Photochem Rev 13:169–189

Neri G, Walsh JJ, Teobaldi G, Donaldson PM, Cowan AJ (2018) Detection of catalytic intermediates at an electrode surface during carbon dioxide reduction by an earth-abundant catalyst. Nat Catal 1:952–959

Nie X, Esopi MR, Janik MJ, Asthagiri A (2013) Selectivity of CO2 reduction on copper electrodes: the role of the kinetics of elementary steps. Angew Chemie Int Ed 52:2459–2462

Omer AM (2008) Green energies and the environment. Renew Sust Energ Rev 12:1789–1821

Pahlevanpour G, Bashiri H (2022) Kinetic Monte Carlo simulation of hydrogen production from photocatalytic water splitting in the presence of methanol by 1 wt% Au/TiO2. Int J Hydrog Energy 47:12975–12987

Pérez-Gallent E, Figueiredo MC, Calle-Vallejo F, Koper MTM (2017) Spectroscopic observation of a hydrogenated CO dimer intermediate during CO reduction on Cu(100) electrodes. Angew Chemie Int Ed. 56:3621–3624

Ren T, Patel MK (2009) Basic petrochemicals from natural gas, coal and biomass: energy use and CO2 emissions. Resour Conserv Recycl 53:513–528

Ren D, Fong J, Yeo BS (2018) The effects of currents and potentials on the selectivities of copper toward carbon dioxide electroreduction. Nat Commun 9:925

Sahara G, Ishitani O (2015) Efficient photocatalysts for CO2 reduction. Inorg Chem 54:5096–5104

Samokhvalov A (2017) Hydrogen by photocatalysis with nitrogen codoped titanium dioxide. Renew Sust Energ Rev 72:981–1000
Schouten KJP, Qin Z, Gallent EP, Koper MTM (2012) Two pathways for the formation of ethylene in CO reduction on single-crystal copper electrodes. J Am Chem Soc 134:9864–9867
Shah AH et al (2018) Revisiting electrochemical reduction of CO2 on cu electrode: where do we stand about the intermediates? J Phys Chem C 122:18528–18536
Sharma V, Dhiman R, Samanta S, Karwal S, Mahajan A (2025) PANI/MXene/MnO2 ternary heterostructure with multi-redox active sites and enhanced interfacial charge transfer for efficient hydrogen production via electro/photocatalytic water splitting. Int J Hydrog Energy 97:532–548
Shkrob IA, Marin TW, He H, Zapol P (2012) Photoredox reactions and the catalytic cycle for carbon dioxide fixation and methanogenesis on metal oxides. J Phys Chem C 116:9450–9460
Singh V, Beltran IJC, Ribot JC, Nagpal P (2014) Photocatalysis deconstructed: design of a new selective catalyst for artificial photosynthesis. Nano Lett 14:597–603
Sorcar S et al (2019) CO2, water, and sunlight to hydrocarbon fuels: a sustained sunlight to fuel (Joule-to-Joule) photoconversion efficiency of 1%. Energy Environ Sci 12:2685–2696
Tanaka K, Ooyama D (2002) Multi-electron reduction of CO2 via Ru–CO2, –C(O)OH, –CO, –CHO, and –CH2OH species. Coord Chem Rev 226:211–218
Ting LRL et al (2020) Enhancing CO2 electroreduction to ethanol on copper-silver composites by opening an alternative catalytic pathway. ACS Catal 10:4059–4069
Todorova TK, Schreiber MW, Fontecave M (2020) Mechanistic understanding of CO2 reduction reaction (CO2RR) toward multicarbon products by heterogeneous copper-based catalysts. ACS Catal 10:1754–1768
Tu W, Zhou Y, Zou Z (2014) Photocatalytic conversion of CO2 into renewable hydrocarbon fuels: state-of-the-art accomplishment, challenges, and prospects. Adv Mater 26:4607–4626
Ursúa A, Gandía LM, Sanchis P (2012) Hydrogen production from water electrolysis: current status and future trends. Proc IEEE 100:410–426
Vasileff A et al (2019) Selectivity control for electrochemical CO2 reduction by charge redistribution on the surface of copper alloys. ACS Catal 9:9411–9417
Wang W, Huang G, Yu JC, Wong PK (2015) Advances in photocatalytic disinfection of bacteria: development of photocatalysts and mechanisms. J Environ Sci 34:232–247
Wang JW et al (2022) Co-facial π-πInteraction expedites sensitizer-to-catalyst electron transfer for high-performance CO2 photoreduction. JACS Au 2:1359–1374
Xue L et al (2022) Unveiling the reaction pathway on Cu/CeO2 catalyst for electrocatalytic CO2 reduction to CH4. Appl Catal B Environ 304:120951
Yaemsunthorn K, Macyk W, Ortyl J (2024) Semiconductor photocatalysts in photopolymerization processes: mechanistic insights, recent advances, and future prospects. Prog Polym Sci 158:101891
Yan Q et al (2025) CuBi2O4/AgIO3 photocatalyst based on the synergistic effects of oxygen vacancies and S-scheme heterojunctions for enhanced tetracycline degradation. J Water Process Eng 70:107150
Yang Y et al (2020) Insight into the formation and transfer process of the first intermediate of CO2 reduction over Ag-decorated dendritic Cu. Chem A Eur J 26:4080–4089
Yang H et al (2021) Predictions of chemical shifts for reactive intermediates in CO2 reduction under operando conditions. ACS Appl Mater Interfaces 13:31554–31560
Yang L et al (2023) Intermediates and their conversion into highly selective multicarbons in photo/electrocatalytic CO2 reduction reactions. J Mater Chem A 11:19172–19194
Yang H et al (2025) Dual-driven charge transport enabled by S-scheme heterojunction and solid solution in CdS@N-NiCoO photocatalysts for enhanced hydrogen evolution. Sep Purif Technol 362:131819
Yi J et al (2020) Highly selective CO2 electroreduction to CH 4 by in situ generated Cu2O single-type sites on a conductive MOF: stabilizing key intermediates with hydrogen bonding. Angew Chemie 132:23849–23856

Yin S et al (2021) Enhanced electron–hole separation in SnS2/Au/g-C3N4 embedded structure for efficient CO2 photoreduction. Chem Eng J 406:126776
Yin S et al (2023) Nanocluster-mediated electron–hole separation for efficient CO2 photoreduction. Chem Eng J 477:147292
Yu S, Wilson AJ, Heo J, Jain PK (2018) Plasmonic control of multi-electron transfer and C-C coupling in visible-light-driven CO2 reduction on Au nanoparticles. Nano Lett 18:2189–2194
Zhang T, Lin W (2014) Metal-organic frameworks for artificial photosynthesis and photocatalysis. Chem Soc Rev 43:5982–5993
Zhang H, Goddard WA, Lu Q, Cheng MJ (2018) The importance of grand-canonical quantum mechanical methods to describe the effect of electrode potential on the stability of intermediates involved in both electrochemical CO2 reduction and hydrogen evolution. Phys Chem Chem Phys 20:2549–2557
Zhang S, Fan Q, Xia R, Meyer TJ (2020) CO2 reduction: from homogeneous to heterogeneous Electrocatalysis. Acc Chem Res 53:255–264
Zhang B et al (2021) Single-atom Electrocatalysts for multi-electron reduction of CO2. Small 17:1–17
Zhang N et al (2022) Visible-light driven boosting electron-hole separation in CsPbBr3 QDs@ 2D Cu-TCPP heterojunction and the efficient photoreduction of CO2. J Colloid Interface Sci 608:3192–3203
Zhang J et al (2024) Cascade electrocatalytic and thermocatalytic reduction of CO2 to propionaldehyde. Angew Chemie 136:1–6
Zhang Y, Qiu J, Sun Z (2025) Development of a stable and high-performance Z-scheme In2O3/TiO2 heterojunction photocatalyst for tetracycline degradation. Appl Surf Sci 687:162231
Zheng Y, Grant R, Hu W, Marek E, Scott SA (2019a) H2 production from partial oxidation of CH4 by Fe2O3-supported Ni-based catalysts in a plasma-assisted packed bed reactor. Proc Combust Inst 37:5481–5488
Zheng Y et al (2019b) Understanding the roadmap for electrochemical reduction of CO2 to multi-carbon oxygenates and hydrocarbons on copper-based catalysts. J Am Chem Soc 141:7646–7659
Zhi X, Jiao Y, Zheng Y, Vasileff A, Qiao SZ (2020) Selectivity roadmap for electrochemical CO2 reduction on copper-based alloy catalysts. Nano Energy 71:104601
Zhou B et al (2020) Highly efficient binary copper? Iron catalyst for photoelectrochemical carbon dioxide reduction toward methane. Proc Natl Acad Sci USA 117:1330–1338
Zhou G et al (2024) Concentrated C2+ alcohol production enabled by post-intermediate modulation and augmented CO adsorption in CO electrolysis. J Am Chem Soc 146. https://doi.org/10.1021/jacs.4c10629
Zhu Z, Ban Y, Tang L, Li H (2025) Construction of S-scheme heterojunction CeCuO3/g-C3N4 photocatalyst for degradation of tetracycline under visible light. Diam Relat Mater 152:111888
Zou X et al (2017) How nitrogen-doped graphene quantum dots catalyze electroreduction of CO2 to hydrocarbons and oxygenates. ACS Catal 7:6245–6250
Zuo W et al (2021) Degradation of organic pollutants by intimately coupling photocatalytic materials with microbes: a review. Crit Rev Biotechnol 41:273–299

Chapter 9
Green Photocatalytic Energy Conversion

M. Muthulakshmi, T. S. Subha, and M. Muthukumaran

Abstract Green photocatalytic energy conversion is attracting increasing attention from researchers, owing to its potential to solve the global energy crisis and environmental problems caused by the overconsumption of fossil fuels. A solar-based green photocatalytic energy conversion system is efficient and satisfactory for removing pollutants by decomposition. The manufacture of green and sustainable materials in the field of energy and environmental sectors has received global demand in the form of reuse or reprocessing in a biocompatible manner. Photocatalysis is a green technology that demonstrates artificial photosynthesis in which eco-friendly solar energy is converted and stored into high-energy chemicals. Green photocatalytic energy conversion technologies are low-cost, robust, biocompatible, and can tackle high-concentration toxic organic pollutants. Moreover, it is considered to be a perspective technology to resolve environmental pollution and control CO_2 discharge. This chapter deals with green energy, photocatalysts, photosynthesis, green photocatalysis, and recent advances in green photocatalytic energy conversion.

Keywords Photocatalysis · Toxic · Green photocatalyst · Organic pollutants · Green energy · Environmental pollution

---

The original version of the chapter has been revised. A correction to this chapter can be found at https://doi.org/10.1007/978-981-95-1618-6_13

M. Muthulakshmi (✉)
Ethiraj College for Women (Autonomous), Chennai, India

T. S. Subha
PG and Research Department of Botany, Bharathi Women's College, Chennai, India

M. Muthukumaran
PG and Research Department of Botany, Ramakrishna Mission Vivekananda College (Autonomous), Chennai, India

C. M. Hussain, G. Hussain (eds.), *Green Nanostructured Photocatalysts for Clean Energy*, Clean Energy Production Technologies,
https://doi.org/10.1007/978-981-95-1618-6_9

9.1 Introduction

In the twenty-first century, the rapid growth of population with socioeconomic development inevitably leads to continuously increasing energy demand and environmental crisis. The synthesis of green sustainable photocatalysts has received global recognition in recent years. According to Dittmar 2014, the sustainable development and green economy are "Growth that helps the desire of present generation, to convene their own needs to next generation without compromising anything." In general, green photocatalysis is the process of using natural resources, biomaterials, and biological extracts to prepare various photocatalysts (Fox and Dulay 1993). These precursors are bio-dependent, biocompatible, and can convert the light energy to chemical energy over the semiconductor materials, for instance, in the treatment of dye wastewater, reduction of toxic molecules (Herrmann 1999, 2005), and production of energy (Lin et al. 2018). Currently, 65% of the world's electric power is generated by the combustion of fossil fuels (Cover et al. 2016). Figure 9.1 shows that the world's total final energy consumption has seen an increasing percentage of geothermal, solar, biomass, and wind energy sources in recent years. As of now, the usage of wind, nuclear, and hydroelectric energies has emitted the lowest carbon dioxide levels. So, the hazardous waste produced in power generation can be termed as "green and clean." By using green photocatalytic energy

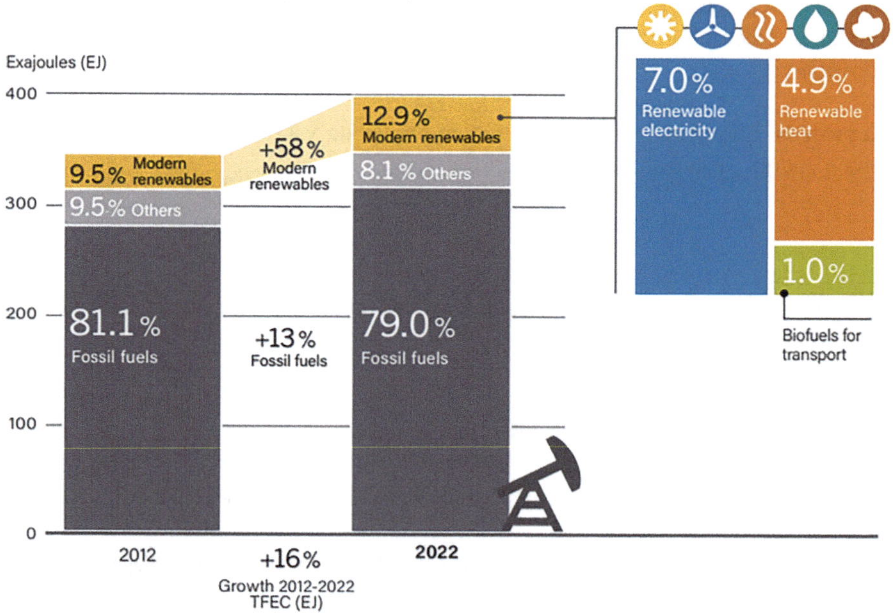

Fig. 9.1 Total final energy consumption by Source, 2012 and 2022. (Source: International Energy Agency, REN21)

converters, it is possible to increase the cycling efficiency and reduce carbon dioxide emissions.

Excessive consumerism and human activities release toxic materials, causing severe environmental degradation which releases greenhouse gases like carbon dioxide (CO_2), carbon monoxide (CO), nitrogen oxides (NO_x), and methane and sulfur oxides (SO_2, SO_3). These greenhouse gases cause a significant threat on a planetary scale by making adverse effects (Chantara et al. 2019). The devastating environmental effects were created by combustion of fossil fuels and the disposal and release of urban, industrial, electronic, agricultural, and plastic wastes (Khan et al. 2022a, b; Joshi et al. 2021). In addition to the new ecological problems, uneven distribution of fossil fuel resources and non-renewability are challenging problems since abundant products are shaped using compounds derived from crude oil (Gani 2021; Khan et al. 2022a, b; Liu et al. 2021; Regin et al. 2008). The amount of energy consumed by humanity was 580×10^{18} J in 2018 (Hassani et al. 2021). Based on the report provided by the "World Energy Outlook 2014" published by the International Energy Agency (IEA) from 2011 to 2035, the energy consumption of the world will increase by a third. It is assessed that a 20% increase in fossil fuel-based energy production capacity will release 37.2 gigatons of CO_2. According to the report of "World Energy Outlook 2021," the goal of net-zero emissions is achieved by 2050. This net-zero emission will be achieved by reducing the dependence of factories on fossil fuels and developing production methods that employ renewable and clean energy technologies (Berdysheva and Ikonnikova 2021). Solar energy is one of the dominant market shares of the renewable energy production industry. Solar energy is converted into chemical energy which is utilized in numerous applications such as photocatalytic, photovoltaic, and photothermal approaches (Yaghoubi et al. 2024).

Photocatalytic systems of energy conversion for environmental remediation are a promising solution for various environmental crises, including the efficient decomposition of pollutants. The inevitable broad utilization of photocatalysts in the conversion of solar energy is applied in various fields (Nordin et al. 2022). In photocatalytic processes, the solar energy is renewed to chemical energy through artificial photosynthesis, photocatalytic synthesis, and photodegradation reactions (Yaghoubi et al. 2024). Manufacture of photocatalytic systems from green processes or green chemistry is inevitable for future generations. Natural-based materials are available for disinfection of hazardous pollutants, pathogens, and toxic chemical waste in drinking water (Naushad et al. 2017; Ghasemi et al. 2014; Alqadami et al. 2017). Natural bio-based molecules are abundantly present in our ecosystem. It possesses great biodegradable and biological activities. The natural bio-macromolecules are formed by applying Bamboo wood under pyrolysis and carbonization (Yahya et al. 2015). These natural methods are highly efficient than chemical-based ones, and their recycling capability is also highly reproducible.

9.2 Nanotechnology in Energy Systems

Nanotechnology is defined as the design of materials, components, or systems at near near-atomic level or nanometer scale between 1 and 100 nm (Cao and Wang 2011). This field mainly concentrates on synthesizing, measuring, imaging, modeling, and manipulating matter at the nano level. In the past two decades, significant progress has been observed in designing, analyzing, and synthesizing nano materials and devices, and this tendency will continue for a few more decades in fundamental studies, research, and developmental fields (Rogers et al. 2011).

Nanomaterials are the chief component of solar energy conversion devices and have been applied in the following three ways.

(a) The assembly of molecular groups of oxidizing and reducing components imitating photosynthesis.
(b) Semiconductor-assisted photocatalysis in the production of solar fuel.
(c) By employing the nanostructured semiconductors in solar cells.

Among the solar energy conversion devices, binary and ternary metal oxides are the most widely used nanostructured systems which have a promising future in this field (Nuraje et al. 2012, 2013; Lei et al. 2015). Although numerous books have been published on solar cells, renewable energy, energy conversion, and photocatalysis, very few books only published on green photo-active nanomaterials and their applications. Most of the books elaborated on a broad spectrum of photocatalysts like metal and non-metal oxides. However, this chapter introduces and summarizes green photocatalysis, green nanostructured photoactive materials for energy conversion, the role of microorganisms in energy mitigation, and health and environmental aspects of green photoactive nanomaterials up until the present time.

9.3 Green Photocatalysis

Photocatalysis is an advanced technique by semiconductors, molecules, or ions in water hydrolysis to produce hydrogen as fuel and to recover polluted effluents (Ravelli et al. 2009). Even though it is used in energy conversion technology, its practical usage has been constrained due to its poor efficacy and inadequate light absorption capability, and it is highly active only under UV light. To overcome these limitations, various technologies like the manufacture of green photocatalysts or biogenic catalysts have been implemented leading to some fruitful findings (Lang et al. 2014; Dong et al. 2015; Moniz et al. 2015; Fagan et al. 2016). Currently, researchers have fabricated metal- and non-metal-doped TiO_2 to engineer the native properties of TiO_2. This kind of materialistic modulation of TiO_2 could improve the photocatalytic performance (Etacheri et al. 2015; Jaiswal et al. 2015). TiO_2 is an n-type white hue semiconductor which has drawn huge attention because of its numerous applications across various sectors. During the synthesis of physical or chemical processes, the materials produce toxic or hazardous chemical precursors that could create pollutants. To overcome these drawbacks, researchers move towards green photocatalysts/biogenic photocatalysts which are simpler, sustainable, and cost-effective. During biogenic or green photocatalyst synthesis, the biological molecules present in plants and microorganisms facilitate the bio-reduction and capping processes. The natural reducing agents in living organisms such as proteins, enzymes, and phytochemicals are employed in the green photocatalyst synthesis process helps in oxidation and reduction reactions of photocatalytic activity (Verma et al. 2022).

9.4 Green Nanostructured Photoactive Materials

With the help of light energy and natural resources, the green photocatalysts are used to create fossil fuels. For example, zero-carbon emission transportation fuels such as hydrogen fuel and biodiesel manufactured from biomass (Demirbas 2008, 2011). The production of high energy through artificial photosynthesis by employing green nanostructured photocatalysts offers hope for meeting the world's energy demand. The green nano-based photocatalysts act as the base surface for green hydrogen production. The integration of multi-component nanostructures like metal-free polymeric graphitic carbon nitrides can be more appropriate in catalytic activity than a single-component photocatalysis (Sahu et al. 2024). The synthesis of functionalized polymeric nanocomposites is prepared and employed in the energy conversion process (Raja and Mattoso 2019). Bayahia designed a highly efficient green nanostructured $ZnFe_2O_4$ photocatalyst by using *Ziziphus mauritiana* and *Salvadora persica* extracts for the degradation of crystal violet in sunlight (Bayahia 2023). The development of green ZnO-nanoparticles synthesized by using *Carica papaya* leaves has been used as a photo-anode in energy production (Yashni et al. 2020; Rathnasamy et al. 2017; Ahmad et al. 2022).

9.5 Green Nanostructured Photocatalysts for Environmental Consideration

The air, soil, and water quality are worsening due to industrialization which has significantly raised the suspended particulate concentration and emission of greenhouse gases, and these will likely continue for the next decades, further affecting human life severely (Asmatulu et al. 2015). Greenhouse gases like methane, carbon dioxide (CO_2), and nitrogen oxides (NO_X) are the primary sources for global warming and climate change which has now become a worldwide challenge. Thus, we have to reduce this emission by addressing microorganisms such as microalgae, viruses, bacteria, fungi, and molds which is an effective way. Nanostructured photocatalysts can also offer structural features for reducing the emission of CO_2 and other toxic gases in an eco-friendly manner (Babu et al. 2013).

9.6 Role of Microorganisms in Energy Mitigation Through Nanostructures

Recent studies indicate the potential for improvement of biological biomass production by applying green nanostructured photocatalytic materials. This technology can pointedly raise the biomass conversion rate and biodiesel production. In addition to that, it improves enzyme loading efficiency and mobilization, lipid accumulation and extraction, nanoscale catalytic activity, the rate of separation and purification of liquids from other liquids and solids, and the design of bioreactors and their applications (Rajamehala et al. 2022; Babu et al. 2013; Zhang et al. 2013; Yang et al. 2013; Kassim et al. 2020). In order to assemble the components for separating water molecules into H_2 and O_2, harmless bacterial viruses were modified genetically. This genetically engineered virus yielded a fourfold boost in production efficiency. This novel process mimics plant photosynthesis. In this research study, a biological scaffold was made with a bioengineered virus equipped to split a water molecule (Zhang et al. 2013; Asmatulu et al. 2015). To extract electrical energy, nanoprobes tap into algae and bacteria which leads to the increasing efficacy of fuel conversion rates. A single-cell alga, *Chlamydomonas reinhardtii*, is an ideal cell for high-energy electron harvesting during the step prior to the downstream process. However, the cell will eventually die due to leakage in the membrane where the nano-electrodes are inserted into its body (Siva and Marimuthu 2015).

9.7 Conclusion

Fossil fuels like natural gas, coal, and oil have been utilized continuously by many developed and developing countries. Overproduction and consumption of fossil fuel-based energy systems have a drastic impact on human health and the environment causing global warming and pollution in the air, water, and soil. Inspiration from the natural environment like the production of hydrogen energy through photosynthesis can be an option for solving the energy crisis. We can fabricate light-harvesting assemblies, new synthesis methods for fuels, functional materials for water splitting (Hydrolysis) units, pollution controlling devices, and so on. Green photocatalytic nanomaterials are highly reactive (due to high surface area to volume ratio), eco-friendly, and biodegradable, yet at the same time potentially more damaging. To overcome this problem, proper studies on green nanostructured materials should be analyzed prior to photocatalytic applications.

References

Ahmad W, Singh V, Ahmed S, Nur-e-Alam M (2022) A comprehensive study on antibacterial antioxidant and photocatalytic activity of Achyranthes aspera mediated biosynthesized Fe2O3 nanoparticles. Res Eng 14:100450

Asmatulu R, Nuraje N, Mul G (2015) Green photo-active nanomaterials: sustainable energy and environmental remediation. The Royal Society of Chemistry, pp 1–12

Alqadami AA, Naushad Mu, Abdalla MA, Ahamad T, ALOthman ZA, Alshehri SM, Ghfar AA (2017) Efficient removal of toxic metal ions from wastewater using a recyclable nanocomposite: A study of adsorption parameters and interaction mechanism, J Clean Prod, 156:426–436, ISSN 0959-6526, https://doi.org/10.1016/j.jclepro.2017.04.085

Arise AK, Malomo SA, Omar SA, Arubuola OV, Arise RO (2025) Chemical composition, sensory attributes and in vivo anti-diabetic effects of wheat- walnut composite flour-based croissants on streptozotocin-induced diabetic Wistar rats, Food Chemistry Advances 6:100929,ISSN 2772–753X, https://doi.org/10.1016/j.focha.2025.100929

Babu V, Thapliyal A, Patel GK (2013) Biofuels production. Wiley

Bayahia H (2023) The highly efficient green synthesis of nanostructured ZnFe2O4 photocatalyst by using *Ziziphus mauritiana* and *Salvadora persica* extracts for the photocatalytic degradation of crystal violet in sunlight. J King Saud Univ Sci 35(4):102584. Various plant extracts were used to synthesis $M^{2+}Fe_2O_4$green nanoparticles where M^{2+} can be cobalt (Co), nickel (Ni), gold (Au), copper (Cu), or silver (Ag) among other metals (Elumalai et al. 2021; Jaast and Grewal 2021)

Berdysheva S, Ikonnikova S (2021) The energy transition and shifts in fossil fuel use: the study of international energy trade and energy security dynamics. Energies 14(17):5396

Cao G, Wang Y (2011) Nanostructures and nanomaterials: synthesis, properties and applications, 2nd edn. World Scientific

Chantara S, Thepnuan D, Wiriya W, Prawan S, Tsai YI (2019) Emission of pollutant gases, fine particulate matters and their significant tracers from biomass burning in an open-system combustion chamber. Chemosphere 224:407–416

Cover KS, van Schijndel RA, Bosco P, Damangir S, Redolfi A (2018) Can measuring hippocampal atrophy with a fully automatic method be substantially less noisy than manual segmentation over both 1 and 3 years? Psychiatry Research: Neuroimaging, 280:39–47, ISSN 0925-4927, https://doi.org/10.1016/j.pscychresns.2018.06.011.

Demirbas A (2008) Biofuels sources, biofuel policy, biofuel economy and global biofuel projections. Energy Convers Manag 49(8):2106–2116

Demirbas A (2011) Competitive liquid biofuels from biomass. Appl Energy 88(1):17–28

Dittmar H, Bond R, Hurst M, Kasser T (2014) The relationship between materialism and personal wellbeing: A meta-analysis. J Pers Soc Psychol 107(5):879–924. https://doi.org/10.1037/a0037409

Dong S, Feng J, Fan M, Pi Y, Hu L, Han X, Liu M, Sun J, Sun J (2015) Recent developments in heterogeneous photocatalytic water treatment using visible light-responsive photocatalysts: a review. RSC Adv 5(19):14610–14630

Dwivedi MK, Pandey SK, Singh PK (2022) Larvicidal activity of green synthesized zinc oxide nanoparticles from Carica papaya leaf extract, inorganic and nano-metal, vol 53. Chemistry, pp 1–11

Elumalai N, Prabhu S, Selvaraj M, Shanavas S, Navaneethan M, Harish S, Ramu P, Ramesh R (2021) Investigation on synergistic effect of rGO and carbon quantum dots-embedded ZnO hollow spheres for improved photocatalytic aqueous pollutant removal process. J Mater Sci Mater Electron 32(24):28633–28647

Etacheri V, Valentin CD, Schneider J, Bahnemann D, Pillai SC (2015) Visible-light activation of TiO2 photocatalysts: Advances in theory and experiments, J Photochem Photobiol C: Photochem Rev, 25:1–29, ISSN 1389-5567, https://doi.org/10.1016/j.jphotochemrev.2015.08.003.

Fagan R, McCormack DE, Dionysiou DD, Pillai SC (2016) A review of solar and visible light active TiO2 photocatalysis for treating bacteria, cyanotoxins and contaminants of emerging concern. Mater Sci Semicond Process 42:2–14

Fox MA, Dulay MT (1993) Heterogeneous photocatalysis. Chem Rev 93(1):341–357

Gani A (2021) Fossil fuel energy and environmental performance in an extended STIRPAT model. J Clean Prod 297:126526

Hassani A, Krishnan S, Scaria J, Eghbali P, Nidesh PV (2021) Z- scheme photocatalyst for visible-light- driven pollutants degradation: a review on recent advancements. Curr Opin Solid State Mater Sci 25(5):100941

Herrmann J-M (1999) Heterogeneous photocatalysis: fundamentals and applications to the removal of various types of aqueous pollutants. Catal Today 53(1):115–129

Herrmann J-M (2005) Heterogeneous photocatalysis: state of the art and present applications in honor of Pr. RL Burwell Jr (1912–2003), former Head of Ipatieff Laboratories, Northwestern University, Evanston (Ill). Top Catal 34(1–4):49–65

Jaast S, Grewal A (2021) Green synthesis of silver nanoparticles, characterization and evaluation of their photocatalytic dye degradation activity. Curr Res Green Sustain Chem 4

Jaiswal M, Dudhe R, Sharma PK (2015) Nanoemulsion: an advanced mode of drug delivery system. 3 Biotech 5, 123–127. https://doi.org/10.1007/s13205-014-0214-0

Joshi A, Kaur M, Kaur R, Grover A, Nash D, El-Mohandes A (2021) Predictors of COVID-19 Vaccine Acceptance, Intention, and Hesitancy: A Scoping Review. Front Public Health 13(9):698111. https://doi.org/10.3389/fpubh.2021.698111. PMID: 34485229; PMCID: PMC8414566

Kassim MA, Meng TK, Serri NA, Yusoff SB, Shahrin NAM, Seng KY, Bakar MHA, Keong LC (2020) Sustainable biorefinery concept for industrial bioprocessing. Biorefinery production. Technol Chem Energy. https://doi.org/10.1002/9781119593065.ch2

Khan M, Assal ME, Tahir MN, Ashraf M, Hatshan MR, Khan M, Varala R, Badawi NM, Adil SF (2022a) Graphene/inorganic nanocomposites: evolving photocatalysts for solar energy conversion for remediation. J Saudi Chem Soc 26(6):101544

Khan I, Zakari A, Dagar V, Singh S (2022b) World energy trilemma and transformative energy developments as determinants of economic growth and environmental sustainability. Energy Econ 108:105884

Lang X, Chen X, Zhao J (2014) Heterogeneous visible light photocatalysis for selective organic transformations. Chem Soc Rev 43(1):473–486

Lei Y, Asmatulu R, Nuraje N (2015) Photo-active metal oxide nanomaterials for water splitting. ScienceJet 4:169–173

Lin Z, Ye M, Wang M (eds) (2018) Multifunctional photocatalytic materials for energy. Woodhead Publishing

Liu Y, Lin B, Xu B (2021) Modeling the impact of energy abundance on economic growth and CO_2 emissions by quantile regression: evidence from China. Energy 227:120416

Moniz T, Arntfield S, Miller K, Lingard L, Watling C, Regehr G (2015) Considerations in the use of reflective writing for student assessment: issues of reliability and validity. Med Educ 49(9):901–8. https://doi.org/10.1111/medu.12771. PMID: 26296406.

Naushad M, Ahamad T, Al-Maswari BM, Alqadami AA, Alshehri SM (2017) Nickel ferrite bearing nitrogen-doped mesoporous carbon as efficient adsorbent for the removal of highly toxic metal ion from aqueous medium. Chemical Engineering Journal. 330. https://doi.org/10.1016/j.cej.2017.08.079.

Nordin NA, Mohamed MA, Salehmin MN, Yusoff SFM (2022) Photocatalytic active metal-organic frame work and its derivatives for solar-driven environmental remediation and renewable energy. Coord Chem Rev 468:214639

Nuraje N, Asmatulu R, Kudaibergenov S (2012) Metal oxide-based functional materials for solar energy conversion: a review. Curr Inorg Chem 2:124–146

Nuraje N, Kudaibergenov S, Asmatulu R (2013) In: Luque R, Balu AM (eds) Production of fuels using nanomaterials. Taylor and Francis, pp 95–117

Raja S, Mattoso LHC, Moreira FKV (2019) Biomass-Derived Nanomaterials. In: Rajendran S, Naushad M, Balakumar S (eds) Nanostructured Materials for Energy Related Applications. Environmental Chemistry for a Sustainable World, vol 24. Springer, Cham. https://doi.org/10.1007/978-3-030-04500-5_10
Raja S, Mattoso LHC (2020) Functionalized polymer-based composite photocatalysts. In: Naushad M, Rajendran S, Lichtfouse E (eds) Green photocatalysts. Environmental chemistry for a sustainable world, vol 34. Springer, Cham. https://doi.org/10.1007/978-3-030-15608-4_7
Rajamehala M, Renugaa S, Gopalakrishnan B, Muthukumarapandian A, Vijay Pradap Singh M (2022) Boundaries and opening of biorefineries towards sustainable biofuel production. Biofuels Bioenergy:3–22
Rathnasamy R, Thangasamy P, Thangamuthu R, Sampath S, Alagan V (2017) Green synthesis of ZnO nanoparticles using Carica papaya leaf extracts for photocatalytic and photovoltaic applications. J Mater Sci Mater Electron 28:10374–10381
Ravelli D, Dondi D, Fagnoni M, Albini A (2009) Photocatalysis. A multi-faceted concept for green chemistry. Chem Soc Rev 38:1999–2011
Regin AF, Solanki SC, Saini JS (2008) Heat transfer characteristics of thermal energy storage system using PCM capsules: a review. Renew Sust Energ Rev 12(9):2438–2458
Rogers B, Adams B, Pennathur S (2011) Nanotechnology understanding small systems, 2nd edn. CRC Press
Sahu PK, Champati A, Pradhan A, Naik B (2024) Design and development of nanostructured photocatalysts for large scale solar green hydrogen generation. Sustain Energy Fuels 8:1872–1917
Siva S, Marimuthu C (2015) Production of biodiesel by transesterification of algae oil with an assistance of nano-CaO catalyst derived from egg shell. S Int J Chem Tech Res 74:2112–2116
Verma V, Dossari MA, Singh J, Rawat M, Kordy MGM, Shaban M (2022) A review on green synthesis of TiO_2 NPs: photocatalysis and antimicrobial applications. Polymers 14(7):1444
Yaghoubi S, Mousavi SM, Babapoor A, Binazadeh M, Lai CW, Althomali RH, Rahman MM, Chiang WH (2024) Photocatalysts for solar energy conversion: recent advances and environmental applications. Renew Sust Energ Rev 200:114538
Yahya MA, Al-Qodah Z, Ngah CZ (2015) Agricultural Bio-Waste Materials as Potential Sustainable Precursors Used for Activated Carbon Production: A Review. Renewable and Sustainable Energy Reviews, 46:218–235. https://doi.org/10.1016/j.rser.2015.02.051
Yang ST, El-Ensashy H, Thongchul N (2013) Bioprocessing technologies in biorefinery for sustainable production of fuels, chemicals, and polymers. Wiley
Yashni G, Willy K, Al-Gheethi A, Mohamed R, Salleh SM, Hashim MA (2020) A review on green synthesis of ZnO nanoparticles using Coriandrum sativum leaf extract for degrading dyes in textile wastewater: a prospect towards green chemistry. In: IOP conference series: materials science and engineering. IOP Publishing. 042003
Zhang XL, Yana S, Tyagia RD, Surampalli RY (2013) Biodiesel production from heterotrophic microalgae through transesterification and nanotechnology application in the production. Renew Sustain Energy Rev 26:216–223

Part IV
Scale-Up and Commercialization of Green Nanostructured Photocatalysts

Chapter 10
Environmental and Safety Considerations

Muhammad Ibrahim, Martins Lateef Yusuf, Muhammad Yousuf, Hui Li, Shaoguang Li, and Fan Xia

Abstract Nanostructure photocatalysts have recently gained attention as effective materials for environmental remediation, offering potential in the degradation of various pollutants. These materials, including metal oxides, precious metals, and carbon-based nanostructures, demonstrate significant capacity in breaking down contaminants. However, alongside their performance, there are increasing concerns about their long-term environmental impact, particularly regarding safety and sustainability. This chapter provides an overview of prominent nanostructure photocatalysts utilized in pollutant removal, examining their mechanisms and effectiveness in various applications. Importantly, we look into the potential environmental consequences of their use, as improper disposal or degradation could exacerbate pollution problems. To address this, we emphasize a lifecycle perspective, assessing the role these materials play from synthesis to disposal, and discussing the environmental implications throughout their lifecycle. For the future perspective, we also briefly describe the methods for the safe synthesis and sustainability of photocatalystuse, via adopting eco-friendly practices, improved structural stability, and functional adjustments. This approach contributes to a more responsible utilization of nanostructured materials, aligning their effectiveness with environmental stewardship and minimizing potential adverse effects.

Keywords Environment · Life cycle assessment · Pollutants · Green synthesis · Photocatalyst

---

M. Ibrahim · M. L. Yusuf · H. Li · S. Li (✉) · F. Xia
State Key Laboratory of Geomicrobiology and Environmental Changes, Engineering Research Center of Nano-Geomaterials of the Ministry of Education, Faculty of Materials Science and Chemistry, China University of Geosciences, Wuhan, China
e-mail: lisg@cug.edu.cn

M. Yousuf
School of Environmental Studies, China University of Geosciences, Wuhan, China

C. M. Hussain, G. Hussain (eds.), *Green Nanostructured Photocatalysts for Clean Energy*, Clean Energy Production Technologies,
https://doi.org/10.1007/978-981-95-1618-6_10

10.1 Introduction

10.1.1 Environmental and Safety Problems in Nano-Based Technology

The environment is a sum of living and non-living elements surrounding us and contains chemical, biological, and physical factors that influence the life of an organism (Tahir et al. 2022). There are two types of environments: natural environment and human environment. Interaction among these factors is constant, and any imbalance or alteration can cause problems (Shetty et al. 2023; Ponizovskiy 2022). Pollutants are mainly released into the environment (air, water, and soil) by anthropogenic or natural sources. The common pollutants in the air are particulate matter, ozone, sulfur oxide (SOX), and nitrogen oxide (NOX) (Shi and Yu 2021), while those in water are heavy metals, organic pollutants, and industrial waste. Additionally, soil pollutants include microplastics, pesticides, and herbicides (Amoatey et al. 2019). To address these environmental issues, scientists have been researching active materials capable of reducing or eliminating these pollutants. Nanostructured photocatalysts are one of them which are synthesized at the nanoscale, act as semiconductors, and absorb light (Sivaraman et al. 2022).

Although nanostructured photocatalysts are being used as environmental cleaning agent, they still have adverse effects on the environment because they are emitted during their synthesis, application, and disposal (Hassaan et al. 2023). Once released into the environment, these nanostructured photocatalysts can enter the air, soil, water, and ultimately living organisms, and they have the potential to persist in bodies for a long time and cause health issues to living organisms (Adochite and Andronic 2020). These nanoparticles can enter the human body through three different pathways, including inhalation, absorption, and the nasal route. Inhalation is the most common way because it directly enters the respiratory system and then into the bloodstream (Liu et al. 2022). Through the nasal cavity, it can reach the brain leading to severe health conditions. For example, nanoparticles like titanium dioxide (TiO_2) are less reactive or non-reactive in bulk form because of low surface area but exhibit high reactivity at the nanoscale. The increased reactivity of nanoparticles enables them to penetrate living organisms quite easily, unlike larger particles that cannot pass into cells (Thoma et al. 2024).

Nanotechnology is developing rapidly, but currently, regulatory frameworks are lagging behind the race of nanotechnology development, leading to creating a gap that could cause unforeseen health risks and environmental consequences (Foo et al. 2022; Miron-Barroso et al. 2021). More importantly, the safety of researchers and industrial workers should be considered (Hassan Raza et al. 2024). Overexposure of workers in industries where nano photocatalysts are synthesized may lead to severe diseases. Some nanoparticles are flammable or reactive, raising concerns about fire hazards in various applications. All these concerns prompt an urgent need to implement safety measures to lessen the risks (Gondal 2023). The purpose of this chapter is to communicate the environmental and safety concerns of photocatalysts based on life cycle assessment, emissions, and green synthesis.

10.1.2 Overview of Photocatalyst Safety and Environmental Implications

There are different types of nanostructured photocatalysts including bismuth-based nanostructure photocatalysts, metal oxide nanostructures (Fe_2O_3, WO_3, CeO_2, and ZnO), metal sulfide nanostructures (CdS and ZnS), and carbon-based nanostructures (graphene-based composites and carbon quantum dots). These nanostructured photocatalysts provide promising solutions for ecological remediations (Oladipo and Mustafa 2023; Farhan et al. 2024). For example, TiO_2 is a widely used nanostructured photocatalyst that offers benefits for environmental remediation (Dharma et al. 2022). When used extensively, there is growing concern regarding its release into the environment, its accumulation in the ecosystem, and its possible toxicity to organisms (Sengul and Asmatulu 2020).

However, it offers an environmental advantage by degrading the organic pollutants, dyes, antibiotics, and contaminants in air and water. During application, the emission into the environment causes toxic issues. In addition, certain nanostructured photocatalysts, such as carbon nanotubes, cause pulmonary toxicity that damages the lungs and systemic toxicity. This is because these particles cross biological barriers, and the organism's immune system becomes weak. More so, during the synthesis of nanostructured photocatalyst, incomplete pollutant oxidations can produce byproducts from the BTEX (Benzene, toluene, ethylbenzene, xylenes) compound.

The current rules and regulations regarding nanoparticles are insufficient to address the long-term environmental impacts. So, further studies might provide a better understanding of the long-term environmental and safety implications thoroughly. Despite their advantages, risk management is necessary. While photocatalysts offer significant environmental benefits, their use, especially in nanostructured forms, requires careful consideration of potential risks and implementation of appropriate safety measures (Compton et al. 2019).

10.2 Environmental Impact of Nanostructured Photocatalysts

10.2.1 Life Cycle Assessment of Photocatalytic Materials

The life cycle assessment (LCA) of nanostructured photocatalysts evaluates their environmental impact from production to disposal. By studying each stage, LCA identifies key areas of environmental concern, guiding improvements for sustainability (Nizam et al. 2021). Mohamed A. Hassaan and his team highlighted that the use of nanostructured photocatalysts can yield both positive and negative environmental effects (Hassaan et al. 2023). On the positive side, these photocatalysts are effective in degrading pollutants and facilitating sustainable energy production. On the other hand, numerous health and environmental hazards have been discussed earlier (Logozzi et al. 2022). Therefore, a life cycle assessment is a valuable tool for evaluating the potential benefits and risks associated with the use, production, and disposal of photocatalysts (Fernandes et al. 2020).

Nano photocatalysts' life cycle assessment (LCA) is a thorough analytical method used to assess the environmental effects of each stage of the catalyst's life cycle, from the extraction of raw materials through materials processing, manufacturing, distribution, use, repair and maintenance, and disposal or recycling (Tahir et al. 2022). The importance of photocatalyst life cycle assessment (LCA) is found in its ability to thoroughly analyze these materials' environmental effects throughout their life cycle. Energy consumption, resource depletion, and emissions from the extraction of raw materials during synthesis, usage, and end-of-life are all factors that LCA helps to measure in the overall ecological footprint of photocatalysts (Jolliet et al. 2014).

Using a comprehensive approach, industries and researchers can analyze various synthesis techniques or kinds of photocatalysts, pinpoint environmental hotspots in the production process, and direct advancements toward more environmentally friendly solutions. LCA ensures that photocatalysts' ecological costs and efficacy are balanced by considering both functional performance and environmental implications. This is especially crucial for nanoparticles like TiO_2, where possible ecological hazards must be balanced against the advantages of their photocatalytic capabilities. Finally, the photocatalyst's life cycle assessment (LCA) promotes informed decision-making, encourages innovation in environmentally friendly synthesis techniques, and helps create more sustainable nanotechnology applications in environmental remediation and other domains (Wu et al. 2019).

The comprehensive LCA of photocatalysts has been rarely conducted as most researchers studied only a portion of the life cycle assessment (McCartney et al. 2024). For example, some studied the synthesis portion, and some studies concentrated on the applications or use. Additionally, some researchers studied the end-of-life of nanostructure photocatalysts assessing how these materials behave once they are no longer useful and their potential effects on waste management systems (Pesqueira et al. 2024). There are different nanostructured photocatalyst synthesis methods, while some researchers used physio-chemical methods, others deployed green methods to synthesize the photocatalyst. TiO_2 is one of the photocatalysts that can be synthesized by both chemical and green methods. Both green and chemical methods possess an impact on the environment in terms of climate, oxidant formation, and toxicity (Goncalves et al. 2021).

Sonia et al. performed the life cycle assessment of a mono-doped TiO_2 photocatalyst. Particularly, the doping was done with Fe and Co. They compared the doped photocatalyst with the undoped form. The performance of the doped photocatalyst was significantly improved making it possible to utilize a minute quantity of the photocatalyst. Moreover, they also showed that using a specific solvent like isopropyl during synthesis enhances the performance in comparison to ordinary ethanol. The chemical isopropyl substantially contributed to the environmental impact, except in mineral resources scarcity and land use classes. The use of ethanol instead of isopropyl may improve sustainability (Fernandes et al. 2020).

Maria et al. studied the two traditional chemical methods, chloride and sulfate, for synthesizing TiO_2. The green method always has an advantage over chemical methods; they studied the life cycle of TiO_2 synthesis, and they used aqueous extracts of *Cymbopogon citratus* (lemongrass). They also studied the LCA of TiO_2-functionalized coated float glass, which shows that it has the highest impact on the environment during production (Rodríguez-Rojas et al. 2024).

Wu and co-workers performed the life cycle assessment of TiO_2. They used different synthesis processes and reagents and investigated that the use of various processes and reagents in the synthesis of TiO_2 significantly impacts the environment as shown in Fig. 10.1a–f (Wu et al. 2019). They used six different synthetic methods and different reagents and noticed the impacts of these various methods and reagents on the environment.

Hydrothermal route and sol-gel method: In these processes, the precursor material influences the environmental impact by contributing from 48.8% to 79.4% for hydrothermal and 60.7% to 96.7% for the sol-gel method. The use of reagents and electricity is minimal, reducing the environmental burden. Isopropanol is mainly used in the sol-gel method. In contrast, acetic acid is used in hydrothermal synthesis, so the use of acetic acid in hydrothermal increases the environmental burden compared to the sol-gel process (Wu et al. 2019).

Microemulsion and biological methods: In these processes, surfactants (organic reagents) and solvents play crucial roles, contributing 70% of the environmental burden. Because organic surfactants are used in microemulsion, it has a notable environmental impact. Biological synthesis: bacterial culture media play significant roles in this process, which 91% impact on the environment. In this process, the impact is high because a bulk of bacteria (2500 kg) is required for 1 kg of TiO_2 NPs.

FSP and RF plasma: Two main contributors to these physical routes are supporting gases and electricity. The supporting gases (methane, argon, and oxygen), as well as the combustion of these gases, cause global warming potential, acidification potential, and photochemical smog.

María del Pilar Rodríguez-Rojasa and his team studied that during the green synthesis of TiO_2, some non-methane volatile organic compounds (NMVOCs) were

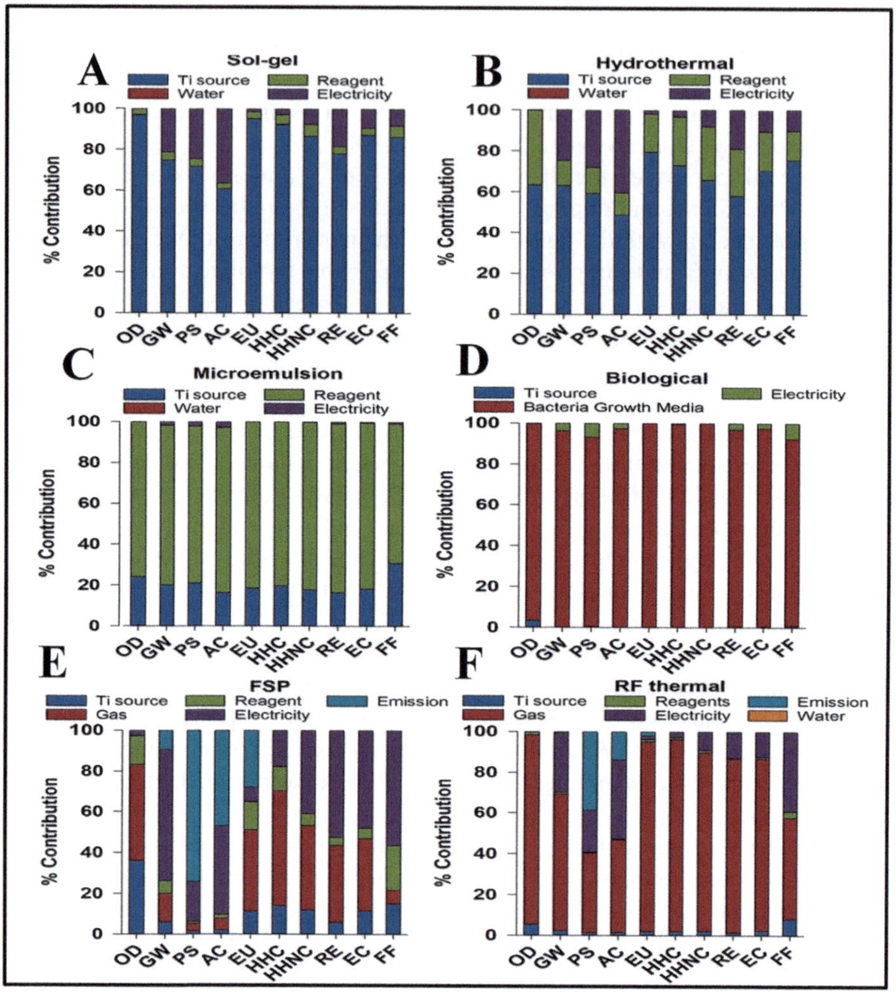

Fig. 10.1 (**a–f**) A comparison of the environmental impact of reagents and processes in six different methods to synthesize TiO2 NPs. (ACS 2019 copyright (Wu et al. 2019))

emitted. These compounds contribute (Fig. 10.2a, b) to the formation of photochemical oxidants. The maximum release of these organic compounds is shown from the synthesis route, particularly leaching and filtration. The figure illustrates the diverse environmental effects of each stage of the synthesis process (Rodríguez-Rojas et al. 2024).

Another example is the ecotoxicity profile for the green manufacture of titanium dioxide (TiO_2) nanoparticles, as measured in kilograms of 1,4-dichlorobenzene equivalents (1,4-DB eq). The leaching process is highlighted as the primary contributor to ecotoxicity since it involves extracting bioactive chemicals, such as flavonoids, from lemongrass leaves, which can release organic wastes into the

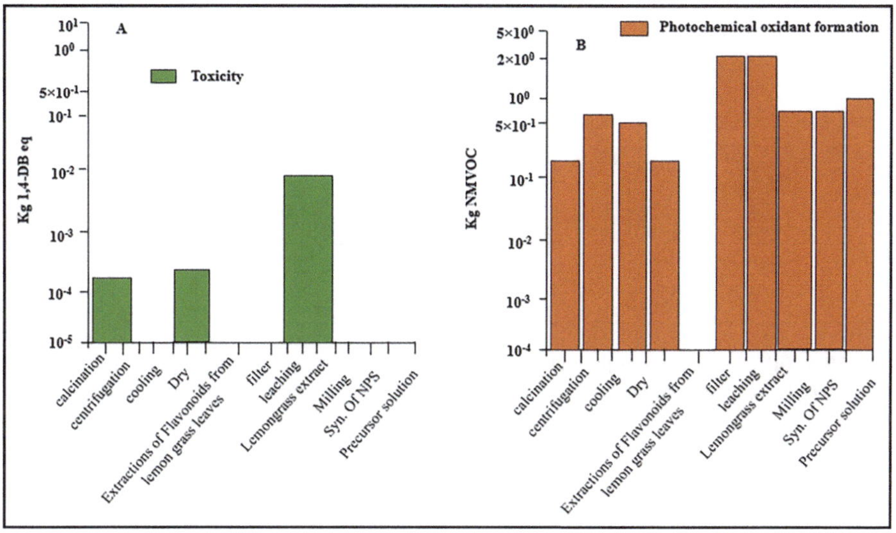

Fig. 10.2 Photochemical oxidant formation during the green synthesis of TiO_2 (B) and ecotoxicity of TiO_2 NPs during green synthesis(A). (Reproduced under the license (https://creativecommons.org/licenses/by/4.0/) (Rodríguez-Rojas et al. 2024))

environment if not controlled effectively. These leftovers may contaminate water and soil, leading to acidification and eutrophication. The filtration process also adds to ecotoxicity due to the possibility of TiO_2 nanoparticles and organic compounds in the waste. Other methods, such as nanoparticle manufacturing, cooling, and milling, have little or no effect on ecotoxicity, indicating that they do not leave primary harmful residues. This figure emphasizes the crucial stages in the synthesis process that contribute to environmental concerns.

Sonia Fernandes and co-worker synthesized a mono-doped photocatalyst (TiO_2 doped with Fe and Co) under visible light (VL) and ultraviolet light (UV) and compared the impact of each mono-doped photocatalyst on human health, ecosystem, and resources. UV-A cobalt-doped TiO_2 has the highest impact on human health, ecosystem, and resources (Fig. 10.3a), while UV-A iron-doped photocatalyst has the least impact indicating iron-doped TiO_2 is sustainable to the environment. They also synthesized a mono-doped photocatalyst (TiO_2 doped with Ni, Co, and Mn) under visible light (VL) and ultraviolet light (UV) (Fig. 10.3b) and compared the impact of each mono-doped photocatalyst on human health, ecosystem, and resources. VL-Mn-TiO_2 and UV-A-Mn-TiO_2 have the highest impact on the environment. So, the overall performance of the TiO_2 photocatalyst can be improved by mono-doping of iron (Fe).

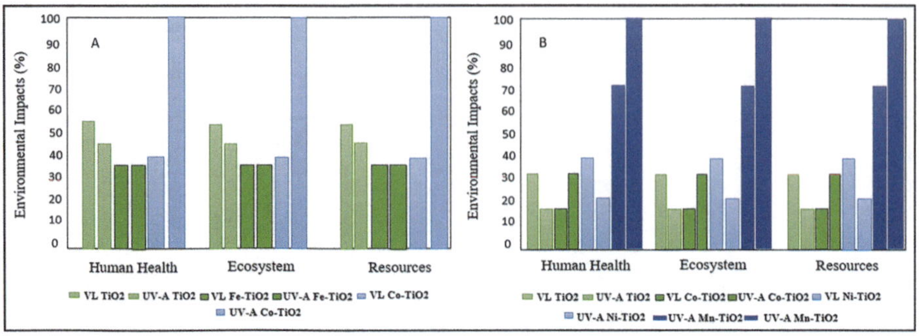

Fig. 10.3 (**a**) Comparison of the environmental impact of Fe and Co doped-TiO_2 under VL and UV-A. (**b**) Comparison of the environmental impact of Co, Ni, and Mn-doped TiO2 under VL and UV-A (Fernandes et al. 2020). (Reproduced under the license http://creativecommons.org/licenses/by/4.0/)

10.3 Emission of Nanoparticles into the Environment

As we discussed in previous sections, nanostructured photocatalyst has a significant role in reducing pollutants from the environment, but the release of these materials poses a risk to the environment and health. Engineered nanoparticles are used nowadays in different products such as in paints and coatings and are considered important applications. The purpose of using nanoparticles in paints is for protection against microbes, resistance to chemicals, and improving physical deterioration.

Ahmed Al-Kattan and his team investigated the release of TiO_2 from paints. The author illustrates the emission of nanoparticles (TiO_2) from two different paints under environmental weathering conditions (Fig. 10.4a); they show both the size and amount of nanoparticles (TiO_2) (Al-Kattan et al. 2013). They compare the release of TiO_2 from two different paint formulations. A1 paint contains both pigment and nano-TiO_2, and A2 contains pigment-TiO_2. A minimal amount of TiO2, quantified at 1.5 mg/L, was released over time from both paint formulations, A1 and A2. TiO_2 from both paints is well embedded in a matrix of paints, limiting the amount of TiO_2 released into the environment and also focuses on the size fractionation of TiO_2 from paint A1, which contains both pigment-TiO_2 and nano-TiO_2. This shows that the maximum particles emitted are 100 nm or under 100 nm in size. Large particles and particles under 100 nm are released at the start of weathering. As weathering continues, the release of large particles larger than 100 nm decreases.

This figure shows how these nanoparticles (TiO_2) can be released from consumer products and enter the environment through natural weathering processes. Paints release both large-sized and small-sized particles, and small-sized nanoparticle emissions are considered a risk to the aquatic ecosystem due to their high reactivity and ability to penetrate membranes. Even though the paint matrix helps limit the emission of nanoparticles, there is still the potential for the release of nanoparticles from paints into the environment through the natural weathering process. These

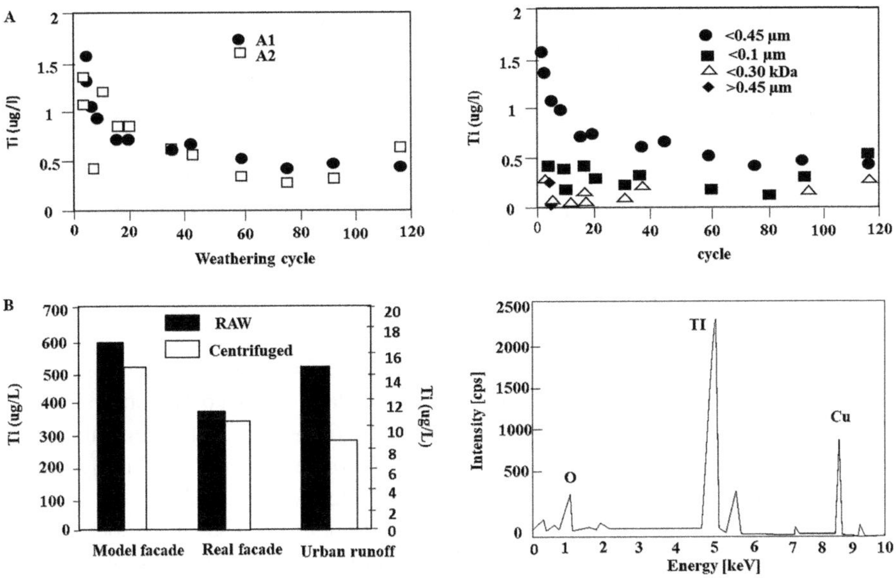

Fig. 10.4 (**a**) Emission of TiO2 from paints A1 and A2, and size fractionation of TiO2 released from paint A1. (**b**) Ti concentrations were measured by ICP-MS in surface runoff before and after centrifugations, and TEM-EDX of Ti was detected in urban runoff samples. (**a**) RSC 2013 copyrights, (**b**) (Al-Kattan et al. 2013). (Elsevier 2008 copyrights (Kaegi et al. 2008))

figures exemplify the potential for photocatalysts such as (TiO_2) in natural ecosystems through common material paints (Armaković et al. 2022).

Kaegi et al. showed that TiO_2 particles are emitted from buildings into aquatic environments. These authors showed that nanoparticles are emitted from the external parts of buildings into water. By using different analytical tools, they were able to find the concentration and types of nanoparticles. It has been noticed that nanoparticles with the pigment TiO_2 are emitted from buildings even though they do not contain nanoparticles. Almost 30% of the pigment TiO_2 contains nanoparticles of a size lower than 100 nm. It is proved from experiments that the sources of nanoparticles (TiO_2) are not only the use of nanoparticles in paints and pigment particles (Kaegi et al. 2008).

Another study illustrates the mass concentrations of titanium in surface runoff samples measured using ICP-MS. (Fig. 10.4b) This study compared the raw and centrifuged samples from three different sources: model facade, actual facade, and urban runoff. The results reveal a large amount of titanium in all runoff samples, showing the presence of titanium dioxide (TiO_2) nanoparticles. This data indicates that most of the nanoparticles in the runoff sample range from 20 to 300 nm. The higher concentrations of Ti in large particles compared to facade runoff suggest that there may be an additional source of TiO_2 in urban environments, such as paints and other coatings. This study also shows the spectrum of TEM-EDX of titanium dioxide (TiO_2) present in urban runoff samples. It shows that titanium dioxide nanoparticles were detected in water samples collected from urban runoff before they reached surface water. The spectrum confirms the titanium dioxide nanoparticles by detecting the titanium (Ti) with other elements such as copper and oxygen. This picture shows that synthetic nanoparticles from different sources, such as building facades, can be present in urban water runoff.

In another study, the author investigated the release of nanoparticles (Ag-NPs) from coated pints on the exterior of a building. They exposed the exterior to ambient weather for 1 year and investigated the release of nanoparticles. It is noticed that 30% of nanoparticles (Ag-NPs) are released into the environment as colloids with organic binders. After releasing these particles into the atmosphere, they are converted into a less toxic form (Reinsch et al. 2012).

Jimenez et al. investigated the leaching of TiO_2 and nitrates from nano-functionalized construction materials. Over 800 days of extensive outdoor exposure, the researcher noticed that titanium consistently leached into rainwater runoff from almost all tested materials, with the highest concentrations of 60 ug/L, raising concerns about water contamination (Jimenez-Relinque et al. 2020).

Josune Olabarrieta and his team illustrated the release of TiO_2 from coated material in aging water (H_2O) at various experimental conditions containing UVA and dark, and various solutions such as deionized water, sodium chloride, humic acid and humic acid, and salt. The author and his team explained the emission of TiO_2 from coating materials under different conditions (Fig. 10.5a, b). The highest release of TiO_2 was found in salt and salt with humic acid. So overall, the use of UVA and water conditions can increase the emission of titanium dioxide (TiO_2) (Olabarrieta et al. 2012). These studies indicate that metal oxide photocatalysts, when exposed to weathering or disposed of after use, can gradually release particles into the environment.

This release poses a potential threat to living organisms, as these materials may accumulate in ecosystems and cause harmful effects. Consequently, it is essential to carefully consider the environmental impact of metal oxide photocatalysts and prioritize sustainable practices to mitigate potential risks. Enhanced focus on lifecycle analysis, safe disposal, and environmentally friendly alternatives can help reduce

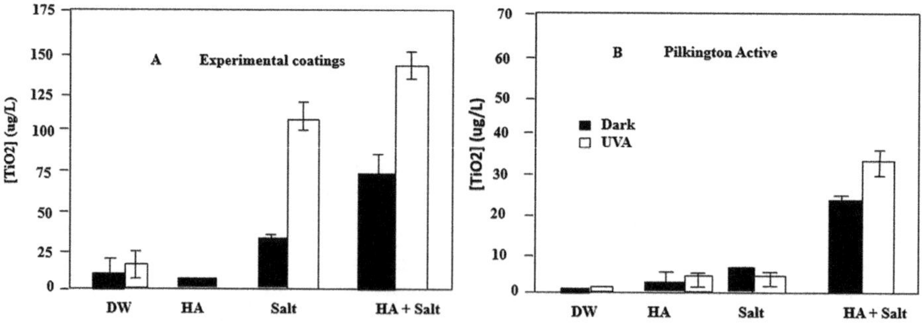

Fig. 10.5 TiO2 concentrations in aging water, measured by ICP-OES, (**a**) coating materials and (**b**) Pilkington active surface. (Elsevier 2012 copyrights) (Olabarrieta et al. 2012)

these adverse effects, to ensure safer use of these materials while maintaining their efficacy in industrial and environmental applications.

10.4 Human Health and Safety Considerations

As discussed earlier, when nanostructured photocatalysts are emitted into the environment, their small particle size and large surface area allow them to penetrate and enter the human body through various pathways: inhalation, absorption through the nasal cavity, and skin contact (Ganguly et al. 2018). There are various types of nanostructured photocatalysts such as TiO_2, ZnO, CuO, and Ag (Assadian et al. 2018; Saravanan et al. 2022). TiO_2 is the most studied nanostructured photocatalyst. Like TiO_2, ZnO also promotes cytotoxicity in cells, CuO is highly toxic in both vivo and vitro, and Ag nanoparticle also damages the DNA by mutations (Ramakrishna and Rao 2011; Zhang et al. 2022; Saravanan et al. 2022). Once inside, they have the potential to cause toxicity in various organs, affecting systems such as the respiratory, cardiovascular, liver, kidney, and nervous systems (Rajput et al. 2020). The mechanism of toxicity of nanostructured photocatalyst includes inflammation, the production of reactive oxygen species, and mitochondrial damage. When nanostructured photocatalysts interact with immune cells, they can cause acute inflammations leading to cell damage. Some nanoparticles damage the mitochondria, leading to cell death and reducing the production of cellular energy (Rajput et al. 2020). Over long-term exposure to human beings, some nanostructure photocatalysts (metal-based) can cause cancer, by inducing mutations in cells and damaging the DNA.

Nanostructured photocatalysts exhibit complex interactions with cellular components due to their unique properties, resulting in biological effects. The author outlines three primary mechanisms: direct nanoparticle contacts with cellular components, oxidative stress induction, and ion release (Fig. 10.6a). These interactions affect several cellular components, including proteins, DNA, and cell membranes, resulting in protein interactions, DNA damage, and membrane disruption, ultimately contributing to cellular toxicity (Kang et al. 2008).

They also explained the production of reactive oxygen species by nanostructured photocatalyst and their effect on cellular systems (Fig. 10.6b). The nanoparticles become activated and generate reactive oxygen species whenever exposed to ultraviolet or visible light. This is the common property of photocatalytic materials. This figure also shows that there are different types of reactive oxygen species produced during the activation process such as single oxygen, superoxide ion, and hydroxyl radical. After the production of different reactive oxygen species, they interact with DNA, proteins, and lipids and cause oxidative damage. In the end, ROS generations lead to inflammation, genotoxicity, and cell death. Ilina et al. studied the toxic effect of iron-doped titanium dioxide (TiO_2-Fe3) on human beings.

They compared the effect of iron-based nanostructured photocatalysts in vitro and in vivo. Some previous studies show that iron-doped materials do not show a

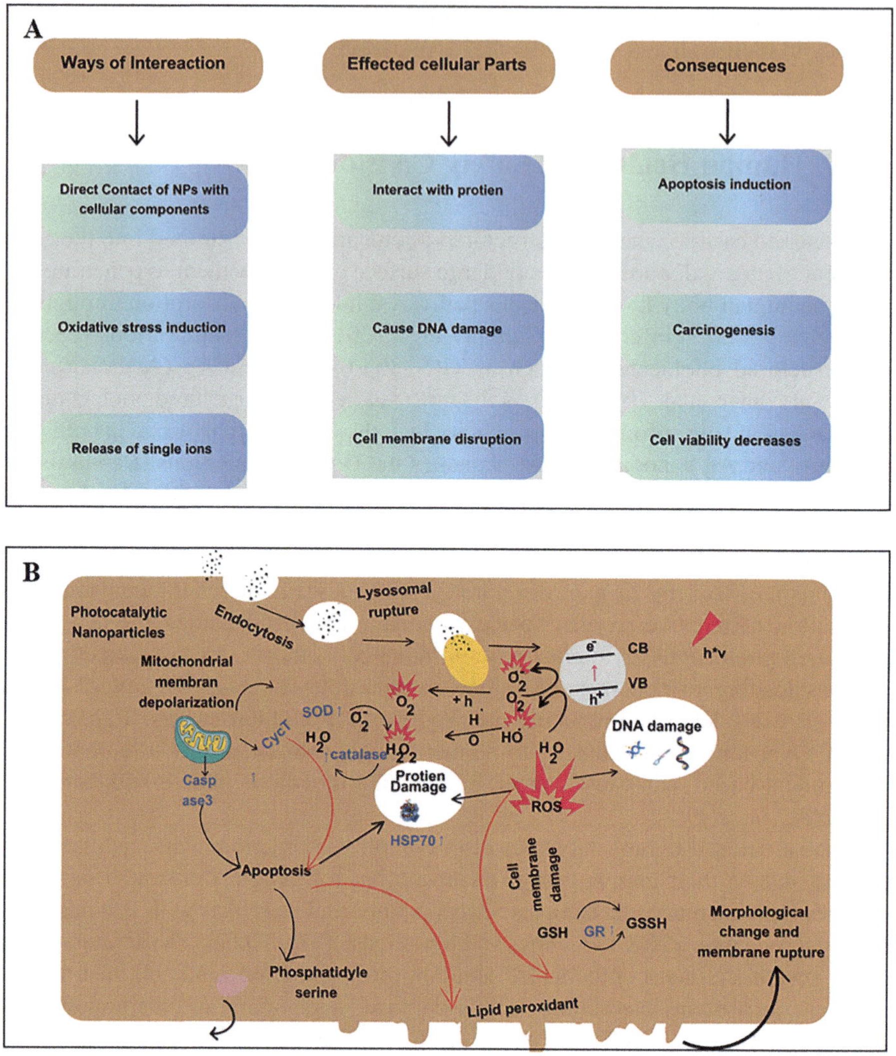

Fig. 10.6 (**a**) Complex interactions with cellular components. (**b**) Productions of reactive oxygen species. (Elsevier 2016 copyrights (Friehs et al. 2016))

significant effect on the red blood cells of human beings. An in vitro study shows that the destruction of red blood cells is less compared to an in vivo study. But in in vivo evaluations, it is indicated that these nanoparticles promote problems such as systemic and pulmonary inflammations, hypo-toxicity, and oxidative stress (Medina-Ramírez et al. 2014). Rajnihs Gupta et al. illustrated the toxicity of nano-structured photocatalysts (CdS) (Choudhary et al. 2023). Reactive oxidative stress and cytotoxicity are induced by cadmium (Cd) ions, released from nanostructured photocatalysts (International Journal of Environment, Agriculture and Biotechnology n.d.). Therefore, the precautions from exposure to nanostructured photocatalysts (CdS NPs) are important. If these nanoparticles are exposed to human beings, they cause biochemical alterations and damage the liver and kidneys. The toxic effect of this nanostructured photocatalyst on human beings raises concerns about environmental safety (Gupta 2019). Shan Yuju et al. experimented to see the toxicity of tungsten trioxide photocatalyst in rats through direct administration and inhalation. The purpose of the study was to see the persistence of nanoparticles in the lungs after exposure to tungsten trioxide. The tungsten trioxide increases the neutrophils and shows short-lived inflammatory marks instead of long-lasting inflammations, and no development of any tumor was found, indicating its less toxicity at 2 mg/m^3 (Yuju et al. 2023).

Although some scientists agree with this statement that ZrO_2 is safe for biomedical use but some studies show safety concerns about the use of ZrO_2 (Hichem et al. 2022). Studies show that the main toxicity of ZrO_2 involves the formation of reactive oxygen species. The production of reactive oxygen species causes oxidative stress, changes in the biochemical system, and even cell death (Hichem et al. 2022). Hridoy and his team said that SnO_2 (tin oxide) photocatalyst is commonly not absorbed by the skin and gastrointestinal tract; this property of SnO_2 makes it less toxic to human beings. However, long-term exposure can cause health problems, particularly inflammation in the lungs, coughing, and respiratory issues. So overall, it is not considered highly toxic to human beings (Roy et al. 2023).

Gurunathan et al. investigated graphene oxide which is generally considered toxic to human cells (embryonic kidney). Graphene oxide induces toxicity by increasing the reactive oxygen species generation, decreasing energy unit synthesis (ATP), decreasing antioxidant levels, and disrupting the mitochondrial membrane potential. This nanostructured photocatalyst also induces apoptosis via mitochondrial impairment and oxidative stress (Gurunathan et al. 2019).

In another study, Lin Wu et al. stated that Fe_3O_4 is highly toxic due to its small size ranging from 2.3 to 4.2 nm. Through hydroxyl radicals and reactive oxygen species, it produces significant oxidative stress and may cause heart failure. The small size of nanoparticles is important because particles with large sizes do not have such toxicity. This study suggests that toxicity induced in organs is because of iron concentration as well as size (Wu et al. 2022).

Yu Hang Leung's co-worker suggests that CeO_2 can produce oxidative stress and damage the cell membrane by generating reactive oxygen species, but some studies are vice versa. They illustrated that CeO_2 particles can act as antioxidants and reduce oxidative stress. Therefore, the toxicity of CeO_2 depends upon its chemical and physical properties (Leung et al. 2015).

The various types of photocatalysts including TiO_2, CdS, ZnO, CuO, TiO_2-Fe3, CuO, ZrO_2, and CeO_2 are focusing on their interactions and toxicity to human beings. Each photocatalyst produces specific reactive oxygen species which damage the cell. For example, CdS damages the liver and kidneys, ZrO_2 produces oxidative stress, and TiO_2-Fe_3 decreases red blood cells (RBC). Additionally, CeO_2 shows the properties of an oxidant and an antioxidant while Fe_3O_4 may cause heart problems or failure. These findings revealed the significant health risks posed by photocatalysts, showing that they are unsuitable for promoting human health.

10.5 Sustainable Design and Synthesis of Nanostructured Photocatalyst

Nanostructured photocatalysis plays a vital role in environmental remediation, but it is crucial to minimize the environmental impact of the synthesis process (Panchal et al. 2021). Titanium dioxide (TiO_2), a widely used nanocrystalline photocatalyst, is favored for its UV-activated catalytic performance. However, TiO_2 has a band gap of 3.2 eV, limiting its absorption to the UV spectrum, which may require excessive amounts of material to achieve optimal effectiveness (Mahy et al. 2021). This overuse can potentially result in environmental consequences. To enhance the photocatalytic performance of TiO_2, modification approaches, such as metal or non-metal doping and polymer capping, are often used (Wang et al. 2014). They use combinations of materials to produce heterojunctions and Z-scheme systems, combining machine learning tools, and most focus area nowadays is hydrogen production. These methods, though effective, involve multiple steps that need careful consideration of their environmental impact. For example, Jiang et al. demonstrated a modification of TiO_2 using a reaction between the hydroxyl groups on TiO_2 nanoparticles and the –NCO group on toluene diisocyanate (TDI) (Jiang et al. 2007). While this modification enhanced the photocatalyst's performance, it is important to note that TDI is a hazardous organic compound. Toluene diisocyanate poses significant environmental risks due to its toxicity. It can cause severe respiratory issues, skin irritation, and long-term environmental damage. Once released into the environment, TDI can persist, contributing to air and water pollution and potentially harming aquatic life. Similarly, Margan and his team doped TiO_2 with cadmium to improve its photocatalytic properties (Margan and Haghighi 2016). While cadmium doping enhanced performance, cadmium itself is a highly toxic heavy metal with severe environmental consequences. Cadmium can accumulate in ecosystems and is harmful to aquatic and terrestrial organisms. Prolonged exposure to cadmium can lead to bioaccumulation, disrupting food chains and causing kidney damage and cancer in humans (Choudhary et al. 2023). Its release into the environment must be carefully controlled to prevent contamination.

These examples highlight the need for caution when modifying photocatalysts to prevent exacerbating environmental pollution. Although modifications can significantly improve photocatalytic efficiency, the environmental impact of the materials used in these processes must be critically assessed. Safer alternatives, such as green synthesis methods, should be explored to balance performance enhancement with environmental sustainability.

10.5.1 Green Chemistry Principles and Nano-Material Design

One effective way to ensure the safe modification of photocatalysts is through the adoption of green chemistry principles (Goncalves et al. 2021). Green synthesis reduces material pollution and minimizes the environmental impact during production (Li et al. 2022). One promising approach involves leveraging carbon-based materials from biodegradable sources, which has gained significant attention. For example, Liu et al. demonstrated a green synthesis method using rice hulls, a by-product of rice, to create a photocatalyst composite in a single-step process (Liu et al. 2024). This material not only enhanced the degradation of oxytetracycline under sunlight at low dosage and energy but also exemplified an environmentally conscious synthesis method. Similarly, Jansanthea et al. utilized grape extract to harmonize copper acetate, ferric nitrate, and zinc acetate, forming a $CuO/Fe_2O_3/ZnO$ composite (Jansanthea et al. 2024). The grape extract served as a reducing agent for the metallic nanoparticles, improving the photocatalyst's properties while showcasing an impactful green synthesis route. Further research in this field aims to refine these eco-friendly approaches (Zinatloo-Ajabshir et al. 2019). For instance, Mostafa and colleagues employed carbohydrates such as glucose, lactose, and starch to cap $NaEuTiO_2$ metal oxide nanoparticles (Hosseinpour-Mashkani and Sobhani-Nasab 2016). These biodegradable materials acted as reducing agents and prevented nanoparticle agglomeration, contributing to a greener and safer synthesis process. These studies illustrate the potential of green chemistry to improve photocatalyst performance while minimizing environmental risks. Exploring biodegradable and non-toxic materials, researchers are developing sustainable routes for photocatalyst modification, ensuring safer environmental practices in nanomaterial production (Li et al. 2022).

10.5.2 Waste Reduction in Synthesis and Manufacturing

While it is essential to use eco-friendly materials and adopt green routes for photocatalyst synthesis, fully achieving a green process can be challenging (Cheng et al. 2022). This is particularly true for metal oxide photocatalysts, such as titanium dioxide (TiO_2), which carry inherent environmental risks (Adochite and Andronic 2020). Thus, minimizing waste during synthesis and managing it effectively becomes crucial to reduce environmental impact. For instance, in TiO_2 nanoparticle synthesis, water is often used as the primary solvent (Muniandy et al. 2017). Proper management of wastewater discharge is vital to ensure minimal environmental harm, as metal oxide photocatalysts can have detrimental effects on aquatic organisms. The interaction of metal oxides in water is complex and influenced by various factors, making it difficult to predict their behavior. Morelli et al. studied the interaction between algae and TiO_2 nanoparticles and found that, rather than forming the usual agglomerates (Lin et al. 2015), the nanoparticles remained suspended and

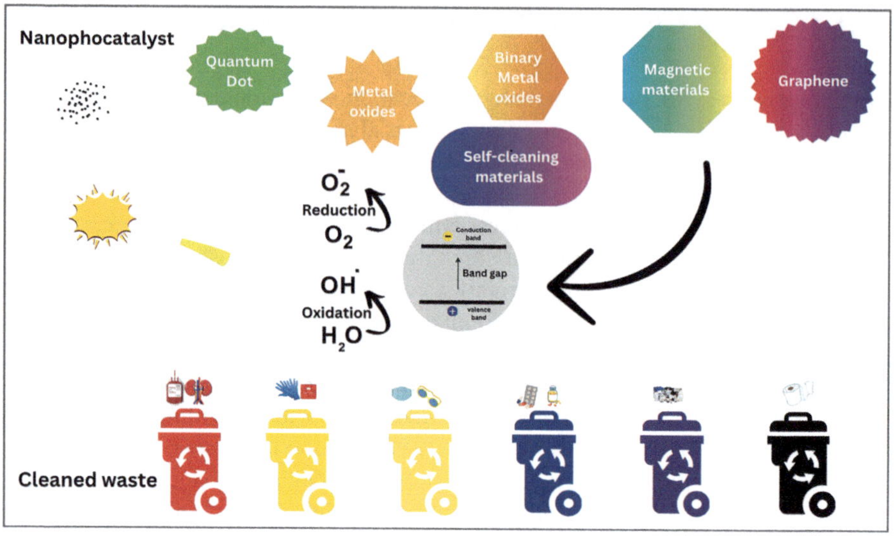

Fig. 10.7 A schematic highlighting key nano-photocatalysts, including quantum dots, metal oxides, binary metal oxides, self-cleaning materials, magnetic materials, and graphene. The figure illustrates how light triggers these nanomaterials to generate reactive radicals (O2⁻ and OH·) through reduction and oxidation reactions involving oxygen and water. These radicals then degrade various types of biomedical waste—such as kidney-related waste, fibers, masks, medicines, blood-contaminated waste, and tissues—transforming them into non-toxic, cleaned waste. (Reproduced under the license http://creativecommons.org/licenses/by/4.0/ (Hooshmand et al. 2020))

stabilized for hours, leading to prolonged retention in the water under specific conditions such as ionic strength and pH (Morelli et al. 2018).

Titanium dioxide (TiO_2) can also form hetero agglomerates with algae cells complicating the prediction of how photocatalysts behave in aquatic environments (Ma et al. 2015). Given these challenges, reducing waste during synthesis to the lowest possible levels is critical to minimizing environmental risks. Managing discharge and recycling solvents or materials can help mitigate the impact of metal oxide photocatalysts on ecosystems. As synthesis processes advance, ensuring proper waste treatment and developing methods to control and minimize environmental contamination will play a key role in making photocatalyst production more sustainable. Figure 10.7 illustrates how various types of nanostructured photocatalysts produce different reactive oxygen species by triggering the light and cleaning biomedical waste (Hooshmand et al. 2020).

10.5.3 Alternative Photocatalytical Nanoparticles

Graphene-based materials have been widely used to enhance the properties of metal oxide photocatalysts, and recent research suggests that graphene itself holds promise as a photocatalyst (Chen et al. 2022). With its wide band gap and semiconductor-like properties, graphene offers a potential alternative to traditional metal oxides. What makes graphene particularly appealing is its ability to be synthesized from biodegradable materials, making it a more environmentally friendly option (Zinatloo-Ajabshir et al. 2019). Given these advantages, future studies could focus on further developing graphene as a standalone photocatalyst, exploring its potential to replace metal oxides in applications ranging from environmental remediation to energy conversion. Graphene's green synthesis route and promising photocatalytic properties make it a valuable candidate for sustainable and effective photocatalysis (Hurtado et al. 2017).

10.6 Conclusion and Outlook

In this chapter, we discussed the role of nanostructured photocatalysts in the environment. Of note is their enormous importance towards remediation and pollutant degradation, but there might be environmental risk as well to their usage. There are numerous ongoing methods of ensuring they are more environmentally friendly like the synthesis approaches. Future work should focus more on ways to better manage photocatalysts for effective use. In this regard, scientists are developing new approaches to synthesize the nanostructured photocatalyst with optimized conditions to get a more efficient nanostructured photocatalyst (Arif et al. 2024; Bazgir et al. 2023). They use combinations of materials to produce heterojunctions and a Z-scheme system, which enhances charge separations and absorptions of light (Sanoja-López et al. 2024). Researchers are also trying to use multiple complementary tools such as machine learning to optimize the synthesis of photocatalyst and their performance. For water treatment, nanostructured photocatalysts are used nowadays instead of conventional methods. Hydrogen production from water splitting is the most focused area nowadays. With time, researchers are trying to improve the efficiency of materials used for hydrogen production like $TiO_2/Fe_2O_3@Ti_3C_2$ MXene heterojunctions (Sanoja-López et al. 2024). It is trying to develop a more efficient and cost-effective photocatalyst that can work in different environmental conditions such as acidic environments (Sanoja-López et al. 2024).

The safety measurement of nanostructured photocatalysts is essential as these have environmental applications. There is a potential impact on the aquatic environment and human life during the synthesis and applications of nanostructured photocatalysts, so a comprehensive study is essential. There should be a deep study to understand the emission and long-term stability of nanostructured photocatalysts. As with time, there is advancement in this field, so there should be regulations:

(Nasrollahzadeh et al. 2021) Guidelines for the use, handling, and disposal of nanostructured photocatalysts and standardized protocols to find the efficiency and safety of these materials (Arif et al. 2024).

The primary objective of this advancement is to boost the performance of photocatalysts without compromising environmental safety. The development of more sustainable strategies that address environmental and energy concerns is essential to ensure that technological progress aligns with the imperative of ecological protection.

Acknowledgments This work was supported by the National Natural Science Foundation of China (22274144, 22474130, 22090050), the National Key Research and Development Program of China (2021YFA1200403), the Joint NSFC-ISF Research Grant Program (Grant No: 22161142020), the Natural Science Foundation of Shenzhen (JCYJ20220530162406014).

References

Adochite C, Andronic L (2020) Aquatic toxicity of photocatalyst nanoparticles to green microalgae chlorella vulgaris. Water 13:77

Al-Kattan A, Wichser A, Vonbank R, Brunner S, Ulrich A, Zuin S, Nowack B (2013) Release of Tio2 from paints containing pigment-Tio2 or Nano-Tio2 by weathering. Environ Sci Process Impacts 15:2186–2193

Amoatey P, Omidvarborna H, Baawain MS, Al-Mamun A (2019) Emissions and exposure assessments of Sox, Nox, Pm10/2.5 and trace metals from oil industries: a review study (2000–2018). Process Saf Environ Prot 123:215–228

Arif MD, Hoque ME, Rahman MZ, Shafoyat MU (2024) Emerging directions in green nanomaterials: synthesis, physicochemical properties and applications. Mater Today Commun 40:109335

Armaković SJ, Savanović MM, Armaković S (2022) Titanium dioxide as the most used photocatalyst for water purification: an overview. Catalysts 13:26

Assadian E, Zarei MH, Gilani AG, Farshin M, Degampanah H, Pourahmad J (2018) Toxicity of copper oxide (Cuo) nanoparticles on human blood lymphocytes. Biol Trace Elem Res 184:350–357

Bazgir H, Issaabadi Z, Arabi H (2023) Green nanomaterials as photocatalyst/catalyst: exploration of properties. In: Handbook of green and sustainable nanotechnology. Springer Nature, pp 973–1003

Chen Z, Sun S, Prakash J (2022) Design and engineering of graphene nanostructures as independent solar-driven photocatalysts for emerging applications in the field of energy and environment. Mol Syst Des Eng 7:213–238

Cheng S, Zhao S, Xing B, Shi C, Meng W, Zhang C, Bo Z (2022) Facile one-pot green synthesis of magnetic separation photocatalyst-adsorbent and its application. J Water Process Eng 47:102802

Choudhary P, Sharma P, Kaur S, Randhawa J, Borse L (2023) A comprehensive review on the deleterious effects of heavy metal bioaccumulation on the gills and other tissues of freshwater fishes. Biosci Biotechnol Res Asia 20:395–405

Compton JS, Peterson CA, Dervishogullari D, Sharpe LR (2019) Spray pyrolysis as a combinatorial method for the generation of photocatalyst libraries. ACS Comb Sci 21:489–499

Dharma HNC, Jaafar J, Widiastuti N, Matsuyama H, Rajabsadeh S, Othman MHD, Rahman MA, Jafri NNM, Suhaimin NS, Nasir AM, Alias NH (2022) A review of titanium dioxide (Tio(2))-based photocatalyst for oilfield-produced water treatment. Membranes (Basel) 12(3):345

Falciglia, Pietro Paolo (2020) International Journal of Environment, Agriculture and Biotechnology.

Farhan S, Hassan Raza A, Yang S, Yu Z, Wu Y (2024) Boosted photocatalytic hydrogen evolution of S-Scheme N-doped Ceo(2-Delta)@Znin(2)S(4) heterostructure photocatalyst. J Colloid Interface Sci 669:430–443

Fernandes S, Esteves da Silva JCG, Pinto da Silva L (2020) Life cycle assessment of the sustainability of enhancing the photodegradation activity of Tio(2) with metal-doping. Materials (Basel) 13(7):1487

Foo JJ, Ng S-F, Ong W-J (2022) Dimensional heterojunction design: the rising star of 2d bismuth-based nanostructured photocatalysts for solar-to-chemical conversion. Nano Res 16:4310–4364

Friehs E, AlSalka Y, Jonczyk R, Lavrentieva A, Jochums A, Walter J-G, Stahl F, Scheper T, Bahnemann D (2016) Toxicity, phototoxicity and biocidal activity of nanoparticles employed in photocatalysis. J Photochem Photobiol C: Photochem Rev 29:1–28

Ganguly P, Breen A, Pillai SC (2018) Toxicity of nanomaterials: exposure, pathways, assessment, and recent advances. ACS Biomater Sci Eng 4:2237–2275

Goncalves RA, Toledo RP, Joshi N, Berengue OM (2021) Green synthesis and applications of Zno and Tio(2) nanostructures. Molecules 26:2236

Gondal AH (2023) Nanotechnology advancement in the elimination of chemical toxins from air spectrums. Int J Environ Sci Technol 20:12775–12792

Gupta R (2019) Cadmium nanoparticles and its toxicity. J Crit Rev 6(5):1–7

Gurunathan S, Arsalan Iqbal M, Qasim M, Park CH, Yoo H, Hwang JH, Uhm SJ, Song H, Park C, Do JT, Choi Y, Kim JH, Hong K (2019) Evaluation of graphene oxide induced cellular toxicity and transcriptome analysis in human embryonic kidney cells. Nanomaterials (Basel) 9:969

Hassaan MA, El-Nemr MA, Elkatory MR, Ragab S, Niculescu VC, El Nemr A (2023) Principles of photocatalysts and their different applications: a review. Top Curr Chem (Cham) 381:31

Hassan Raza A, Farhan S, Yu Z, Wu Y (2024) Double S-Scheme Zns/Zno/Cds heterostructure photocatalyst for efficient hydrogen production. Acta Phys-Chim Sin 11:2406020

Hichem N, Hadjer Z, Fateh S, Feriel L, Wang Z (2022) The potential exposure and hazards of zirconia nanoparticles: a review. Ecotoxicol Environ Contam 17:1–21

Hooshmand S, Kargozar S, Ghorbani A, Darroudi M, Keshavarz M, Baino F, Kim HW (2020) Biomedical waste management by using nanophotocatalysts: the need for new options. Materials (Basel) 13:3511

Hosseinpour-Mashkani SM, Sobhani-Nasab A (2016) Green synthesis and characterization of Naeuti2o6 nanoparticles and its photocatalyst application. J Mater Sci: Mater Electron 28:4345–4350

Hurtado RB, Calderon-Ayala G, Cortez-Valadez M, Ramírez-Rodríguez LP, Flores-Acosta M (2017) Green synthesis of metallic and carbon nanostructures. In: Nanomechanics. https://doi.org/10.5772/intechopen.68483

Jansanthea P, Inyai N, Chomkitichai W, Ketwaraporn J, Ubolsook P, Wansao C, Wanaek A, Wannawek A, Kuimalee S, Pookmanee P (2024) Green synthesis of Cuo/Fe(2)O(3)/Zno ternary composite photocatalyst using grape extract for enhanced photodegradation of environmental organic pollutant. Chemosphere 351:141212

Jiang D, Xu Y, Hou B, Wu D, Sun Y (2007) Synthesis of visible light-activated Tio2 photocatalyst via surface organic modification. J Solid State Chem 180:1787–1791

Jimenez-Relinque E, Grande M, Duran T, Castillo A, Castellote M (2020) Environmental impact of nano-functionalized construction materials: leaching of titanium and nitrates from photocatalytic pavements under outdoor conditions. Sci Total Environ 744:140817

Jolliet O, Frischknecht R, Bare J, Boulay A-M, Bulle C, Fantke P, Gheewala S, Hauschild M, Itsubo N, Margni M, McKone TE, Canals LM, Posthuma L, Prado-Lopez V, Ridoutt B, Sonnemann G, Rosenbaum RK, Seager T, Struijs J, van Zelm R, Vigon B, Weisbrod A (2014) Erratum to: global guidance on environmental life cycle impact assessment indicators: findings of the scoping phase. Int J Life Cycle Assess 19:1566–1566

Kaegi R, Ulrich A, Sinnet B, Vonbank R, Wichser A, Zuleeg S, Simmler H, Brunner S, Vonmont H, Burkhardt M, Boller M (2008) Synthetic Tio2 nanoparticle emission from exterior facades into the aquatic environment. Environ Pollut 156:233–239

Kang SJ, Kim BM, Lee YJ, Chung HW (2008) Titanium dioxide nanoparticles trigger P53-mediated damage response in peripheral blood lymphocytes. Environ Mol Mutagen 49:399–405
Leung YH, Yung MM, Ng AM, Ma AP, Wong SW, Chan CM, Ng YH, Djurisic AB, Guo M, Wong MT, Leung FC, Chan WK, Leung KM, Lee HK (2015) Toxicity of Ceo2 nanoparticles – the effect of nanoparticle properties. J Photochem Photobiol B 145:48–59
Li Y, Zhang D, Qiao W, Xiang H, Besenbacher F, Li Y, Su R (2022) Nanostructured heterogeneous photocatalyst materials for green synthesis of valuable chemicals. Chem Synth 2:9
Lin M, Tseng YH, Huang C-P (2015) Interactions between nano-Tio2 particles and algal cells at moderate particle concentration. Front Chem Eng 9:242–257
Liu YY, Liu J, Wu H, Zhang Q, Tang XR, Li D, Li CS, Liu Y, Cao A, Wang H (2022) Endocytosis, distribution, and exocytosis of polystyrene nanoparticles in human lung cells. Nanomaterials (Basel) 13:84
Liu Z, Zhang D, Mou R (2024) Rice hull-derived biochar with photocatalytic activity for efficient degradation of oxytetracycline in water under sunlight. Process Saf Environ Prot 188:1550–1561
Logozzi M, Di Raimo R, Mizzoni D, Fais S (2022) The potentiality of plant-derived nanovesicles in human health-a comparison with human exosomes and artificial nanoparticles. Int J Mol Sci 23(9):4919
Ma S, Zhou K, Yang K, Lin D (2015) Heteroagglomeration of oxide nanoparticles with algal cells: effects of particle type, ionic strength and Ph. Environ Sci Technol 49:932–939
Mahy JG, Lejeune L, Haynes T, Lambert SD, Marcilli RHM, Fustin C-A, Hermans S (2021) Eco-friendly colloidal aqueous sol-gel process for Tio2 synthesis: the peptization method to obtain crystalline and photoactive materials at low temperature. Catalysts 11(2):768
Margan P, Haghighi M (2016) Hydrothermal-assisted sol–gel synthesis of Cd-doped TiO2 nano-photocatalyst for removal of acid orange from wastewater. J Sol-Gel Sci Technol 81:556–569
McCartney JD, Ferrara ML, Jayakrishnan PV, Brachman RM (2024) Comparative life cycle assessment study for erosion control application with conventional shotcrete and geomat solutions. E3S Web Conf 569:09003
Medina-Ramírez I, Liu JL, Hernández-Ramírez A, Romo-Bernal C, Pedroza-Herrera G, Jáuregui-Rincón J, Gracia-Pinilla MA (2014) Synthesis, characterization, photocatalytic evaluation, and toxicity studies of Tio2–Fe3+ nanocatalyst. J Mater Sci 49:5309–5323
Miron-Barroso S, Domenech EB, Trigueros S (2021) Nanotechnology-based strategies to overcome current barriers in gene delivery. Int J Mol Sci 22(6):8537
Morelli E, Gabellieri E, Bonomini A, Tognotti D, Grassi G, Corsi I (2018) Tio(2) nanoparticles in seawater: aggregation and interactions with the green alga Dunaliella Tertiolecta. Ecotoxicol Environ Saf 148:184–193
Ponizovskiy MR (2022) Interactions environmental processes and development of human organisms activities via mechanisms of mental activity from the point of views of biophysics, biochemistry and thermodynamics. J Psychi Res Rev Rep 138:1–15
Muniandy SS, Kaus NH, Jiang ZT, Altarawneh M, Lee HL (2017) Green synthesis of mesoporous anatase Tio2 nanoparticles and their photocatalytic activities. RSC Adv 7:48083–48094
Nasrollahzadeh M, Sajjadi M, Iravani S, Varma RS (2021) Green-synthesized nanocatalysts and nanomaterials for water treatment: current challenges and future perspectives. J Hazard Mater 401:123401
Nizam NUM, Hanafiah MM, Woon KS (2021) A content review of life cycle assessment of nanomaterials: current practices, challenges, and future prospects. Nanomaterials (Basel) 11(12):3324
Olabarrieta J, Zorita S, Peña I, Rioja N, Monzón O, Benguria P, Scifo L (2012) Aging of photocatalytic coatings under a water flow: long run performance and Tio2 nanoparticles release. Appl Catal B: Environ 123–124:182–192
Oladipo AA, Mustafa FS (2023) Bismuth-based nanostructured photocatalysts for the remediation of antibiotics and organic dyes. Beilstein J Nanotechnol 14:291–321
Panchal P, Meena P, Nehra SP (2021) A rapid green synthesis of ag/Agcl-Nc photocatalyst for environmental applications. Environ Sci Pollut Res Int 28:3972–3982

Pesqueira JFJR, Pereira MFR, Silva AMT (2024) Carbon-based composites in advanced wastewater treatment: a life cycle assessment of Tio2 and go-Tio2 solar photocatalysis. J Clean Prod 444(1):140845

Rajput V, Minkina T, Sushkova S, Behal A, Maksimov A, Blicharska E, Ghazaryan K, Movsesyan H, Barsova N (2020) Zno and Cuo nanoparticles: a threat to soil organisms, plants, and human health. Environ Geochem Health 42:147–158

Ramakrishna D, Rao P (2011) Nanoparticles: is toxicity a concern? J Int Fed Clin Chem Lab Med 22:92

Reinsch BC, Levard C, Li Z, Ma R, Wise A, Gregory KB, Brown GE Jr, Lowry GV (2012) Sulfidation of silver nanoparticles decreases Escherichia coli growth inhibition. Environ Sci Technol 46:6992–7000

Rodríguez-Rojas MD, Bustos-Terrones V, Díaz-Cárdenas MY, Vázquez-Vélez E, Martínez H (2024) Life cycle assessment of green synthesis of Tio2 nanoparticles vs. chemical synthesis. Sustainability 16:7751

Roy H, Rahman TU, Khan MAJR, Al-Mamun MR, Islam SZ, Khaleque MA, Hossain MI, Khan MZH, Islam MS, Marwani HM, Islam A, Hasan MM, Awual MR (2023) Toxic dye removal, remediation, and mechanism with doped Sno2-based nanocomposite photocatalysts: a critical review. J Water Process Eng 54:104069

Sanoja-López KA, Loor-Molina NS, Luque R (2024) An overview of photocatalyst eco-design and development for green hydrogen production. Catal Commun 187:106859

Saravanan A, Kumar PS, Jeevanantham S, Anubha M, Jayashree S (2022) Degradation of toxic agrochemicals and pharmaceutical pollutants: effective and alternative approaches toward photocatalysis. Environ Pollut 298:118844

Sengul AB, Asmatulu E (2020) Toxicity of metal and metal oxide nanoparticles: a review. Environ Chem Lett 18:1659–1683

Shetty SS, Deepthi D, Harshitha S, Sonkusare S, Naik PB, Madhyastha H (2023) Environmental pollutants and their effects on human health. Heliyon 9:e19496

Shi Y, Yu M (2021) Assessing the environmental impact and cost of the tourism-induced Co2, Nox, Sox emission in China. Sustainability 13:604

Sivaraman C, Vijayalakshmi S, Leonard E, Sagadevan S, Jambulingam R (2022) Current developments in the effective removal of environmental pollutants through photocatalytic degradation using nanomaterials. Catalysts 12(5):544

Tahir MB, Sohaib M, Sagir M, Rafique M (2022) Role of nanotechnology in photocatalysis. In: Encyclopedia of smart materials. Elsevier, pp 578–589

Thoma T, Ma-Hock L, Schneider S, Honarvar N, Treumann S, Groeters S, Strauss V, Marxfeld H, Funk-Weyer D, Seiffert S, Wohlleben W, Dammann M, Wiench K, Lombaert N, Spirlet C, Vasquez M, Dewhurst N, Landsiedel R (2024) Toxicological inhalation studies in rats to substantiate grouping of zinc oxide nanoforms. Part Fibre Toxicol 21:24

Wang Y, He Y, Lai Q, Fan M (2014) Review of the progress in preparing nano Tio2: an important environmental engineering material. J Environ Sci (China) 26:2139–2177

Wu F, Zhou Z, Hicks AL (2019) Life cycle impact of titanium dioxide nanoparticle synthesis through physical, chemical, and biological routes. Environ Sci Technol 53:4078–4087

Wu L, Wen W, Wang X, Huang D, Cao J, Qi X, Shen S (2022) Ultrasmall iron oxide nanoparticles cause significant toxicity by specifically inducing acute oxidative stress to multiple organs. Part Fibre Toxicol 19:24

Yuju S, Xiujuan T, Dongsheng S, Zhiruo Z, Meizhen W (2023) A review of tungsten trioxide (Wo(3))-based materials for antibiotics removal via photocatalysis. Ecotoxicol Environ Saf 259:114988

Zhang N, Xiong G, Liu Z (2022) Toxicity of metal-based nanoparticles: challenges in the nano era. Front Bioeng Biotechnol 10:1001572

Zinatloo-Ajabshir S, Morassaei MS, Salavati-Niasari M (2019) Eco-friendly synthesis of Nd2sn2o7–based nanostructure materials using grape juice as green fuel as photocatalyst for the degradation of erythrosine. Compos Part B Eng 167:643–653

Chapter 11
Future Perspectives and Emerging Trends in Green Nanostructured Photocatalysis

Mohd Mazhar, Junaid Aman, Radha Khirwar, Mthokozisi Dladla, Soumya Ghosh, Gözde Koşarsoy Ağçeli, Sıla Hoşoğlu, Upasana Rana, and Kanika Dulta

Abstract Photocatalysis showed great potential as a solution to fossil fuel depletion and other clean energy source limitations. By employing photocatalysis for water splitting, hydrogen can be produced which can be stored and used for different applications, using photocatalysts and light irradiation. Similarly, to solve this problem, green nanomaterials (both as a broad class of materials and, in specific instances, as excellent photocatalysts owing to their semiconducting nature and narrow band gap) are becoming trending with well-efficient characteristics for sensing and concomitant destruction of harmful chemical contaminants from water and soil. The conventional (physical and chemical) methods of synthesis have some

---

M. Mazhar
Taqgene Training and Research Institute, Dehradun, Uttarakhand, India

J. Aman (✉) · R. Khirwar
School of Applied and Life Sciences, Uttaranchal University, Dehradun, Uttarakhand, India
e-mail: Junaidaman@uumail.in

M. Dladla
University of the Free State, Bloemfontein 9301, South Africa

Human Molecular Biology Unit (School of Biomedical Sciences), Faculty of Health Sciences, University of the Free State, Bloemfontein 9300, South Africa
e-mail: 2017383189@ufs4life.ac.za

S. Ghosh
University of the Free State, Bloemfontein 9301, South Africa

Natural and Medical Sciences Research Center, University of Nizwa, Nizwa 616, Oman

G. K. Ağçeli · S. Hoşoğlu
Department of Biology/Biotechnology, Hacettepe University, Ankara, Turkey

U. Rana
Uttaranchal School of Computing Sciences, Uttaranchal University, Dehradun, India

K. Dulta (✉)
Center of Advanced Innovation Technologies, VŠB-Technical University of Ostrava, 708 00, Ostrava-Por, Czechia, Ostrava, Czech Republic

C. M. Hussain, G. Hussain (eds.), *Green Nanostructured Photocatalysts for Clean Energy*, Clean Energy Production Technologies,
https://doi.org/10.1007/978-981-95-1618-6_11

limitations of toxicity and energy consumption. Therefore, green innovation (environmentally amicable, low-operating cost, and high efficiency) is required to be developed on account of ecological issues. A comprehensive overview has been presented about the different types of green-synthesized nanoparticles including plant extracts, polymer-based metal oxide, ionic liquids metal oxides, their applications in photocatalytic activities, and their mechanism of work under UV-visible light or sunlight illumination.

Keywords Nanomaterials · Photocatalyst · Sustainable · Environment · Process

11.1 Introduction

Energy consumption has increased in recent decades due to population growth and resource demands. Green nanostructure photocatalysis is an emerging cutting-edge technology used for clean energy that helps in sustainable development. Innovations in material synthesis, increased catalytic efficiency, and sustainable applications will contribute to major improvements in green nanostructured photocatalysis in the future. New developments show a move toward the use of environmentally friendly materials and techniques, which enhance photocatalytic performance and support worldwide sustainability goals. Nanoparticles (NPs) showing semiconducting properties, including light absorption, charge transfer, and a desirable electronic structure, are known as photocatalysts. Green synthesis in photocatalysis refers to the use of an environmentally friendly process to create photocatalysis. In the past, we used traditional methods that involved toxic chemicals and energy-intensive processes that impacted the environment in a negative manner. Green synthetic techniques, like employing microbial and plant extracts, are becoming more popular for creating environmentally friendly nanostructured photocatalysts like silver nanoparticles (AgNPs) and titanium oxide (TiO_2). Materials having single-atom catalysts have enhanced photocatalytic reaction efficiency, especially for CO_2 conversion and pollutant degradation. Many of the green chemistry principles are in line with photocatalysis, a technique that uses light energy to propel chemical reactions. Green chemistry reduces or mitigates the risks associated with chemical feedstocks, reagents, solvents, and products, thereby diminishing pollution at its source. Green chemistry technology involves various technologies that reduce or eliminate perilous chemicals used to clean up environmental toxicity. Green chemistry comprises 12 principles such as prevent waste, maximize atom economy, design safer chemicals and products, increase energy efficiency, etc. Green nanostructure photocatalysis focuses on multiple principles of green chemistry like using safer solvents and reaction conditions, using renewable feedstocks, designing less hazardous chemical synthesis, and using catalysts rather than stoichiometric reagents. There are various green synthesis methods to create photocatalysts like plants, microbes, solar energy, microwave (MW), ultrasound (US), mild reducing agents, and mild reaction conditions. To develop a new green nano-catalyst, a thermal technique was used from the sepals of waste tomato (WT) plants. This approach is renowned for being economical and environmentally friendly, adhering to the concepts of green chemistry. Mechanochemistry (grinding), photochemical reactions, ultrasonic or sonochemical techniques, microwave irradiation, and solvent-free conventional thermal heating are examples of environmentally friendly synthesis techniques. By emphasizing sustainable procedures, these methods minimize the amount of hazardous materials used in the production of chemicals while maintaining economic viability and the welfare of society. Using tamarind fruit extract as a chelating agent, this study describes the ecologically friendly sol-gel synthesis of magnetic photo-Fenton catalysts based on CuFe2O4 at various annealing temperatures (700, 800, and 900 °C). X-ray diffraction, scanning electron microscopy, nitrogen adsorption, Fourier-transform infrared spectroscopy, and vibrating sample magnetometry were used to characterize the catalysts. The environmentally friendly properties of the 5-hydroxyisoxazolidine synthesis process make it significant. A key advantage in contemporary chemical procedures is that they make use of non-toxic and non-hazardous reagents. This method is a sustainable choice for chemical synthesis since it produces as little non-toxic waste as possible and doesn't require metal catalysts. The high-entropy oxide TiZrHfNbTaO11 is being studied as a photocatalyst for hydrogen generation. Its bandgap is comparable to that of TiO_2 polymorphs, and the synergistic effects of cations with different electronegativities result in increased water splitting efficiency.

The main advantage of photocatalytic treatment is its ability to break down pollutants into mineral end products without moving them from one phase to another, which is typically the case with traditional treatment techniques. Photocatalysis must compete with other Advanced oxidation process (AOP) technologies in order to upgrade water treatment systems. One viable alternative for renewable energy sources like solar or wind is photocatalysis; the improvement of photocatalysts' characteristics by functionalization raises the possibility that they will eventually consist of multiple types of materials. Growing use of plant extracts and microbial techniques to create nanoparticles, improve crystallinity, and avoid agglomeration are examples of emerging trends. Photocatalysis has a promising future, particularly when domains are combined to improve environmentally friendly chemical processes. For the enhancement of photocatalysis to industrial applications, we have a main focus on understanding of mechanistic and reaction quantum. In order to create more effective systems and modify and transform energy storage, waste treatment, and synthetic chemistry, future research can expand our knowledge of photocatalytic systems and optimize their performance and reproducibility.

11.2 Green Synthetic Routes for Nanostructured Photocatalysts

The "green" synthetic routes towards these nanostructured photocatalysts aim for environmentally friendly and sustainable approaches providing low-energy paths and biogenic precursors with reduced eco-toxicological footprint. Techniques like hydrothermal and solvothermal synthesis utilize water or green solvents under controlled conditions to produce nanostructures without toxic by-products. On the other hand, biological methods utilize plants, algae, and microorganisms as bio-reducing and bio-stabilizing agents. Plant extracts containing polyphenols and flavonoids below a certain concentration reduce metal ions and act as a stabilizing agent for nanoparticles; algae and microorganisms synthesize nanoparticles using biopolymers and enzymes. These approaches produce biocompatible, stable, and functional photocatalysts applicable for water purification and solar energy conversion. Green synthesis methods allow for minimizing environmental impact by reducing energy consumption, avoiding the use of toxic solvents, and using renewable resources (Jadoun et al. 2021). Solvent-free and low-energy synthesis methods in mechanochemical synthesis are the main environmentally friendly ways to create nanostructured photocatalysts. To create chemical reactions without the need for solvents, solid reactants can be ground using ball milling or other mechanical forces. TiO_2 nanoparticle synthesis for photocatalysis in the presence of visible light, for example. Benefits include trash reduction, energy efficiency, and the use of no solvents. In decomposition by heat to create nanostructures, solid precursors are heated in air or under inert conditions. Study of a case: Zinc precursors are converted into ZnO photocatalysts in ambient circumstances. Hydrothermal synthesis involves producing controlled-morphology nanoparticles through reactions in water at high temperatures and pressures. $BiVO_4$ nanorods serve as efficient visible-light photocatalysts, utilizing water as an environmentally safe solvent, with tunable sizes and shapes that enhance their photocatalytic performance. Similar to hydrothermal synthesis, solvothermal synthesis makes use of less hazardous and biodegradable non-aqueous solvents like ethanol or ethylene glycol. Solvothermal production of TiO_2 nanospheres in ethanol for dye degradation is the case study. Biological synthesis uses microorganisms, plants, and algae. In plant-based synthesis, metal oxide nanoparticles are synthesized using plant extracts as stabilizing and reducing agents, e.g., ZnO nanoparticle synthesis employing leaf extract from *Azadirachta indica* (neem). Benefits include low cost, non-toxicity, and use of renewable resources. Because rich in proteins and polysaccharides, algae-mediated synthesis makes it easier to create nanoparticles (Khan et al. 2019). In microbial synthesis, microbials like *Aspergillus niger* and *Bacillus subtilis* use enzymes to create nanoparticles. TiO_2 nanoparticles produced for photocatalysis by *Aspergillus flavus* are the subject of this case study. *Chlorella vulgaris* is used in the green production of TiO_2 nanoparticles (Singh et al. 2018).

11.2.1 Solvent-Free and Low-Energy Synthesis Methods

These methods focus on eliminating the use of solvents and minimizing energy requirements during the synthesis of nanostructured photocatalysts, aligning with green chemistry principles. In mechanochemical synthesis, solid precursors are mechanically ground in a ball mill or other such apparatus. Nanoparticles are created when chemical reactions are triggered by mechanical energy. Benefits include high efficiency, simplicity and scalability, avoidance of hazardous solvents, and decreased waste and energy usage. Synthesis applications include perovskite-based photocatalysts, ZnO, and TiO_2. For example, zinc acetate and oxalic acid are used in the mechanochemical production of ZnO nanoparticles. The solid-state process reactions are induced by directly heating solid precursors to high temperatures. Liquid-phase synthesis is avoided with this technique. Benefits include ease of use, being solvent-free, and suitability for producing oxides and sulfides in large quantities. This is used in the manufacturing of photocatalysts such as FeO_3 or CuO_3. For example, bismuth-based precursors can undergo thermal breakdown to produce $BiVO_4$ for visible light photocatalysis. Green combustion synthesis creates nanoparticles without the need for solvents by inducing an exothermic reaction with less fuel and oxidants. Benefits include quick and efficient production of highly crystalline materials with applications of ZnO or CeO_2 nanoparticle synthesis for environmental cleanup. These methods demonstrate how solvent-free and low-energy synthesis aligns with green chemistry principles by reducing environmental impact while producing efficient photocatalytic materials (Chandrakala et al. 2022).

11.2.2 Hydrothermal and Solvothermal Green Synthesis Approaches

Hydrothermal and solvothermal synthesis are widely regarded as green methods for producing nanostructured photocatalysts. They employ water or green solvents under controlled temperature and pressure to create materials with precise morphologies and properties as shown in Table 11.1 (Faisal et al. 2021).

11.2.3 Combining Solvothermal and Hydrothermal Processes with Green Additives

By using green additives in solvothermal and hydrothermal processes including biopolymers, plant extracts, or other ecologically friendly components, these techniques raise the syntheses' green credentials. For example, TiO_2 nanoparticles are hydrothermally synthesized using aloe vera extract (Rui et al. 2024).

Table 11.1 Hydrothermal and solvothermal green synthesis approaches

Synthesis	Description	Advantages	Applications	Example	References
1. Hydrothermal synthesis	Aqueous-based synthesis conducted at elevated temperatures and pressures in a sealed system, often an autoclave. It uses water as the solvent and reaction medium	Water is a green and abundant solvent Allows control over particle size, shape, and crystallinity Energy-efficient and scalable	Fabrication of photocatalysts like TiO_2, $BiVO_2$, and ZnO Environmental applications such as pollutant degradation under visible light	Hydrothermal synthesis of $BiVO_2$ nanoparticles for visible-light-driven photocatalytic degradation of organic pollutants	Wang et al. (2015)
2. Solvothermal synthesis	Similar to hydrothermal synthesis but uses organic solvents such as ethanol, glycerol, or ethylene glycol instead of water. These solvents can provide better solubility for certain precursors and additional control over product morphology	Organic solvents can act as reducing and capping agents, reducing the need for additional chemicals. Tailored to produce unique structures like nanowires, nanorods, and hollow spheres Solvents like ethanol and glycerol are biodegradable and considered green	Synthesis of graphene oxide composites with metal oxides for enhanced photocatalytic performance Production of TiO_2 or ZnO nanoparticles for dye degradation	Solvothermal synthesis of TiO_2 nanoparticles using ethanol as the solvent for photocatalytic applications	Faisal et al. (2021)

11.2.4 Comparison of Hydrothermal and Solvothermal Methods

The comparison of hydrothermal and solvothermal procedures is very flexible and in line with the principles of green chemistry since it reduces the use of hazardous chemicals and enhances reaction conditions as shown in Table 11.2 (Li et al. 2021).

Table 11.2 Comparison of hydrothermal and solvothermal methods

Feature	Hydrothermal	Solvothermal
Solvent	Water	Organic solvents (e.g., ethanol)
Environmental impact	Very green (uses water)	Green if biodegradable solvents used
Morphological control	Moderate	High due to solvent properties
Applications	Oxides, sulfides, layered materials	Composites, doped materials

11.2.5 Biological Synthesis of Nanostructured Photocatalysts

Utilizing environmentally safe and sustainable methods, the biological production of nanostructured photocatalysts uses microbes, plants, and algae. But these biological systems act as natural reducing and stabilizing agents, thus eliminating the need for the use of toxic chemicals. The bioactive compounds (polyphenols, flavonoids, and alkaloids) present in plants help in metal ion reduction and act as capping agents for the stable formation of nanoparticles. Bio-capping has been hypothesized to lead towards enhanced photocatalytic activity because of stabilization and further coupling of several bi-molecules into nanoparticles that will help them degrade via the process of energy transfer. For instance, algae such as Chlorella vulgaris utilize their polysaccharides and proteins to facilitate the synthesis of nanostructures that offer stability and distinctive coatings with functional properties to improve photocatalytic performance. Lastly, microorganisms (bacteria and fungi) reduce metal ions via enzymatic activity, e.g., the action of nitrate reductases, while secreting different proteins, which act as capping agents to stabilize the nanoparticle formed and prevent aggregation. Not only does this method manufacture biocompatible photocatalysts, but it also provides adherence to green chemistry by effectively reducing the environmental footprint (Iravani 2011).

11.3 Mechanisms of Green Nanostructure Formation

Green synthesis techniques create nanostructures with little effect on the environment by using biological agents, low-energy procedures, or environmentally benign precursors. Natural or benign chemicals aid in the nucleation, growth, and stabilization processes that result in the development of nanostructures. Green nanostructure generation strategies rely on environmentally friendly methods that use natural stabilizing and reducing agents to create nanoparticles without the use of hazardous chemicals (Bai et al. 2018). In chemical methods, it is possible to isolate from plants the active compounds such as polyphenols, flavonoids, and alkaloids that act as reducing agents where they provide their electrons to metal ions to form nanoparticles while functional groups ensure the stability of the structures thus preventing aggregation. In hydrothermal and solvothermal processes, water or green solvents are used as the reaction solvent at a specific temperature and pressure for the nucleation and growth of nanostructures. In biological processes, bacteria, fungi, and algae are employed which contain an enzymatic system like nitrate reductase to reduce the metal ions and release proteins or polysaccharides for the stabilization of the nanoparticles. These bio-capping agents improve performance and, in conjunction, the structural integrity of the nanostructures. These mechanisms reduce environmental risks while developing high-performance nanostructures for catalysis, energy, and environmental applications. Green synthesis of nanostructures involves the use of extracts, microorganisms, renewable resource agents such as plants, and energy-efficient processes. They aim at reducing the waste and toxicity while at the same time having a high level of control over the morphology and functionality of the nanostructures (Pinto et al. 2020).

11.3.1 Chemical and Biological Mechanisms of Nanostructure Formation During Green Synthesis

The green synthesis of nanostructures takes advantage of chemical and biological processes which do not pose harm to the environment and are sustainable. Chemical mechanisms on the other hand are made of the reduction of metal ions to nanoparticles using green reducing agents, for example, plant extracts, sugars, or polyols. These substances are abundant in functional groups like hydroxyls, flavonoids, and phenols, which let metal ions receive electrons, forming nuclei and promoting the development of nanostructures. For instance, the polyphenols found in plant extracts are utilized in the synthesis of silver nanoparticles as a stabilizing agent and reductant to prevent the nanoparticles from adhering to one another. Similarly, safe processes like hydrothermal or solvothermal techniques and green solvents like water or ethanol provide a controlled environment devoid of hazardous consequences (Chen et al. 2021).

Since living organisms including bacteria, fungi, algae, and plants produce nanoparticles, they are vital to biological processes. Metal ion stability and reduction are mediated by bioactive compounds, including proteins, terpenoids, and alkaloids, which are present in plant extracts. Although bacteria such as *Bacillus subtilis* and *Aspergillus niger* use enzymes like nitrate reductase to enzymatically reduce metal ions, which transfer electrons for the nucleation of nanoparticles, algae use their proteins and polysaccharides as natural templates for the production of nanoparticles. In addition to simplifying the manufacturing process, these biological systems provide the nanoparticles with biocompatible and functional coatings. For example, the production of TiO_2 nanoparticles by algae shows enhanced stability and photocatalytic efficacy due to the bio-capture of algal components (Chaudhary et al. 2020).

11.3.2 Role of Bioactive Molecules in Reduction and Stabilization Processes

Bioactive chemicals are crucial for stabilizing and reducing metal ions during the ecologically friendly formation of nanostructures. These molecules, which are typically found in plant extracts, algae, and microbes, are responsible for stabilizing metal ions by encasing them to prevent them from aggregating or growing out of control in addition to reducing them into nanoparticles. Bioactive compounds such as proteins, polyphenols, flavonoids, terpenoids, and polysaccharides provide functional groups to stabilize the nanostructures and work as reducing agents as shown in Table 11.3 (Mandal et al. 2016).

Key Mechanisms Involved
Decreased with their reduction to their zero-valent state by the electrons provided by bioactive molecules, metal ions (such as Ag^+ and Au^{3+}) nucleate and form nanoparticles. By preserving stability, functional substances from bioactive molecules, like amino, carboxyl, and hydroxyl groups, stick to the nanoparticles' surface to increase stability and prevent agglomeration. These bioactive materials help to produce nanoparticles in an environmentally responsible manner and guarantee that the final nanostructures are biocompatible, environmentally benign, and have improved functionality for a range of applications, such as environmental remediation and catalysis (Jain 2022).

11.4 Case Studies: Green Synthesized Photocatalysts

11.4.1 Successful Examples of Plant-Based Syntheses

Green synthesis approaches offer a major innovation in the production of photocatalysts with environmentally friendly and sustainable methods (Mousa et al. 2024; Taghavi Fardood et al. 2024). Plant extracts are an effective tool in the reduction of metal ions and stabilization of nanoparticles with their rich content in natural compounds. Metal oxides and metal nanoparticles such as titanium dioxide (TiO_2), zinc oxide (ZnO), silver (Ag), and gold (Au) can be produced with this method (Das et al. 2020; Joseph and Vijayanandan 2021; Mirzaeifard and Shariatinia 2024; Rani et al. 2024).

The synthesis process usually consists of three basic steps. First, the herbal extract is prepared; for this, the plant leaves, fruits, flowers, roots, or barks are cleaned, and extraction is carried out with solvents such as water, ethanol, or methanol. During this process, phenols, flavonoids, tannins, and other phytochemicals in the extract are dissolved and contribute to the formation of nanoparticles. In the second stage, the prepared herbal extract is mixed with a solution containing metal ions (e.g., titanium nitrate, zinc nitrate, or silver nitrate). This mixture is kept at a

Table 11.3 Role of bioactive molecules in reduction and stabilization processes

Bioactive molecule	Source	Role in reduction	Role in stabilization	Examples of nanostructures synthesized	References
Polyphenols	Plant extracts (e.g., tea, neem)	Donate electrons to metal ions, reducing them to nanoparticles	Form a capping layer on the nanoparticle surface, preventing aggregation	Silver (Ag), gold (Au), copper (Cu) nanoparticles	Iravani (2011)
Flavonoids	Plants (e.g., citrus, onion)	Reduce metal ions through electron transfer, facilitating nanoparticle formation	Stabilize nanoparticles through functional groups such as hydroxyl groups, preventing growth	Silver (Ag), zinc oxide (ZnO) nanoparticles	Chan et al. (2022)
Proteins	Microorganisms (e.g., *Bacillus subtilis*)	Reduce metal ions through enzymatic activity, often involving nitrate reductase	Serve as natural stabilizing agents, preventing nanoparticle aggregation and enhancing biocompatibility	Silver (Ag), gold (Au), platinum (Pt) nanoparticles	Alprol et al. (2023)
Polysaccharides	Algae (e.g., *Chlorella vulgaris*)	Reduce metal ions through complexation and electron transfer mechanisms	Capping agent that stabilizes nanoparticles and prevent agglomeration	Zinc oxide (ZnO), Titanium dioxide (TiO_2) nanoparticles	Wang et al. (2015)
Terpenoids	Plant extracts (e.g., eucalyptus)	Reduce metal ions to nanoparticles by providing electrons	Prevent aggregation by acting as a stabilizing agent on the nanoparticle surface	Silver (Ag), copper (Cu) nanoparticles	Khan et al. (2022)

certain temperature and pH value to reduce the metal ions and form nanoparticles. In the last stage, the formed nanoparticles are filtered, washed, and generally stabilized by a heat treatment called calcination to develop their crystal structures (Khadka et al. 2024; Krishna et al. 2022) (Fig. 11.1).

Examples from different studies include the synthesis of zinc oxide (ZnO) nanoparticles with Neem (*Azadirachta indica*) leaf extract and the photocatalytic degradation of textile dyes (Rani et al. 2024). Similarly, titanium dioxide (TiO2) nanoparticles were produced using aloe vera extract and found to be effective in the

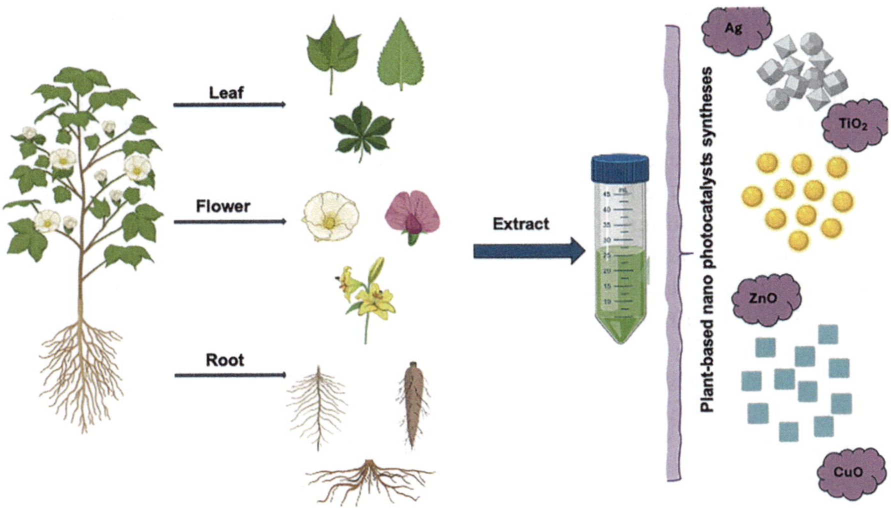

Fig. 11.1 Plant-based nano photocatalyst syntheses

degradation of organic pollutants in industrial waste under UV light (Dülger et al. 2024). Silver nanoparticles (AgNPs) were synthesized from ginger (*Zingiber officinale*) (Mehrotra et al. 2024) and tea plant (*Camellia sinensis*) (Banumathi et al. 2017) extracts, and these particles were used in water disinfection. In addition, nanoparticles produced with *Moringa oleifera* extract were found to be effective in heavy metal removal, while ZnO nanoparticles synthesized with lemon grass (*Cymbopogon citratus*) showed photocatalytic activity under UV light and antimicrobial activity against food pathogens (Ramzan et al. 2022). Iron oxide (Fe_2O_2) nanoparticles were produced using tea plant (*Camellia sinensis*) extract, and these nanoparticles were found to be effective in the removal of organic pollutants in water treatment and photocatalytic applications (Anwer et al. 2022). Copper oxide (CuO) nanoparticles were synthesized with *Moringa oleifera* extract, and these photocatalysts were successful in environmental applications such as water treatment and heavy metal removal (Das et al. 2020). Titanium dioxide (TiO_2) nanoparticles were synthesized with lavender (*Lavandula angustifolia*) extract and were effective in air purification applications (Rashed et al. 2019). Finally, silver nanoparticles (AgNPs) were produced with clove (*Syzygium aromaticum*) extract, and these nanoparticles were used in water disinfection with their antibacterial and antifungal properties (Makky et al. 2023). These studies show that plant extracts offer an environmentally friendly and economical solution for photocatalyst synthesis. These studies show that plant-based syntheses offer both environmental and economic advantages.

11.4.2 Biopolymer and Metal Nanoparticle Combinations

Composites of biopolymers and metal nanoparticles are attracting attention as environmentally friendly, biodegradable, and highly efficient materials in photocatalytic applications. These combinations combine the photocatalytic activity of metal nanoparticles with the biocompatibility and environmentally friendly properties of biopolymers, providing effective solutions in areas such as the removal of environmental pollutants and water treatment. For example, the combination of chitosan and silver nanoparticles (AgNPs) has achieved successful results in water disinfection and pathogen removal applications with its antibacterial properties (Badawy et al. 2019; Korkut et al. 2024; Rezazadeh et al. 2020). Alginate and zinc oxide (ZnO) nanoparticles have been found to be effective in water treatment and photocatalytic degradation of organic pollutants, and alginate increased the photocatalytic activity of ZnO nanoparticles by providing stabilization (Hmmam et al. 2023; Siddiqui et al. 2022; Yao et al. 2024). In addition, composites formed by chitosan and titanium dioxide (TiO_2) nanoparticles provide a biocompatible environment in photocatalytic degradation processes and have been used especially in water treatment and elimination of environmental pollution (Castillo et al. 2022; Dhanya and Aparna 2016; Halawa et al. 2022; Spoială et al. 2022). Agar and zinc nanoparticles (ZnNPs) have been successful in microbial disinfection and pollutant removal applications by accelerating photocatalytic reactions (Kumar et al. 2019). *Moringa oleifera* extract and copper oxide (CuO) nanoparticles have been effective in heavy metal removal and also provided environmentally friendly and biodegradable composites (Jadhav et al. 2022). The combination of cellulose and silver nanoparticles (AgNPs) has achieved successful results in water disinfection and photocatalytic degradation of organic pollutants (Hanif et al. 2021; Jadhav et al. 2022; Madivoli et al. 2022; Zeng et al. 2023). Such biopolymer-metal nanoparticle composites increase the efficiency of photocatalytic processes and offer environmentally friendly solutions thanks to their biodegradability.

11.4.3 Nanostructures Obtained from Environmentally Friendly Synthetic Processes

Green synthesis photocatalysts and nanostructures obtained from environmentally friendly synthetic processes are an important area of research in terms of developing more efficient and non-toxic photocatalytic materials by minimizing environmental impacts. While traditional photocatalysts usually require hazardous chemicals and high energy input, green synthesis methods reduce environmental impact by using renewable resources such as plant extracts, microorganisms, and natural polymers. These green synthesis nanostructures have unique properties such as high surface area, tunable morphology, and stability and are highly effective in applications such as environmental remediation, energy production, and catalysis.

Bimetallic nanostructures developed by green synthesis methods are also quite promising. For example, silver-gold (Ag-Au) alloy nanoparticles were synthesized using cinnamon bark extract. Cinnamon extract reduces both silver and gold ions and provides stabilizing agents, which ensure homogeneous sizing and high surface area of the nanoparticles (Anjum et al. 2019; Granja Alvear et al. 2024; Premkumar et al. 2018). These bimetallic nanoparticles increase photocatalytic activity thanks to the interaction between silver and gold and show high performance even under visible light. With these properties, they are extremely suitable for energy production and environmental cleaning applications.

Another remarkable example is the green synthesis of iron oxide (Fe_2O_3) nanostructures using bamboo leaf extract. Bamboo extract contains polyphenolic components, which reduce iron salts and form highly stable Fe_2O_2 nanoparticles. Fe_2O_2 nanoparticles can be effective in water purification, especially in the removal of dyes and heavy metals. Iron oxide nanoparticles also have magnetic properties, which allow them to be easily separated from the reaction medium and make them suitable for continuous use in large-scale applications (Fatimah et al. 2019; Salgado et al. 2024).

Photocatalysts and nanostructures obtained by green synthesis offer many advantages, such as cost-effectiveness, easy synthesizability, and the use of environmentally friendly materials. In addition, these materials are generally more biocompatible and safer than those synthesized by traditional chemical methods (Aliannezhadi et al. 2024; Ying et al. 2022). Therefore, green synthesis photocatalysts and nanostructures have great potential in a wide range of applications such as environmental pollution removal, energy production, and sustainable industrial processes. In the future, optimization of the properties of these materials by controlled synthesis, further exploration of different plant-based and natural resources, and development of hybrid materials consisting of a combination of organic and inorganic components can further improve their performance in this field (Abuzeid et al. 2023; Chakroborty et al. 2022; Osman et al. 2024).

11.5 Advantages of Green Synthesis in Photocatalysis

The conduction of green synthesis in photocatalysis is both a sustainable approach and an environmentally friendly way, and provides transformed advantages to both chemical production and environmental remediation. This approach uses renewable reagents (plant extract) and proceeds under mild conditions that consume minimal energy (Mancuso and Iervolino 2022). Green photocatalysis is a unique field of synthesis that utilizes the principles of green chemistry to promote reactions by using very few toxic reagents and generating little waste compared to traditional synthesis. A recent study suggests its utility for synthesizing photocatalysts such as titanium dioxide (TiO_2) and silver-based nanomaterials, exhibiting enhanced capabilities for pollutant degradation and antimicrobial activity (Verma et al. 2022). Also, the economic advantages of green synthesis are significant. These processes

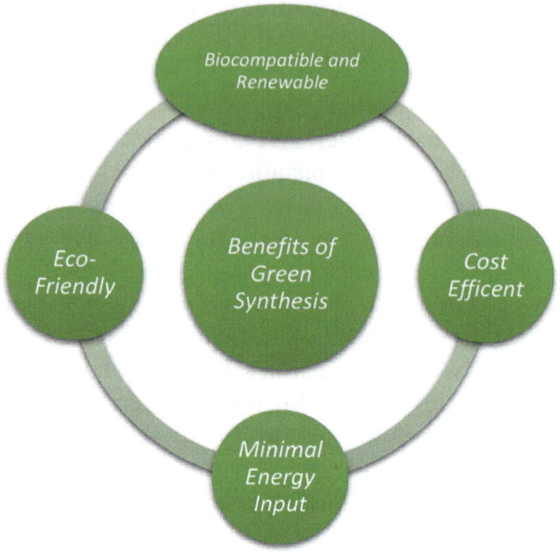

Fig. 11.2 Energy and resource savings compared to traditional synthesis methods

use abundant and renewable biological feedstock including (but not limited to) plant biomass and energy-efficient processes leading to lower production costs coupled with compatibility with industrial scales of production (Fig. 11.2).

Ultimately, green synthesis in photocatalysis should be viewed not simply as an alternative route to traditional approaches, but rather a quantifiable change in the pursuit of industrial-scale, energy-capturing, pollution-reducing sustainable methods applicable to real-world environmental processes. This field is still in development, and if pursued, there is great potential for alleviating global sustainability objectives.

The green synthesis plays a paradigm shift in photocatalysis towards a more sustainable approach than conventional methods from the point of view of environmental, economic, and energy perspectives. Traditional synthesis approaches face significant sustainability challenges, in large part because of the high process mass intensity (PMI), which is a measure of how much resources are used compared to the yield of the final product. Much of this stems from the use of organic solvents in both the work-up and purification steps. These solvents are typically associated with a high environmental impact such as toxicity and waste production. Many photocatalyst syntheses are energy-intensive and also involve rare and expensive metals, adding to the environmental and economic burden. In addition, the waste produced during photocatalyst synthesis also has a significant impact on the total sustainability of catalytic protocols, despite the low catalyst loading typically employed. Catalysts with high PMI numbers can wildly escalate the waste burden of a reaction. For example, solvent use, specifically in post-synthesis purification, is recognized as the greatest waste producer (Quintavalla et al. 2024). Green synthesis addresses these drawbacks, utilizing renewable materials like plant extracts, biomolecules, or non-toxic precursors at mild reaction conditions that dramatically lower the environmental impact.

Example of direct use of eco-friendly precursors like plant extracts in the synthesis of photocatalyst, e.g., TiO_2 (Mancuso and Iervolino 2022). They are used as reducing agents to eliminate toxic solvents and enhance the efficiency of photocatalysts under visible light irradiation. Such an approach is consistent with the principles of green chemistry, especially waste prevention and using safer, renewable feedstocks.

Moreover, a new study emphasizes the more sustainable ability of green synthesis methods to optimize the performance of photocatalysts while avoiding significant resource wastage. These also guarantee scalability without loss of efficiency through the use of low-energy processes, like hydrothermal synthesis or solvent-free grinding. The researchers also mention that green-synthesized photocatalysts have higher stability and reusability than those synthesized via traditional routes, which adds to their sustainability (Mancuso and Iervolino 2022). Another study notes that bio-template–developed doped photocatalysts exhibited higher degradation efficiency of pollutants like pharmaceuticals and dyes, under ambient conditions (Aba-Guevara et al. 2017).

There are huge economic profits from green synthesis as well. This method reduces production costs by avoiding expensive and hazardous chemicals, and its scalability ensures the suitability of the method for industrial applications (Lin and Wang 2019). Importantly, synthesis via renewable resources (as agricultural waste or plant derivatives) promotes circular economy approaches, where waste material is transformed into high-value products.

11.5.1 Reduction of Use of Toxic Chemicals

Photocatalysis is a potent method for conducting green and sustainable chemical transformations. It has the potential to cause a major reduction in the usage of toxic chemicals when used in conjunction with green synthesis approaches. Traditional chemical synthesis typically starts and speeds reactions using toxic solvents, hazardous reagents, and heavy metals. They not only pollute our environment but also affect human health. In comparison, photocatalysis facilitates chemical conversions stimulated by light which is considered much safer and environmentally friendly (Briffa et al. 2020). In this light, photocatalysis serves as a good model of the "green chemistry" ideas because it avoids toxic chemicals.

Photocatalysis-driven reactions make use of light energy as a driving force for either the chemical reduction or oxidation of a synthetic substance. These processes can be done without solvents or toxic chemicals. Photocatalytic reactions are usually initiated using abundant and biocompatible reagents such as water and oxygen. This does not require toxic reagents to degrade organic pollutants and other toxic compounds and, as such, is highly advantageous for the environment. Traditional chemical reactions use solvents and harmful substances; on the other hand, photocatalysis depends less on these by-products (Szczepanik 2017). TiO_2 (titanium dioxide), as an example of a photocatalyst, can photo catalytically degrade organic pollutants under light conditions, significantly reducing hazardous chemical usage and waste generation. Due to the environmentally friendly feature of TiO_2, these reactions are much safer and efficient (Virkutyte and Varma 2013).

One of the most important aspects of photocatalysis is its functionality under low-energy and green conditions. In contrast, photocatalytic reactions can use solar energy, ultraviolet light, or other low-energy light sources instead of relying solely on high-temperature and high-pressure traditional processes. It also reduces the needed energy, as well as the need for other dangerous chemicals and reagents. For instance, photocatalytic organic synthesis can replace toxic solvents and hazardous chemicals with water or biologically compatible materials. This approach minimizes the negative influences on the environment and economic costs (Grätzel 2001).

Nanoparticles are crucial in the enhancement of photocatalytic efficiency, which helps speed up chemical reactions under the influence of light energy and decreases the amount of toxic chemicals required. Nanotechnology advances can make photocatalysts more potent—accelerating reactions and increasing efficiency at lower energy thresholds. For example, metal oxide nanoparticles can harness the energy of light to trigger reactions, eliminating the toxic solvents and reagents commonly employed in classical chemical processes. One of the main targets of such nanoparticles is TiO_2, which is extensively used in photocatalytic applications for the environmentally friendly decomposition of the pollutants that photocatalytic advances should involve substituting harmful chemistries and solvents for more benign alternatives.

Green synthesis combined with photocatalysis works for sustainability in production processes. Conventional chemical synthesis gives rise to toxic materials and pernicious waste, which have posed significant environmental issues. In contrast, reactions that occur via photocatalysis limit the generation of harmful by-products. This minimizes the environmental footprint of the chemical production process. As an example, the generation of harmful by-products and by-products during photocatalytic degradation of organic pollutants is effectively reduced. Photocatalytic cells facilitate the implementation of processes such as energy generation and contaminant degradation in an environmentally benign manner (Grätzel 2001).

11.5.2 Sustainability and Carbon Footprint Assessments

Photocatalysis can use light energy, especially sunlight, to drive chemical reactions, which is one of its major advantages. In contrast to conventional methods operating at high-temperature or high-pressure conditions, photocatalysis functions under ambient conditions, which considerably reduces energy requirements. For example, Zhao et al. 2024 claimed that up to %60 of carbon emissions were saved by the photocatalytic process irradiated by solar energy compared with the traditional methods (Villa et al. 2021; Zhao et al. 2024).

A prime example is water splitting to generate hydrogen via photocatalysis. This approach harnesses light energy to produce hydrogen without emitting greenhouse gases, a step toward carbon-neutral fuel. Grätzel (2001) concluded that hydrogen generated by solar energy can substitute fossil fuels and significantly decrease the global CO_2 emissions.

Another advantage of photocatalysis is its availability for CO_2 reduction. These photocatalytic systems are capable of converting CO_2 to value-added products such as methanol, methane, or formic acid, simultaneously capturing carbon emissions and providing a distinctive source of revenue. Studies pointed out that these types of CO_2 conversion processes could contribute towards a more circular economy by re-routing carbon from the environment to new production cycles instead of being emitted into the atmosphere.

The sustainability of photocatalysis has also been greatly improved through material science innovations. As new materials like g-C_2N_2, doped titanium oxides, or MOFs have developed, they have shown an increased efficiency by absorbing light in an expanded range, incorporating at least some visible light. These developments have been shown to lower energy needs and increase yields while minimizing resource inputs (Meng et al. 2024).

Research comparing traditional synthetic methods with photocatalytic green synthesis demonstrates the advantages of photocatalytic green synthesis for environmental purposes. An LCA study led to the conclusion that photocatalytic approaches or processes coupled with solar energy yield 30%–70% lower CO_2emissions than the respective conventional approaches or processes depending on both the type of reaction and the applied scale. These results illuminate a pathway to incorporate photocatalysis within industrial practices to reach sustainability objectives.

Although photocatalysis offers several advantages, it also poses challenges, including the requirement for efficient catalysts in addition to the scalability issue. These are all things that will need to be addressed, and more research and development are required. Grätzel (2001) indicates that the solar-to-hydrogen conversion efficiency must be increased and the photocatalyst must be less costly for the industrial approach. Moreover, coupling photocatalysis to renewable energy grids would be an extra step to benefit from its sustainability advantages (Michael Grätzel 2001).

11.6 Challenges and Limitations in Green Synthesis

11.6.1 Homogeneity and Controllability Issues

Although green synthesis methods have great potential in the production of environmentally friendly and sustainable photocatalysts, these processes also have some difficulties and limitations. Homogeneity and controllability issues are one of the main obstacles faced by green synthesis. While the sizes and morphologies of nanoparticles can usually be precisely controlled in traditional chemical synthesis methods, biological sources and natural compounds used in green synthesis can make such control processes difficult. For example, biological agents obtained from plant extracts or microorganisms can affect the reduction and growth processes of nanoparticles, but these interactions are usually unpredictable. This can lead to heterogeneity in the size and distribution of the resulting nanomaterials (Amrillah et al. 2022; Venkatesan et al. 2024).

As an example, in the synthesis of silver nanoparticles (AgNPs) using Aloe vera extract, the variety of components of the extract used can cause the size and shape of the nanoparticles to change. Different flavonoids, terpenoids, and phenolic acids in aloe vera react at different speeds during the growth process of nanoparticles, which leads to the formation of heterogeneous structures. Similarly, when titanium dioxide (TiO_2) nanoparticles are synthesized using green tea extract, variations in the composition of the extract can affect the morphology of TiO_2 nanoparticles, and as a result, it can be difficult to obtain a homogeneous material with the desired properties (Dikshit et al. 2021; Kazemi et al. 2023).

Such homogeneity problems can lead to a decrease in the performance of nanostructures in their application areas. For example, in photocatalytic applications, nanoparticles are expected to have a certain size and a smooth surface. Heterogeneous structures can negatively affect photocatalytic processes such as light absorption, electron transfer, and reactive oxygen species (ROS) production. In addition, the inhomogeneity of the size and surface properties of nanoparticles can lead to a decrease in efficiency in environmentally friendly applications (e.g., water treatment) (Kalpana and Devi Rajeswari 2018; Radulescu et al. 2023).

The controllability problem is another important challenge in green synthesis. Especially when plant extracts or biological sources are used, it is difficult to have full control over the synthesis conditions (temperature, pH, time, concentration, etc.). These conditions can vary depending on the content of biological agents and active ingredients used, which can lead to inconsistencies in the production of nanomaterials. For example, when synthesizing zinc oxide (ZnO) nanoparticles using *Moringa oleifera* extract, changes in the extract content (e.g., seasonal factors or harvesting conditions) can affect the morphological properties of the nanoparticles, which can make the process difficult to repeat (Dikshit et al. 2021; Kalpana and Devi Rajeswari 2018; Venkatesan et al. 2024).

In addition, biological agents used in green synthesis usually have slow reduction rates, which can make it difficult to control the size of the nanoparticles. The application of such processes on an industrial scale requires the development of more complex processes to ensure homogeneity while producing large quantities of material. Therefore, the characterization of biological and natural sources used in green synthesis, optimization of synthesis conditions, and development of new methods for the production of homogeneous nanoparticles are among the most important research topics in this field (Amrillah et al. 2022; Dikshit et al. 2021; Kazemi et al. 2023).

11.6.2 Scaling Challenges for High-Yield Production

Although green synthesis methods have great potential to obtain environmentally friendly and sustainable nanomaterials, the scaling-up process for high-throughput production involves significant challenges and limitations. The application of these methods from laboratory scale to industrial scale can make it difficult to increase the efficiency and economic sustainability of certain processes. Since natural materials such as biological sources, plant extracts, or microorganisms are used in green synthesis, the supply, quality, and homogeneity of these materials can be a major obstacle in scaling up the production process (Chakroborty et al. 2022; El-Moslamy et al. 2023; Singh et al. 2023). For example, in the synthesis of silver nanoparticles (AgNPs) using aloe vera extract, the variability of the components in the extract content and biological factors (seasonal changes, harvesting conditions, etc.) can lead to inconsistencies when producing large quantities of material. This can affect the stability of large-scale production.

Another important challenge is that biological agents used in green synthesis usually have low production rates. In syntheses with plant extracts or microorganisms, the degradation processes of nanoparticles are usually slow, which can make it difficult to produce large quantities of material with high yields. For example, when synthesizing zinc oxide (ZnO) nanoparticles using *Moringa oleifera* extract, the slow reaction rate can extend the time for large-scale production, which can negatively affect efficiency. In addition, since the components of plant extracts are naturally variable, it may be difficult to achieve the same quality and efficiency every time with the same extract (Noor Javed et al. n.d.; Sundararajan et al. 2024).

In order for photocatalysts obtained through green synthesis to show high efficiency, especially in photocatalytic applications, the size and morphology of the nanoparticles must be kept within certain limits. However, in large-scale production, it may be difficult to obtain homogeneous particle sizes and shapes. However, controlled conditions applied at the laboratory scale may not be applicable at the industrial scale. For example, when synthesizing titanium dioxide (TiO_2) nanoparticles using green tea extract, while high photocatalytic performance can be achieved in small-scale synthesis, it may be difficult to control the size of the nanoparticles during large-scale production. The inhomogeneity of the nanoparticles may reduce their photocatalytic activity and reduce application efficiency (Amrillah et al. 2022; El-Moslamy et al. 2023; Venkatesan et al. 2024).

In addition, the process optimization required for the application of green synthesis methods on an industrial scale is also a major challenge. The quantity and quality of plant extracts required for nanoparticle production may change over time, which may create problems in the supply of a continuous source. In addition, the cost of processes and materials used in the production of biological agents may also pose a major obstacle on an industrial scale. For example, although syntheses using plant extracts or microorganisms are more economical and environmentally friendly than chemical syntheses, the supply and processing of these biological agents may increase costs in high-volume production (Chakroborty et al. 2022).

As a result, in order to apply green synthesis methods to large-scale production, many factors such as the reliability of biological resources, optimization of synthesis conditions, and acceleration of the production process must be carefully considered. Overcoming these problems is an important step for the efficient and sustainable production of green-synthesized nanomaterials on an industrial scale. At this point, better control of processes, increasing the effectiveness of biological agents, and developing appropriate technologies for large-scale production are the most important research topics in this field (El-Moslamy et al. 2023; Singh et al. 2023; Sundararajan et al. 2024; Thoma et al. 2020).

11.6.3 Sustainable Management of Natural Resources

Although green synthesis methods are an important step in the development of environmentally friendly and sustainable production processes, they face some serious challenges and limitations in the sustainable management of natural resources. In these methods, nanoparticles and other materials are usually synthesized using plant extracts, microorganisms, or other biological agents (Chakroborty et al. 2022; Singh et al. 2023). However, when it is considered that these biological resources must be supplied and managed sustainably, some important problems arise. First, excessive use of natural resources can threaten biodiversity. For example, green synthesis processes using large amounts of plant extracts can cause the depletion of those plants or disrupt the ecosystem balance. High amounts of plants such as aloe vera and *Moringa oleifera* can lead to the destruction of natural habitats and the depletion of these plants. In addition, excessive consumption of soil, water, and other resources required for the growth of these plants can also create environmental pressure. This jeopardizes the sustainability of biological resources and may make it difficult to supply such resources in the future (Sundararajan et al. 2024).

Another important problem is the quality and continuity of supply of biological resources used in green synthesis. Biological agents may vary depending on external factors such as seasonal changes, climatic conditions, and geographical factors. For example, in syntheses with plants such as green tea extract or ginseng, the content of these plants may vary throughout the year, which may lead to fluctuations in the quality of the resulting nanoparticles. The biodiversity and concentration of plant extracts can directly affect the efficiency of synthetic processes, which in turn may negatively affect the sustainability of natural resources (Singh et al. 2023; Thoma et al. 2020).

However, one of the difficulties encountered in the transition of green synthesis methods to large-scale production is the logistical difficulties experienced in the supply of biological resources. The sustainable supply of natural resources is not only an ecological but also an economic problem. In particular, although the acquisition and processing of biological resources is generally easy at the laboratory scale, serious costs and logistical difficulties may be experienced in the supply of these resources in large-scale production. In addition, most of these types of biological resources are limited to certain geographical regions, which leads to limited regional accessibility of resources. This may limit the applicability of green synthesis methods on a global scale and prevent the efficient use of natural resources (Sundararajan et al. 2024).

For the sustainable management of biological agents used in green synthesis, alternative resources and recycling methods need to be developed. For example, utilizing waste biomass or organic waste obtained from the agricultural industry for green synthesis can increase the sustainability of natural resources. However, the biological synthesis capacity of microorganisms can be increased using biotechnological approaches, and more efficient synthesis processes can be developed. However, such alternatives must also be economically viable and environmentally sustainable. In addition, in order to ensure the sustainability of green synthesis methods, it is important to manage the procurement process of biological resources in a local and sustainable manner. This can encourage local communities to use natural resources in harmony with the ecosystem and minimize environmental damage (Debref 2012; Tsibulnikova et al. 2016).

In conclusion, although green synthesis methods promise to use natural resources more efficiently and in an environmentally friendly way, there are serious challenges in the sustainable management of natural resources. In order to overcome these problems, further research and strategies should be developed to sustainably procure biological resources, use alternative resources, and minimize environmental impacts (Chakroborty et al. 2022; Singh et al. 2023).

11.7 Emerging Green Technologies in Nanostructure Synthesis

11.7.1 Microwave and Ultrasound-Assisted Green Synthesis Methods

These innovative methods offer rapid and efficient synthesis of nanostructures using green chemistry principles. Microwave-assisted synthesis utilizes microwave radiation to heat the reaction mixture, leading to uniform and controlled heating, which enhances the reaction rates and product quality (Ren et al. 2022). This method significantly reduces reaction times and energy consumption compared to conventional heating methods. Microwave-assisted synthesis also allows precise control over reaction parameters, resulting in high-purity nanostructures with tailored properties (Kumar et al. 2020). Additionally, microwave-assisted synthesis can be conducted in solvent-free conditions or with minimal use of environmentally benign solvents, further aligning with green chemistry goals (Kumar et al. 2020). Ultrasound-assisted synthesis employs high-frequency sound waves to induce cavitation, promoting the formation of nanostructures through high-energy collisions and localized hot spots (Ahmed and Mohamed 2024). The cavitation process generates extreme temperature and pressure conditions in localized regions, facilitating the formation of unique nanostructures, such as hollow nanoparticles, nanorods, and nanoflowers, that may not be achievable through conventional methods (Wang et al. 2023). This technique enhances mass transfer and mixing, leading to uniform particle size distribution and improved reaction efficiency. Furthermore, ultrasound-assisted synthesis can be performed at ambient temperatures and pressures, reducing the need for harsh reaction conditions and hazardous chemicals, thus minimizing the environmental impact (Behal et al. 2024). Both methods represent a shift towards more sustainable and eco-friendly approaches in nanomaterial synthesis. They not only improve the efficiency and quality of the nanostructures produced but also reduce the ecological footprint of the synthesis process by minimizing the use of toxic reagents and solvents (Adhi et al. 2024). As the field of nanotechnology continues to evolve, these green synthesis methods hold great promise for developing advanced materials with applications ranging from medicine and electronics to environmental remediation and energy storage (Table 11.1).

11.7.2 Plasma-Assisted Environmentally Friendly Processes

Plasma technology provides an eco-friendly alternative for nanostructure synthesis. By generating reactive species at low temperatures, plasma-assisted methods enable the synthesis of nanomaterials with precise control over size, shape, and composition (Shah et al. 2024). These processes reduce the need for hazardous chemicals, such as solvents like benzene and toluene or reagents like sulfuric acid and formaldehyde, and minimize energy consumption, aligning with green chemistry principles (Shah et al. 2024). Traditional synthesis methods often rely on toxic solvents and reagents, which pose environmental and health risks. In contrast, plasma-assisted processes can utilize benign gases such as oxygen, nitrogen, or hydrogen, thereby minimizing the generation of harmful by-products and reducing the overall ecological footprint of the synthesis process (Shah et al. 2024). Plasma-assisted synthesis involves using plasma, an ionized gas consisting of ions, electrons, and neutral species, to drive chemical reactions (Zhang et al. 2023). This method offers several advantages over conventional synthesis techniques. For instance, reactive species in plasma are generated at relatively low temperatures and shorter reaction times than traditional thermal processes (Lamichhane et al. 2022). This efficiency prevents the thermal degradation of sensitive materials. It allows for synthesizing nanostructures that may be unstable at higher temperatures with reduced energy consumption, contributing to cost savings and sustainability (Lamichhane et al. 2022). This low-temperature process is particularly beneficial for producing nanomaterials with high surface areas and enhanced catalytic properties. Additionally, plasma-assisted methods offer precise control over the synthesis environment, including gas composition, pressure, and power input, allowing for fine-tuning nanostructure properties (Li et al. 2023). This level of control enables the production of nanomaterials with uniform size distribution, well-defined shapes, and specific compositions, which are critical for various applications such as catalysis, sensors, and electronic devices (Table 11.1). Furthermore, plasma technology can be integrated with other green synthesis techniques, such as microwave or ultrasound-assisted methods, to enhance reaction rates and material properties (Verma et al. 2024). This synergistic approach can lead to the development of novel nanomaterials with superior performance and environmental benefits.

11.7.3 New Bioactive Agents That Enhance Photocatalytic Performance

Incorporating bioactive agents, such as natural extracts and biopolymers, into nanostructured photocatalysts can significantly enhance their performance (Farajinejad et al. 2024). These bioactive agents can improve the photocatalysts' light absorption, charge separation, and stability, leading to higher efficiency in environmental and energy applications (Farajinejad et al. 2024). Bioactive agents derived from natural sources, including plant extracts, enzymes, and other biological molecules, offer unique properties that can be leveraged to enhance photocatalytic processes. For instance, plant extracts containing flavonoids, tannins, and other phenolic compounds can act as sensitizers, increasing visible light absorption and extending the photocatalysts' photo response range (Khatami and Iravani 2021). This enhanced light absorption enables the utilization of a broader spectrum of sunlight, thereby improving the overall photocatalytic efficiency (Khatami and Iravani 2021). Biopolymers, such as chitosan, cellulose, and alginate, can also play a crucial role in enhancing photocatalytic performance (Sohouli et al. 2022). These biopolymers can form stable composites with photocatalytic materials, providing a supportive matrix that enhances charge separation and prevents recombination of electron-hole pairs. This improved charge separation is essential for increasing the generation of reactive species that drive photocatalytic reactions (Sohouli et al. 2022). Additionally, biopolymers can enhance photocatalysts' structural integrity and durability, making them more resilient under operational conditions (Mittal et al. 2024). Furthermore, bioactive agents can introduce functional groups and active sites that facilitate the adsorption and activation of reactant molecules on the surface of the photocatalysts (Feng et al. 2022). This can lead to more efficient degradation of pollutants in water and air purification applications and improved conversion efficiencies in solar fuel production and other energy-related processes (Table 11.1). The environmental benefits of using bioactive agents in photocatalysis are also significant. These agents are often biodegradable, non-toxic, and derived from renewable sources, aligning with green chemistry and sustainability principles (Omran and Baek 2022). Their use reduces reliance on synthetic chemicals and minimizes the generation of hazardous by-products, contributing to more eco-friendly photocatalytic systems. Moreover, the synergy between bioactive agents and nanostructured photocatalysts can develop hybrid materials with enhanced properties. For example, integrating bioactive molecules with semiconductor nanoparticles can produce photocatalysts with superior light-harvesting capabilities and increased catalytic activity (Mishra et al. 2023). Such hybrid materials have the potential to outperform conventional photocatalysts, offering innovative solutions for addressing environmental challenges and advancing clean energy technologies (Table 11.4).

11.8 Future Directions in Green Synthesis of Photocatalysts

11.8.1 Development of Hybrid Green Methods

The future of green synthesis lies in developing hybrid methods that combine multiple green technologies to optimize the synthesis process. These methods aim to leverage the strengths of different approaches, such as combining microwave-assisted and plasma-assisted techniques, to achieve superior photocatalytic properties. Hybrid green methods represent an innovative frontier in materials science and nanotechnology. Researchers can capitalize on the synergistic effects of combining physical and chemical processes by integrating various environmentally friendly synthesis techniques. For instance, microwave-assisted synthesis provides rapid and uniform heating, which can enhance reaction rates and product uniformity (Goyal et al. 2022). In contrast, plasma-assisted synthesis generates highly reactive species at low temperatures, which can drive reactions that would otherwise require harsh conditions (Joshi and Loganathan 2024).

Microwave-assisted and plasma-assisted hybrid techniques can be compelling. The microwave component ensures efficient and homogeneous heating of the reaction mixture, reducing reaction times and energy consumption. Simultaneously, the plasma component generates reactive ions and radicals that facilitate the formation of nanostructures with precise control over their size, shape, and composition. This combination can produce photocatalysts with enhanced light absorption, improved charge separation, and excellent stability, thereby boosting their overall photocatalytic performance. The integration with other green technologies, such as ultrasound-assisted synthesis or bioactive agent incorporation, can further enhance the effectiveness of hybrid methods. Ultrasound can promote cavitation and improve mass transfer, leading to more uniform particle distribution and enhanced reaction kinetics (Zhang et al. 2022). Incorporating bioactive agents, such as natural extracts or biopolymers, can introduce additional functional groups and active sites, improving the interaction between the photocatalyst and reactant molecules.

Scalability and practical applications of hybrid green methods are also crucial considerations. These methods are designed to be scalable, allowing for the production of nanomaterials in larger quantities without compromising their environmental benefits (Gupta et al. 2023). This scalability is essential for industrial applications, where large-scale production of high-performance photocatalysts can address pressing environmental issues such as water and air purification and renewable energy

Table 11.4 Emerging green technologies in nanostructure synthesis: key benefits and applications

Technology	Description	Key benefits	Applications	References
Microwave and ultrasound-assisted green synthesis methods	*Microwave-assisted synthesis*: Utilizes microwave radiation to uniformly and controllably heat reaction mixtures, enhancing reaction rates and product quality. Reduces reaction times and energy consumption, allows precise control over reaction parameters, and can be solvent-free or use minimal benign solvents. *Ultrasound-assisted synthesis*: Employs high-frequency sound waves to induce cavitation, promoting the formation of nanostructures through high-energy collisions and localized hot spots. Enhances mass transfer and mixing, leading to uniform particle size distribution and improved reaction efficiency	Rapid and efficient synthesis Reduced energy consumption Precise control over reaction parameters Minimal use of hazardous chemicals Ambient temperature and pressure conditions Enhanced product purity and uniformity Potential for continuous flow processing Scalability for industrial applications	Medicine Electronics Environmental remediation Energy storage Pharmaceuticals Food and agriculture Textile industry	Kaur et al. (2023), Adhi et al. (2024), and Behal et al. (2024)
Plasma-assisted environmentally friendly processes	It uses plasma (ionized gas) to generate reactive species at low temperatures, enabling the synthesis of nanomaterials with precise control over size, shape, and composition. Utilizes benign gases (e.g., oxygen, nitrogen, hydrogen) and reduces the need for hazardous chemicals. Offers precise control over synthesis environment parameters (gas composition, pressure, power input)	Low-temperature and short reaction times Reduced energy consumption Minimizes hazardous by-products Produces nanomaterials with high surface areas and enhanced catalytic properties Flexibility in processing various materials (metals, ceramics, polymers) Potential for in-situ functionalization and surface modification Enhances durability and performance of nanomaterials	Catalysis Sensors Electronic devices Surface coatings Biomedical devices Aerospace Automotive industry	Shah et al. (2024), Zhang et al. (2023), and Lamichhane et al. (2022)

(continued)

Table 11.4 (continued)

Technology	Description	Key benefits	Applications	References
New bioactive agents that enhance photocatalytic performance	Incorporates natural extracts and biopolymers into nanostructured photocatalysts to improve light absorption, charge separation, and stability. Bioactive agents (e.g., plant extracts and enzymes) act as sensitizers, enhancing visible light absorption. Biopolymers provide a supportive matrix for improved charge separation and structural integrity	Enhanced photocatalytic efficiency Biodegradable and non-toxic Derived from renewable sources Reduces reliance on synthetic chemicals Minimizes hazardous by-products Enhanced interaction with pollutants Increased surface area and porosity Potential for hybrid nanocomposites with synergistic properties	Environmental purification (water and air) Solar fuel production Clean energy technologies Wastewater treatment Antimicrobial coatings Agricultural pest control Green building materials	Khatami and Iravani (2021), Sohouli et al. (2022), and Feng et al. (2022)

generation. Environmental impact and sustainability are at the core of hybrid green methods (Gupta et al. 2023). By combining multiple green technologies, these methods minimize the use of toxic solvents and hazardous reagents, reduce energy consumption, and lower the overall ecological footprint of the synthesis process. This aligns with the principles of green chemistry, which prioritize reducing waste and using renewable resources (Silvestri et al. 2021). Future research directions in developing hybrid green methods include exploring new combinations of green technologies, optimizing process parameters, and investigating the fundamental mechanisms that drive the synergistic effects observed in hybrid systems. This research may lead to the discovery of novel hybrid techniques that can further enhance the efficiency and sustainability of nanomaterial synthesis.

11.8.2 Nature-Inspired Designs and Biomimetic Approaches

Drawing inspiration from natural processes and biological systems, green synthesis research focuses on biomimetic approaches (Chakraborty et al. 2023). These strategies mimic natural mechanisms, such as photosynthesis and enzymatic reactions, to design efficient and sustainable photocatalysts. Nature-inspired designs seek to emulate biological systems' remarkable efficiency and specificity, which have evolved over millions of years to perform complex chemical reactions under mild conditions. By understanding and replicating these natural processes, scientists may develop innovative materials and methods that are both environmentally friendly and highly effective. Photosynthesis is a prime example of a natural process that can inspire green synthesis. Plants and algae use sunlight to convert carbon dioxide and water into glucose and oxygen through highly efficient reactions (Xie et al. 2022). Researchers can design photocatalysts that harness solar energy to drive chemical reactions by mimicking the key steps of photosynthesis (Lv et al. 2023; Kumar et al., 2024). For instance, artificial photosynthetic systems can produce hydrogen fuel or degrade pollutants, utilizing sunlight as a clean and renewable energy source (Xie et al. 2022; Abas et al. 2020). Enzymatic reactions provide another valuable model for biomimetic approaches. Enzymes are natural catalysts that can accelerate biochemical reactions with high specificity and efficiency under mild conditions (Buller et al. 2023). Researchers can achieve similar catalytic performance by designing photocatalysts that mimic the active sites of enzymes. These biomimetic catalysts can facilitate various reactions, from organic transformations to environmental remediation, without harsh chemicals or extreme conditions (Xu et al. 2021). Bioinspired materials play a crucial role in these approaches. For example, researchers can use natural polymers, such as chitosan or silk fibroin, to create sustainable supports for photocatalysts. These biopolymers provide a biocompatible matrix and contribute to the photocatalytic system's overall stability and functionality (Sohouli et al., 2022). Additionally, incorporating nanostructures inspired by biological surfaces, such as the lotus leaf's superhydrophobic properties or the hierarchical structures of butterfly wings, can enhance photocatalysts' light-harvesting and catalytic capabilities (Wijerathne et al. 2024).

Self-assembly and templating techniques derived from biological processes offer powerful tools for creating nanostructured materials with precise control over their morphology and functionality (Chaouiki et al. 2024). For instance, the self-assembly of proteins or peptides can be used to construct highly ordered nanostructures that mimic the natural organization of cellular components (Levin et al. 2020). These biomimetic nanostructures can be efficient photocatalysts with enhanced surface areas and active sites, improving their catalytic performance. Green chemistry principles are integral to biomimetic approaches. By leveraging natural processes, researchers aim to minimize toxic reagents and solvents, reduce energy consumption, and generate minimal waste (Khosravi et al. 2024). This focus on sustainability aligns with environmental goals and opens up new avenues for developing cost-effective and scalable photocatalytic systems. Additionally, interdisciplinary collaboration is essential for advancing nature-inspired designs and biomimetic approaches. Combining expertise from biology, chemistry, materials science, and engineering can lead to innovative solutions that address complex challenges in photocatalysis. For example, insights from molecular biology can inform the design of enzyme-mimetic catalysts, while advances in nanotechnology can enable the precise fabrication of bioinspired materials. Applications of biomimetic photocatalysts are diverse and impactful. They can be used for solar energy conversion, environmental purification, and sustainable chemical production (Bigham et al. 2024). For instance, photocatalysts inspired by natural photosystems can split water into hydrogen and oxygen, providing a clean and renewable hydrogen fuel source (Yang et al. 2022). Similarly, biomimetic catalysts can degrade organic pollutants in water and air, contributing to environmental cleanup efforts (Jaffri et al. 2021).

11.8.3 The Future of Green Synthesis in Energy and Environmental Applications

The green photocatalyst synthesis will be crucial in addressing global energy and environmental challenges. Future research can explore the integration of green photocatalysts in renewable energy systems, such as solar cells and hydrogen production, and their application in pollution abatement and water purification. Renewable energy systems will greatly benefit from advances in green photocatalysts. For instance, in solar cells, green-synthesized photocatalysts can enhance light absorption and charge separation, leading to higher energy conversion efficiencies (Ahmed et al. 2020). Developing dye-sensitized solar cells (DSSCs) and perovskite solar cells (PSCs) incorporating eco-friendly materials can further drive the adoption of sustainable energy technologies. These solar cells can be manufactured using low-cost and abundant materials, reducing economic and environmental costs.

Significant advancements in pollution abatement will also be seen through the use of green photocatalysts. These materials can degrade organic pollutants and remove harmful substances from air and water. For instance, photocatalysts can be employed in air purification systems to break down volatile organic compounds (VOCs) and other airborne pollutants, leading to cleaner indoor and outdoor air (Talaiekhozani et al. 2021). Additionally, green photocatalysts can be used to treat industrial effluents and agricultural runoff, reducing the release of toxic chemicals into the environment (Saravanan et al. 2022). Water purification is a critical application of green photocatalysts. Photocatalytic processes can effectively degrade organic contaminants, disinfect water by killing bacteria and viruses, and remove heavy metals from contaminated water sources (Soleimani et al. 2021).

11.9 Conclusion

The development of green-synthesized photocatalysts that are both efficient and sustainable can revolutionize water treatment technologies. For example, incorporating bioactive agents or biopolymers into photocatalysts can enhance their performance and stability, making them suitable for long-term use in water purification systems. Additionally, scalability and commercial viability are essential considerations for the future of green synthesis in energy and environmental applications. Future research can focus on developing cost-effective and scalable methods for producing green photocatalysts, ensuring their widespread adoption in various industries. This includes exploring new sources of raw materials, optimizing synthesis processes, and developing robust and durable photocatalytic materials. Interdisciplinary research and collaboration will drive the innovation and application of green photocatalysts. Researchers can develop comprehensive solutions to global challenges by combining expertise from diverse fields. Collaborative efforts can also lead to standardizing green synthesis methods, promoting their acceptance and implementation in industrial practices. Furthermore, policy and regulatory support will play a crucial role in promoting the adoption of green photocatalysts. Governments and regulatory bodies can incentivize research and development in green synthesis, provide funding for sustainable technology projects, and implement policies encouraging the use of eco-friendly materials in energy and environmental applications. Public awareness and education about the benefits of green photocatalysts can also drive consumer demand and support for sustainable technologies.

References

Aba-Guevara CG, Medina-Ramírez IE, Hernández-Ramírez A, Jáuregui-Rincón J, Lozano-Álvarez JA, Rodríguez-López JL (2017) Comparison of two synthesis methods on the preparation of Fe, N-Co-doped TiO2 materials for degradation of pharmaceutical compounds under visible light. Ceram Int 43(6):5068–5079. https://doi.org/10.1016/j.ceramint.2017.01.018

Abas N, Kalair E., Kalair A., Ul Hasan, Q. & Khan, N (2020) Nature inspired artificial photosynthesis technologies for hydrogen production: barriers and challenges. International Journal of Hydrogen Energy, 45:20787–20799

Abuzeid HM, Julien CM, Zhu L, Hashem AM (2023) Green synthesis of nanoparticles and their energy storage, environmental, and biomedical applications. Crystals 13(11). https://doi.org/10.3390/cryst13111576. Multidisciplinary Digital Publishing Institute (MDPI)

Adhi TP, Aqsha A & INDARTO A (2024) Comparative studies on thermal, microwave-assisted, and ultrasound-promoted preparations. Green Chemical Synthesis with Microwaves and Ultrasound, 337–380

Ahmed DS, Mohammed MKA & Majeed SM (2020) Green Synthesis of Eco-Friendly Graphene Quantum Dots for Highly Efficient Perovskite Solar Cells. ACS Applied Energy Materials, 3: 10863–10871

Ahmed MA & Mohamed AA (2024) Advances in ultrasound-assisted synthesis of photocatalysts and sonophotocatalytic processes: A review. Iscience, 27

Aliannezhadi M, Mirsanaee SZ, Jamali M, Shariatmadar Tehrani F (2024) The physical properties and photocatalytic activities of green synthesized ZnO nanostructures using different ginger extract concentrations. Sci Rep 14(1). https://doi.org/10.1038/s41598-024-52455-z

Alprol AE, Mansour AT, Abdelwahab AM, Ashour M (2023) Advances in green synthesis of metal oxide nanoparticles by marine algae for wastewater treatment by adsorption and photocatalysis techniques. Catalysts 13(5):888

Amrillah T, Abdullah CAC, Hermawan A, Sari FNI, Alvani VN (2022) Towards greener and more sustainable synthesis of MXenes: a review. Nanomaterials 12(23). https://doi.org/10.3390/nano12234280. MDPI

Anjum S, Jacob G, Gupta B (2019) Investigation of the herbal synthesis of silver nanoparticles using Cinnamon zeylanicum extract. Emerg Mater 2(1):113–122. https://doi.org/10.1007/s42247-019-00023-x

Anwer Z, Jamali AR, Khan W, Bhatti J, Akhter F, Batool M (2022) Green synthesis of active Fe2O3 nanoparticles using Aloe barbadensis and Camellia sinensis for efficient degradation of malachite green and Congo red dye. Biomass Convers Biorefinery. https://doi.org/10.1007/s13399-022-03626-3

Badawy MEI, Lotfy TMR, Shawir SMS (2019) Preparation and antibacterial activity of chitosan-silver nanoparticles for application in preservation of minced meat. Bull Natl Res Cent 43(1). https://doi.org/10.1186/s42269-019-0124-8

Bai S, Zhang N, Gao C, Xiong Y (2018) Defect engineering in photocatalytic materials. Nano Energy 53:296–336

Banumathi B, Vaseeharan B, Suganya P, Citarasu T, Govindarajan M, Alharbi NS, Kadaikunnan S, Khaled JM, Benelli G (2017) Toxicity of Camellia sinensis-fabricated silver nanoparticles on invertebrate and vertebrate organisms: morphological abnormalities and DNA damages. J Clust Sci 28(4):2027–2040. https://doi.org/10.1007/s10876-017-1201-5

Behal J, Maru MS, Katwal R, Pathak D & Kumar V (2024) Ultrasonic assisted green synthesis approach for nanotechnological materials. Journal of Alloys and Compounds Communications, 100013

Briffa J, Sinagra E, Blundell R (2020) Heavy metal pollution in the environment and their toxicological effects on humans. Heliyon 6(9). https://doi.org/10.1016/j.heliyon.2020.e04691. Elsevier Ltd

Bigham A, Zarepour A, Safarkhani M, Huh Y, Khosravi A, Rabiee N, Iravani S & Zarrabi A (2024) Inspired by nature: Bioinspired and biomimetic photocatalysts for biomedical applications. Nano Materials Science

Buller R, Lutz S, Kaziauskas R, Snajrova R, Moore J & Bornscheuer U (2023) From nature to industry: Harnessing enzymes for biocatalysis. Science, 382, eadh8615

Castillo BE, Prokhorov E, Luna-Bárcenas G, Kovalenko Y (2022) Potential use of chitosan-TiO2 nanocomposites for the electroanalytical detection of imidacloprid. Polymers 14(9). https://doi.org/10.3390/polym14091686

Chakroborty S, Nath N, Pal K (2022) Limitations of nanomaterials insights in green chemistry sustainable route: review on novel applications. Green Process Synth 11(1):951–964. https://doi.org/10.1515/gps-2022-0081. De Gruyter Open Ltd

Chan SS, Low SS, Chew KW, Ling TC, Rinklebe J, Juan JC, Ng EP, Show PL (2022) Prospects and environmental sustainability of phyconanotechnology: a review on algae-mediated metal nanoparticles synthesis and mechanism. Environ Res 212:113140

Chandrakala V, Aruna V, Angajala G (2022) Review on metal nanoparticles as nanocarriers: current challenges and perspectives in drug delivery systems. Emerg Mater 5(6):1593–1615

Chakraborty S, Bera D, Roy L & Ghosh CK (2023) Biomimetic and bioinspired nanostructures: recent developments and applications. Bioinspired and Green Synthesis of Nanostructures: A Sustainable Approach, 353–404

Chaouiki A, Chafiq M & KO Y G (2024) The art of controlled nanoscale lattices: A review on the selfassembly of colloidal metal–organic framework particles and their multifaceted architectures. Materials Science and Engineering: R: Reports, 159,:100785

Chaudhary R, Nawaz K, Khan AK, Hano C, Abbasi BH, Anjum S (2020) An overview of the algae-mediated biosynthesis of nanoparticles and their biomedical applications. Biomolecules 10(11):1498

Chen F, Ma T, Zhang T, Zhang Y, Huang H (2021) Atomic-level charge separation strategies in semiconductor-based photocatalysts. Adv Mater 33(10):2005256

Das PE, Abu-Yousef IA, Majdalawieh AF, Narasimhan S, Poltronieri P (2020) Green synthesis of encapsulated copper nanoparticles using a hydroalcoholic extract of Moringa oleifera leaves and assessment of their antioxidant and antimicrobial activities. Molecules 25(3). https://doi.org/10.3390/molecules25030555

Debref R (2012) The paradoxes of environmental innovations: the case of green chemistry. J Innov Econ Manag 9(1):83–102

Dhanya A, Aparna K (2016) Synthesis and evaluation of TiO2/chitosan based hydrogel for the adsorptional photocatalytic degradation of azo and anthraquinone dye under UV light irradiation. Proc Technol 24:611–618. https://doi.org/10.1016/j.protcy.2016.05.141

Dikshit PK, Kumar J, Das AK, Sadhu S, Sharma S, Singh S, Gupta PK, Kim BS (2021) Green synthesis of metallic nanoparticles: applications and limitations. Catalysts 11(8). https://doi.org/10.3390/catal11080902. MDPI

Dülger B, Özkan G, Angı OS, Özkan G (2024) Green synthesis of TiO2 nanoparticles using Aloe Vera extract as catalyst support material and studies of their catalytic activity in dehydrogenation of Ethylenediamine Bisborane. Int J Hydrog Energy 75:466–474. https://doi.org/10.1016/j.ijhydene.2024.02.223

El-Moslamy SH, Elnouby MS, Rezk AH, El-Fakharany EM (2023) Scaling-up strategies for controllable biosynthetic ZnO NPs using cell free-extract of endophytic Streptomyces albus: characterization, statistical optimization, and biomedical activities evaluation. Sci Rep 13(1). https://doi.org/10.1038/s41598-023-29757-9

Farajinejad Z, Sani IK, Alizadeh M & Amiri S (2024) A review of recent advances in the photocatalytic activity of protein and polysaccharide-based nanocomposite packaging films: antimicrobial, antioxidant, mechanical, and strength properties. Journal of Polymers and the Environment, 32, 3437–3447

Faisal S, Jan H, Shah SA, Shah S, Khan A, Akbar MT, Rizwan M, Jan F, Wajidullah, Akhtar N (2021) Green synthesis of zinc oxide (ZnO) nanoparticles using aqueous fruit extracts of Myristica fragrans: their characterizations and biological and environmental applications. ACS Omega 6(14):9709–9722

Fatimah I, Amaliah SN, Andrian MF, Handayani TP, Nurillahi R, Prakoso NI, Wicaksono WP, Chuenchom L (2019) Iron oxide nanoparticles supported on biogenic silica derived from bamboo leaf ash for rhodamine B photodegradation. Sustain Chem Pharm 13. https://doi.org/10.1016/j.scp.2019.100149

Feng C, Wu ZP, Huang Kw, Ye J & Zhang H (2022) Surface modification of 2D photocatalysts for solar energy conversion. Advanced Materials, 34, 2200180

Goyal H, Chen T.-Y, Chen W & Vlachos DG (2022) A review of microwave-assisted process intensified multiphase reactors. Chemical Engineering Journal, 430, 133183

Granja Alvear A, Pineda-Aguilar N, Lozano P, Lárez-Velázquez C, Suppan G, Galeas S, Debut A, Vizuete K, De Lima L, Saucedo-Vázquez JP, Alexis F, López F (2024) Synergistic antibacterial properties of silver nanoparticles and its reducing agent from cinnamon bark extract. Bioengineering 11(5). https://doi.org/10.3390/bioengineering11050517

Grätzel M (2001) Photoelectrochemical cells. Nature 414:338–344. https://doi.org/10.1038/35104607

Gupta D, Boora A, Thakur A & Gupta TK (2023) Green and sustainable synthesis of nanomaterials: recent advancements and limitations. Environmental Research, 231: 116316

Halawa AA, Elshopakey GE, Elmetwally MA, El-Adl M, Lashen S, Shalaby N, Eldomany E, Farghali A, Sayed-Ahmed MZ, Alam N, Syed NK, Ahmad S, Rezk S (2022) Impact of chitosan administration on titanium dioxide nanoparticles induced testicular dysfunction. Sci Rep 12(1). https://doi.org/10.1038/s41598-022-22044-z

Hanif Z, Khan ZA, Choi D, La M, Park SJ (2021) One-pot synthesis of silver nanoparticle deposited cellulose nanocrystals with high colloidal stability for bacterial contaminated water purification. J Environ Chem Eng 9(4). https://doi.org/10.1016/j.jece.2021.105535

Hmmam I, Ali MAS, Abdellatif A (2023) Alginate-based zinc oxide nanoparticles coating extends storage life and maintains quality parameters of mango fruits "cv. Kiett". Coatings 13(2). https://doi.org/10.3390/coatings13020362

Iravani S (2011) Green synthesis of metal nanoparticles using plants. Green Chem 13(10):2638–2650

Jadhav V, Bhagare A, Ali IH, Dhayagude A, Lokhande D, Aher J, Jameel M, Dutta M (2022) Role of Moringa oleifera on green synthesis of metal/metal oxide nanomaterials. J Nanomater 2022. https://doi.org/10.1155/2022/2147393. Hindawi Limited

Jadoun S, Arif R, Jangid NK, Meena RK (2021) Green synthesis of nanoparticles using plant extracts: a review. Environ Chem Lett 19(1):355–374

Jaffri SB, Ahmad KS, Thebo Kh & Rehman F (2021) Sustainability consolidation via employment of biomimetic ecomaterials with an accentuated photo-catalytic potential: emerging progressions. Reviews in Inorganic Chemistry, 41: 131–150

Jain A (2022) Algae-mediated synthesis of biogenic nanoparticles. Adv Nat Sci Nanosci Nanotechnol 13(4):043001

Joseph A, Vijayanandan A (2021) Photocatalysts synthesized via plant mediated extracts for degradation of organic compounds: a review of formation mechanisms and application in wastewater treatment. Sustain Chem Pharm 22. https://doi.org/10.1016/j.scp.2021.100453

Joshi N & Loganathan S (2024) Cold Plasma Techniques for Sustainable Material Synthesis and Climate Change Mitigation: A Review. Catalysts, 14,:802

Kalpana VN, Devi Rajeswari V (2018) A review on green synthesis, biomedical applications, and toxicity studies of ZnO NPs. Bioinorg Chem Appl 2018. https://doi.org/10.1155/2018/3569758. Hindawi Limited

Kaur N, Singh A & Ahmad W (2023) Microwave assisted green synthesis of silver nanoparticles and its application: A review. Journal of Inorganic and Organometallic Polymers and Materials, 33:663–672

Kazemi S, Hosseingholian A, Gohari SD, Feirahi F, Moammeri F, Mesbahian G, Moghaddam ZS, Ren Q (2023) Recent advances in green synthesized nanoparticles: from production to application. Mater Today Sustain 24. https://doi.org/10.1016/j.mtsust.2023.100500. Elsevier Ltd

Khadka D, Gautam P, Dahal R, Ashie MD, Paudyal H, Ghimire KN, Pant B, Poudel BR, Bastakoti BP, Pokhrel MR (2024) Evaluating the photocatalytic activity of green synthesized iron oxide nanoparticles. Catalysts 14(11):751. https://doi.org/10.3390/catal14110751

Khatami M & Iravani S (2021) Green and eco-friendly synthesis of nanophotocatalysts: an overview. Comments on Inorganic Chemistry, 41:133–187

Khan I, Saeed K, Khan I (2019) Nanoparticles: properties, applications and toxicities. Arab J Chem 12(7):908–931

Khan F, Shahid A, Zhu H, Wang N, Javed MR, Ahmad N, Xu J, Alam MA, Mehmood MA (2022) Prospects of algae-based green synthesis of nanoparticles for environmental applications. Chemosphere 293:133571

Khosravi A, Zarepour A, Iravani S, Varma RS & Zarrabi A (2024) Sustainable synthesis: natural processes shaping the nanocircular economy. Environmental Science: Nano, 11:688–707

Korkut O, Fen AÜ, Enstitüsü B, Kara A, Ozcelik B, Makalesi A (2024) Green synthesis of chitosan-coated silver nanoparticles (Ch-AgNPs): harnessing nature for sustainable and safe nanomaterial production Kitosan Kaplı Gümüş Nanopartiküllerin (Ch-AgNP'ler) Yeşil Sentezi: Sürdürülebilir ve Güvenli Nanomateryal Üretimi için Doğadan Yararlanım. OKU J Inst Sci Technol 7(3):1319–1341

Krishna PG, Chandra Mishra P, Naika MM, Gadewar M, Ananthaswamy PP, Rao S, Boselin Prabhu SR, Yatish KV, Nagendra HG, Moustafa M, Al-Shehri M, Jha SK, Lal B, Stephen Santhakumari SM (2022) Photocatalytic activity induced by metal nanoparticles synthesized by sustainable approaches: a comprehensive review. Front Chem 10. https://doi.org/10.3389/fchem.2022.917831. Frontiers Media S.A

Kumar S, Boro JC, Ray D, Mukherjee A, Dutta J (2019) Bionanocomposite films of agar incorporated with ZnO nanoparticles as an active packaging material for shelf life extension of green grape. Heliyon 5(6). https://doi.org/10.1016/j.heliyon.2019.e01867

Kumar A, Kuang Y, Liang Z & Sun X (2020) Microwave chemistry, recent advancements, and eco-friendly microwave-assisted synthesis of nanoarchitectures and their applications: a review. Materials Today Nano, 11, 100076

Kumar R, Chaudhary M, Al-Ahmed A, Bhattacharyya S, Von Gratowski S & Iqbal J (2024) Photo-to-chemical energy transformation: Pioneering photocatalysts, surface and interface engineering. Materials Research Bulletin, 113046

Lamichhane P, Acharya TR, Kaushik N, Nguyen LN, Lim JS, Hessel V, Kaushik NK & Coi EH (2022) Non-thermal argon plasma jets of various lengths for selective reactive oxygen and nitrogen species production. Journal of Environmental Chemical Engineering, 10, 107782

Levin A, Hakala TA, Schnaider L, Bernardes GJ, Gazit E & Knowles TP (2020) Biomimetic peptide self-assembly for functional materials. Nature Reviews Chemistry, 4: 615–634

Li S-N, Wang R, Ho S-H (2021) Algae-mediated biosystems for metallic nanoparticle production: from synthetic mechanisms to aquatic environmental applications. J Hazard Mater 420:126625

Li C, Zhang G, Lin L, Wu T, Brunner S, Galmarini S, Bi J, Malfait WJ, Zhao S, Ostrikov K (2023) Silica aerogels: from materials research to industrial applications. Int Mater Rev 68(7):862–900

Lv J, Xie J, Mohamed AGA, Zhang, X, Feng Y, Jiao L, Zhou E, Yuan D & Wang Y (2023) Solar utilization beyond photosynthesis. Nature Reviews Chemistry, 7, 91–105

Lin X, Wang J (2019) Green synthesis of well dispersed TiO2/Pt nanoparticles photocatalysts and enhanced photocatalytic activity towards hydrogen production. Int J Hydrog Energy 44(60):31853–31859. https://doi.org/10.1016/j.ijhydene.2019.10.062

Madivoli ES, Kareru PG, Gichuki J, Elbagoury MM (2022) Cellulose nanofibrils and silver nanoparticles enhances the mechanical and antimicrobial properties of polyvinyl alcohol nanocomposite film. Sci Rep 12(1). https://doi.org/10.1038/s41598-022-23305-7

Makky S, Rezk N, Abdelsattar AS, Hussein AH, Eid A, Essam K, Kamel AG, Fayez MS, Azzam M, Agwa MM, El-Shibiny A (2023) Characterization of the biosynthesized Syzygium aromaticum-mediated silver nanoparticles and its antibacterial and antibiofilm activity in combination with bacteriophage. Results Chem 5. https://doi.org/10.1016/j.rechem.2022.100686

Mancuso A, Iervolino G (2022) Synthesis and application of innovative and environmentally friendly photocatalysts: a review. Catalysts 12(10). https://doi.org/10.3390/catal12101074. MDPI

Mandal RP, Sekh S, Sarkar NS, Chattopadhyay D, De S (2016) Algae mediated synthesis of cadmium sulphide nanoparticles and their application in bioremediation. Mater Res Express 3(5):055007

Mehrotra S, Goyal V, Dimkpa CO, Chhokar V (2024) Green synthesis and characterization of ginger-derived silver nanoparticles and evaluation of their antioxidant, antibacterial, and anticancer activities. Plants 13(9). https://doi.org/10.3390/plants13091255

Meng J, Ashry AG, Abou-Elyazed AS, Zhang Z, Li X, Sharara TZ, El-Demerdash SH (2024) Highly efficient visible-light photocatalysts: Bi2O3@TiO2 derived from Ti-MOFs for Eriochrome Black T degradation: a joint experimental and computational study. Catalysts 14(11). https://doi.org/10.3390/catal14110829

Mirzaeifard Z, Shariatinia Z (2024) Economical, one-pot, and green synthesis of plant-based carbon quantum dots for efficient visible-light photocatalytic dye degradation and water purification. J Taiwan Inst Chem Eng 163. https://doi.org/10.1016/j.jtice.2024.105655

Mishra K, Devi N, Siwal SS, Gupta VK & Thakur VK (2023) Hybrid semiconductor photocatalyst nanomaterials for energy and environmental applications: fundamentals, designing, and prospects. Advanced Sustainable Systems, 7, 2300095

Mittal M, Tripathi S & Shin DK (2024) Biopolymeric Nanocomposites for Wastewater Remediation: An Overview on Recent Progress and Challenges. Polymers, 16, 294

Mousa SA, Wissa DA, Hassan HH, Ebnalwaled AA, Khairy SA (2024) Enhanced photocatalytic activity of green synthesized zinc oxide nanoparticles using low-cost plant extracts. Sci Rep 14(1). https://doi.org/10.1038/s41598-024-66975-1

Noor Javed M, Ali Khan Bangash S, Abbas M, Ahmed S, Kaplan A, Iqbal S, Nauman Khan M, Adnan M, Ali A, Zaman F, Wahab S (n.d.) Potential and challenges in green synthesis of nanoparticles: a review, vol 19. http://xisdxjxsu.asia

Omran BA, Baek K-H (2022) Valorization of agro-industrial biowaste to green nanomaterials for wastewater treatment: Approaching green chemistry and circular economy principles. Journal of Environmental Management, 311:114806

Osman AI, Zhang Y, Farghali M, Rashwan AK, Eltaweil AS, Abd El-Monaem EM, Mohamed IMA, Badr MM, Ihara I, Rooney DW, Yap PS (2024) Synthesis of green nanoparticles for energy, biomedical, environmental, agricultural, and food applications: a review. Environ Chem Lett 22(2):841–887. https://doi.org/10.1007/s10311-023-01682-3. Springer Science and Business Media Deutschland GmbH

Pinto I, Silvestre JD, de Brito J, Júlio MF (2020) Environmental impact of the subcritical production of silica aerogels. J Clean Prod 252:119696

Premkumar J, Sudhakar T, Dhakal A, Shrestha JB, Krishnakumar S, Balashanmugam P (2018) Synthesis of silver nanoparticles (AgNPs) from cinnamon against bacterial pathogens. Biocatal Agric Biotechnol 15:311–316. https://doi.org/10.1016/j.bcab.2018.06.005

Quintavalla A, Carboni D, Lombardo M (2024) Green metrics and sustainability in photocatalysis. ChemCatChem 16(14). https://doi.org/10.1002/cctc.202301225. John Wiley and Sons Inc

Radulescu DM, Surdu VA, Ficai A, Ficai D, Grumezescu AM, Andronescu E (2023) Green synthesis of metal and metal oxide nanoparticles: a review of the principles and biomedical applications. Int J Mol Sci 24(20). https://doi.org/10.3390/ijms242015397. Multidisciplinary Digital Publishing Institute (MDPI)

Ramzan M, Ayub F, Shah AA, Naz G, Shah AN, Malik A, Sardar R, Telesiński A, Kalaji HM, Dessoky ES, Elgawad HA (2022) Synergistic effect of zinc oxide nanoparticles and Moringa oleifera leaf extract alleviates cadmium toxicity in Linum usitatissimum: antioxidants and physiochemical studies. Front Plant Sci 13. https://doi.org/10.3389/fpls.2022.900347

Rani N, Yadav S, Mushtaq A, Rani S, Saini M, Rawat S, Gupta K, Saini K, Maity D (2024) Azadirachta indica peel extract-mediated synthesis of ZnO nanoparticles for antimicrobial,

supercapacitor and photocatalytic applications. Chem Pap 78(6):3687–3704. https://doi.org/10.1007/s11696-024-03340-6
Rashed MMA, Zhang C, Ghaleb ADS, Li JP, Nagi A, Majeed H, Bakry AM, Haider J, Xu Z, Tong Q (2019) Techno-functional properties and sustainable application of nanoparticles-based Lavandula angustifolia essential oil fabricated using unsaturated lipid-carrier and biodegradable wall material. Ind Crop Prod 136:66–76. https://doi.org/10.1016/j.indcrop.2019.04.070
Ren X, Ghazani MS, Zhu H, AO W, Zhang, Moreside E, Zhu J, Yang P, Zhong N, Bi X (2022) Challenges and opportunities in microwave-assisted catalytic pyrolysis of biomass: A review. Applied Energy, 315:118970
Rezazadeh NH, Buazar F, Matroodi S (2020) Synergistic effects of combinatorial chitosan and polyphenol biomolecules on enhanced antibacterial activity of biofunctionalaized silver nanoparticles. Sci Rep 10(1). https://doi.org/10.1038/s41598-020-76726-7
Rui K, Xue LI, Han M, Jinling GAO, Jian D, Yongfeng J, Guoxing LIN, Tianjian LU (2024) Ultralight, compact, and load-bearing multifunctional metastructures: concept, design and applications. Appl Math Mech 45(8):949–973. 应用数学和力学
Salgado P, Aedo E, Vidal G (2024) Green synthesis of Fe2O3 nanoparticles using Eucalyptus globulus leaf extract on Pinus radiata sawdust for cationic dye adsorption. Nanomaterials 14(22):1832. https://doi.org/10.3390/nano14221832
Saravanan A, Kumar PS, Jeevanantham S, Anubha M, Jayashree S (2022) Degradation of toxic agrochemicals and pharmaceutical pollutants: Effective and alternative approaches toward photocatalysis. Environmental Pollution, 298:118844
Shah K, Patel J, Kumar S, Pandey R, Maity G, Dubey S (2024) Perspectives on sustainable and efficient routes of nanoparticle synthesis: an exhaustive review on conventional and micro-plasma-assisted techniques. Nanoscale
Siddiqui VU, Ansari A, Ansari MT, Akram MK, Siddiqi WA (2022) Fabrication of a zinc oxide/alginate (ZnO/Alg) bionanocomposite for enhanced dye degradation and its optimization study. RSC Adv 12(12):7210–7228. https://doi.org/10.1039/d1ra08991a
Silvestri C, Silvestri L, Forcina A, Di Bona G, Falcone D (2021) Green chemistry contribution towards more equitable global sustainability and greater circular economy: A systematic literature review. Journal of Cleaner Production, 294:126137
Singh J, Dutta T, Kim K-H, Rawat M, Samddar P, Kumar P (2018) 'Green' synthesis of metals and their oxide nanoparticles: applications for environmental remediation. J Nanobiotechnol 16:1–24
Singh H, Desimone MF, Pandya S, Jasani S, George N, Adnan M, Aldarhami A, Bazaid AS, Alderhami SA (2023) Revisiting the green synthesis of nanoparticles: uncovering influences of plant extracts as reducing agents for enhanced synthesis efficiency and its biomedical applications. Int J Nanomedicine 18:4727–4750. https://doi.org/10.2147/IJN.S419369. Dove Medical Press Ltd
Sohouli E, Irannejad N, Ziarati A, Ehrlich H, Rahimi-nasrabadi M, Ahmadi F, Luque R (2022) Application of polysaccharide-based biopolymers as supports in photocatalytic treatment of water and wastewater: a review. Environmental Chemistry Letters, 20:3789–3809
Soleimani M, Ghasemi JB, Ziarani GM, Karimi-maleh H, Badiei A (2021) Photocatalytic degradation of organic pollutants, viral and bacterial pathogens using titania nanoparticles. Inorganic Chemistry Communications, 130:108688
Spoială A, Ilie CI, Dolete G, Croitoru AM, Surdu VA, Trușcă RD, Motelica L, Oprea OC, Ficai D, Ficai A, Andronescu E, Dițu LM (2022) Preparation and characterization of chitosan/TiO2 composite membranes as adsorbent materials for water purification. Membranes 12(8). https://doi.org/10.3390/membranes12080804
Sundararajan N, Habeebsheriff HS, Dhanabalan K, Cong VH, Wong LS, Rajamani R, Dhar BK (2024) Mitigating global challenges: harnessing green synthesized nanomaterials for sustainable crop production systems. Global Chall 8(1). https://doi.org/10.1002/gch2.202300187. John Wiley and Sons Inc

Szczepanik B (2017) Photocatalytic degradation of organic contaminants over clay-TiO2 nanocomposites: a review. Appl Clay Sci 141:227–239. https://doi.org/10.1016/j.clay.2017.02.029. Elsevier Ltd

Taghavi Fardood S, Ganjkhanlu S, Moradnia F, Ramazani A (2024) Green synthesis, characterization, and photocatalytic activity of superparamagnetic MgFe2O4@ZnAl2O4 nanocomposites. Sci Rep 14(1). https://doi.org/10.1038/s41598-024-67655-w

Talaiekhozani A, Rezania S, Kim K-H, Sanaye R, Amani AM (2021) Recent advances in photocatalytic removal of organic and inorganic pollutants in air. Journal of Cleaner Production, 278:123895

Thoma C, Konnerth J, Sailer-Kronlachner W, Solt P, Rosenau T, van Herwijnen HWG (2020) Current situation of the challenging scale-up development of hydroxymethylfurfural production. ChemSusChem 13(14):3544–3564. https://doi.org/10.1002/cssc.202000581. Wiley-VCH Verlag

Tsibulnikova MR, Pogharnitskaya OV, Adam AM (2016) Challenges and perspectives in sustainable management of natural resources at regional level. http://gubernator.tomsk.ru/words/doklad-gubernatora-na-

Venkatesan K, Sundarababu J, Anandan SS (2024) The recent developments of green and sustainable chemistry in multidimensional way: current trends and challenges. Green Chem Lett Rev 17(1). https://doi.org/10.1080/17518253.2024.2312848. Taylor and Francis Ltd

Verma V, Al-Dossari M, Singh J, Rawat M, Kordy MGM, Shaban M (2022) A review on green synthesis of TiO2 NPs: synthesis and applications in photocatalysis and antimicrobial. Polymers 14(7). https://doi.org/10.3390/polym14071444

Verma DK, Verma C, Fuertes PO (2024) Green chemical synthesis with microwaves and ultrasound, John Wiley & Sons

Villa K, Galán-Mascarós JR, López N, Palomares E (2021) Photocatalytic water splitting: advantages and challenges. Sustain Energy Fuels 5(18):4560–4569. https://doi.org/10.1039/d1se00808k. Royal Society of Chemistry

Virkutyte J, Varma RS (2013) Green synthesis of nanomaterials: environmental aspects. ACS Symp Ser 1124:11–39. https://doi.org/10.1021/bk-2013-1124.ch002

Wang J, Jiang W, Liu D, Wei Z, Zhu Y (2015) Photocatalytic performance enhanced via surface bismuth vacancy of Bi6S2O15 core/shell nanowires. Appl Catal B Environ 176:306–314

Wang S-J, Su D, Zhu Y-F, Lu C-H, Zhang T (2023) The state-of-the-art review on rational design for cavitation assisted photocatalysis. Materials & Design, 112377

Wijerathne B, Liao T, Jiang X, Zhou J, Sun Z (2024) Plant-inspired surfaces and interfaces for sustainable technologies. Materials Futures

Xie Y, Khoo KS, Chew KW, Devadas VV, Phang SJ, Lim HR, Rajendran S, Show PL (2022) Advancement of renewable energy technologies via artificial and microalgae photosynthesis. Bioresource Technology, 363:127830

Xu W. Jiao L, Wu Y, Hu L, Gu W, Zhu C (2021) Metal–organic frameworks enhance biomimetic cascade catalysis for biosensing. Advanced Materials, 33: 2005172

Yang L, Fan D, Li Z, Cheng Y, Yang X, Zhang T (2022) A review on the bioinspired photocatalysts and photocatalytic systems. Advanced Sustainable Systems, 6:2100477

Yao H, Fan Y, Emre EST, Li N, Ge M, Wang J, Wei J (2024) Alginate-modified ZnO anti-planktonic and anti-biofilm nanoparticles for infected wound healing. Int J Biol Macromol 280. https://doi.org/10.1016/j.ijbiomac.2024.135739

Ying S, Guan Z, Ofoegbu PC, Clubb P, Rico C, He F, Hong J (2022) Green synthesis of nanoparticles: current developments and limitations. In: Environmental technology and innovation, vol 26. Elsevier B.V. https://doi.org/10.1016/j.eti.2022.102336

Zeng A, Yang R, Tong Y, Zhao W (2023) Functional bacterial cellulose nanofibrils with silver nanoparticles and its antibacterial application. Int J Biol Macromol 235. https://doi.org/10.1016/j.ijbiomac.2023.123739

Zhang Z, Wang K, Xu C, Zhang Y, Wu W, Lu C, Liu W, Rao Y, Jiang C, Xu C (2022) Ultrasound enhancing the mass transfer of droplet microreactor for the synthesis of AgInS2 nanocrystals. Chemical Engineering Journal, 435:134948

Zhang J, Li X, Zheng J, Du M, Wu X, Song J, Cheng C, Li T, Yang W (2023) Non-thermal plasma-assisted ammonia production: A review. Energy Conversion and Management, 293:117482

Zhao W, Zhao Y, Zheng Y, Chen D, Xin J, Li K, Che H, Li Z, Ma M, Hang Y (2024) Long-term variability in black carbon emissions constrained by gap-filled absorption aerosol optical depth and associated premature mortality in China. Atmos Chem Phys 24(11):6593–6612. https://doi.org/10.5194/acp-24-6593-2024

Chapter 12
Prospective Developments and Emerging Directions in Green Nanostructured Photocatalysis

Kannan Raman, Tharmaraj Vairaperumal, Kalimuthu Karuppanan, Sreeraj Gopi, and Kuppusamy Kanagaraj

Abstract In the rapidly evolving field of green nanostructured photocatalysis (**GNS-PCs**), advanced nanocomposite technology integrates with eco-friendly systems and become promising solution to address many of the environmental and energy dilemmas. This chapter focuses on the new developments and new directions for the field which include how advanced photocatalytic materials capture and utilize solar energy for environmental purification as well as energy production. Building upon the discussion of the generation of biogenic nanoparticles (**BioG-NPs**) and the structural rationality of metal-organic frameworks, into the contemporary applications ranging from water purification to energy conversion. The use of green photocatalysts opens the door not only to the decontamination of water, air, and soil and green hydrogen production but also to industrial applications of photocatalysis, which can fundamentally change the ways many industries work and address their emissions, waste, and energy needs. However, the chapter also enumerates challenges such as scaling up, economic feasibility, and efficiency of using solar energy that remain a challenge in the widespread proliferation of solar energy

K. Raman
Department of Pharmacology, St. John's College of Pharmaceutical Sciences & Research, Kattappana, Idukki, Kerala, India

T. Vairaperumal
Institute of Clinical Medicine, College of Medicine, National Cheng Kung University, Tainan, Taiwan, Republic of China

K. Karuppanan
Department of Biotechnology, School of Bioengineering, SRM Institute of Science and Technology, KTR Campus, Chennai, India

S. Gopi
Molecules Biolabs, Koratty, Kerala, India

K. Kanagaraj (✉)
Center for Supramolecular Chemistry & Catalysis and Department of Chemistry, College of Science, Shanghai University, Shanghai, China

C. M. Hussain, G. Hussain (eds.), *Green Nanostructured Photocatalysts for Clean Energy*, Clean Energy Production Technologies,
https://doi.org/10.1007/978-981-95-1618-6_12

projects. In this way, this chapter tries also to outline the research agenda of investigating emerging technologies and the difficulties and prospects that appear on the horizon and to offer a direction through which these technologies can be further developed toward creating the framework for an efficient vision of the future, pointing for a new course with sustainability and prosperity as the driving force. From a systems perspective, we analyze approaches toward a successful move from "lab to life" for inventions that could potentially redefine the standards of environmental responsibility and efficient energy usage.

Keywords Photocatalysis · Nanostructured photocatalysts · Biogenic nanoparticles · Green chemistry · Clean energy · Environmental remediation · Water treatment · Waste management · Solar energy · Energy storage · Water splitting

Abbreviations

B	Boron
Bi	Bismuth
BioG-NPs	Biogenic nanoparticles
C	Carbon
CB	Conduction band
COD	Chemical oxygen demand
e^-	Electron
EDCs	Endocrine-disrupting chemicals
E_g	Bandgap
ESR	Electron spin resonance
Fe	Iron
$g\text{-}C_3N_4$	Graphitic carbon nitride
GNS-PCs	Green nanostructured photocatalysis
h^+	Hole
$h\gamma$	Photons
MOFs	Metal-organic frameworks
N	Nitrogen
NPs	Nanoparticle
OV	Oxygen vacancies
PCs	Photocatalysis
POPs	Persistent organic pollutants
Pt	Platinum
ROS	Reactive oxygen species
S	Sulfur
SEM	Scanning electron microscopy
Sr	Strontium
TEM	Transmission electron microscopy

TiO_2	Titanium dioxide
UV	Ultraviolet
UV-Vis.	Ultraviolet-visible
VB	Valence band
XRD	X-ray Diffraction
ZnO	Zinc oxide

12.1 Introduction

The detailed investigation of GNS-PCs in the context of a constantly developing nano industry points to a clear potential of nanotechnology's integration with sustainable practices as a way to tackle some of the world's most crucial modern environmental problems. With the rise of global concern and subsequent legislation on emissions and energy efficiency, green nanostructured photocatalysts have become vital (Ikram et al. 2021). The key materials in this area are on the cutting edge of advancement towards the next generation of environmental cleanup technologies while incorporating solar energy to drive pollution reduction and clean energy production for the environment (Thirumalaivasan et al. 2024a). However, in the future, the prospect of developing GNS-PCs is not only highlighted but in fact imperative. Standing before the challenges of climate change, pollution, and energy needs, the emergence and deployment of such technologies might help create a more sustainable and prosperous world. This chapter is intended to introduce and present all modern innovations, as well as to predict the further evolution of these revolutionary technologies to stimulate further development and use of GNS-PCs in the future (Fig. 12.1).

The transition to green nanostructured photocatalysts is an exciting development from orthodox environment and energy solutions that are frequently resource- and chemical-demanding (Kumar et al. 2023). Instead, these advanced materials on the one hand provide the opportunity to reach green and efficient solutions to the problems, using the virtually unlimited source of energy, which is the sun (Baxter et al. 2009). This requires them to function under ambient conditions, without needing additional power inputs apart from solar light energy making them strategic components in the shift towards a low-carbon economy.

These developments have emerged due to the knowledge in material science, nanotechnology, and advanced photocatalytic mechanism. It has to be mentioned that innovators and researchers are working hard, and day by day, new materials are being developed which not only improve the efficiency of the conversion of sun power into useful energy but also provide the material durability for usage in practical real life situations (Green and Bremner 2017). In these aspects, the progress made in enhancing the performance of GNS-PCs is both fast and radical—from synthesizing biogenic nanoparticles called BioG-NPs through eco-friendly processes to designing exquisite metal-organic frameworks that would achieve the best balance of light capture and catalytic turnover.

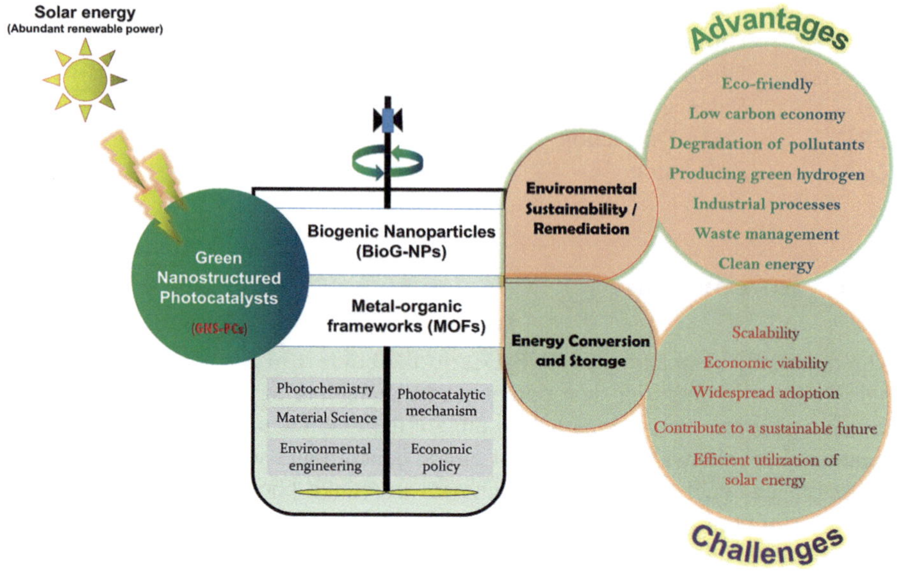

Fig. 12.1 Overview of this chapter on **GNS-PCs** in environmental remediation and energy conversion, highlighting key advantages, challenges, and emerging directions for future research

Trends prescient in this line show the potential of these materials for a plethora of uses, from purifying water and air pollutants to supporting the generation of hydrogen generally known as green hydrogen (Thirumalaivasan et al. 2024b). Work is also being done on incorporating green photocatalysts into presently established systems in industries ranging from manufacturing to waste disposal to energy creation. Such innovations underscore sustainable use of these materials through destroying pollutants so as to promote conservation of the environment and development of sustainable energy sources. Nevertheless, the process of pillars with green nanostructured photocatalysts for wider commercial applications faces some barriers (Ahmad et al. 2024; Şen 2004). The problems of scope, cost, and optimized application of solar energy are still unsolved. Solutions for these challenges would have to involve simultaneous consideration of ideas from chemistry, material sciences, environmental engineering, and economic policy. Thus, this chapter also considers the process of scaling up from laboratory-based analysis to industrial and municipal use, highlighting potential strategies enabling further improvement of such technologies' commercial application and overall effectiveness (García-Serna et al. 2007; Jadhav et al. 2021).

However, in the future, the prospect of developing GNS-PCs is not only highlighted but in fact imperative. Standing before the challenges of climate change, pollution, and energy needs, the emergence and deployment of such technologies might help create a more sustainable and prosperous world (Verma et al. 2024). This chapter is intended to introduce and present all modern innovations, as well as to predict the further evolution of these revolutionary technologies to stimulate further development and use of GNS-PCs in the future (Fig. 12.1).

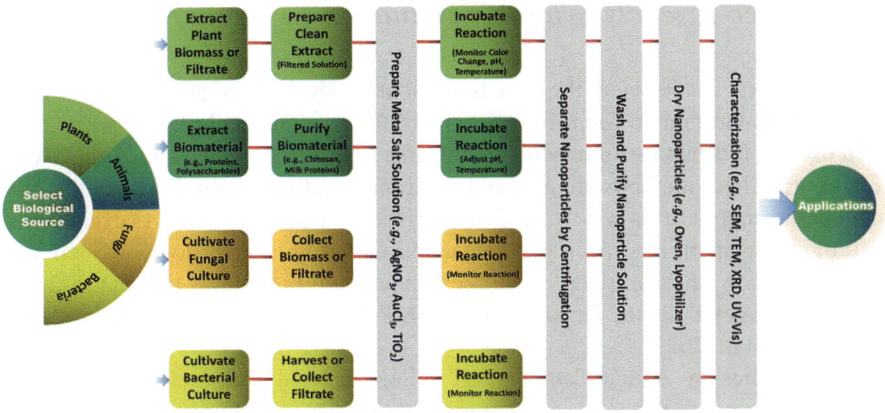

Fig. 12.2 Schematic overview of biogenic nanoparticle (**BioG-NPs**) synthesis pathways using various biological sources. Key steps include extraction or cultivation, reaction with metal salt solution, purification, drying, and characterization, leading to diverse applications

12.2 Advances in Green Nanostructured Materials

12.2.1 Biogenic Nanoparticles (BioG-NPs)

New developments in GNS-PCs have therefore described the enormous benefits that can be derived from BioG-NPs. These NPs are biosynthesized through biological means using microorganisms or plant extracts and are a breakthrough innovation towards the preparation of photocatalytic materials. This section discusses the new techniques for the preparation of these green NPs along with their characteristics and potential uses in an attempt to promote the improvement of the environmental friendliness of industrial processes (Samuel et al. 2022).

Figure 12.2 shows the different NPs synthesized by plants, animals, fungi, and bacteria using a pathway which is somewhat analogous to one another but has distinct intermediate steps. In plant-based synthesis, biomass or filtrate is then extracted, cleaned, and mixed with a metal salt solution. When the NPs are derived from animals, biomaterials such as proteins or chitosan are initially purified before reacting with the metal salts (Manjunathan et al. 2023). Metal biosorption has been achieved using fungal or bacterial synthesis where certain strains are selected and filtrates or biomass are sampled and dispersed in metal salts under controlled conditions such as pH and temperature. As stated earlier, in NP formation, after incubation, the NPs are normally pelleted by centrifugation, washed, and finally dried using methods such as lyophilization. According to characterization techniques such as scanning electron microscopy (SEM), transmission electron microscopy (TEM), X-ray diffraction (XRD), as well as UV-vis spectroscopy, it is revealed that the NPs have appropriate structure, morphology, and optical properties. From this process, the produced NPs have characteristics that are ideal for various uses depending on the biological source as the catalyst and the function that provides the improved performance in environmental, biological, and industrial applications.

12.2.2 Eco-Friendly Synthesis Methods

BioG-NPs are synthesized through biological methods that are friendly to the immediate environment. Microorganisms like fungi, algae, and plant extracts have been used as biosources for producing NPs like titanium dioxide (TiO_2), zinc oxide (ZnO), etc. These methods are not only environmentally friendly but also effective in reducing energy loads which are characteristic of most syntheses. As exemplified by fungi like Aspergillus species and algae like Chlorella, the section will document that they can synthesize metal oxide NPs under mild conditions (Salvadori 2019; Mourdikoudis et al. 2018; Thakur et al. 2024; Pal et al. 2020; Asmat-Campos et al. 2023; Kaur et al. 2023).

Both studies by Asmat-Campos and colleagues and Harpreet Kaur and colleagues show scalability and green synthesis of TiO_2 NPs using plant extracts as reducing and stabilizing agents for photocatalytic applications (Asmat-Campos et al. 2023; Kaur et al. 2023). Asmat-Campos et al. chose to use *Annona muricata L.* extract to prepare TiO_2 NPs with an average size of about 13.41 nm and spherical shape (Asmat-Campos et al. 2023). NPs were successfully characterized using structural, morphological, and optical characterization techniques showing the highly crystalline and monodispersed nature of the synthesized particles (Mourdikoudis et al. 2018). The photocatalytic activity of such samples was examined on the methylene blue dye, where maximum degradation of 59.25% was observed in 30 min. In this study, two aspects worthy of consideration have been demonstrated, namely, the performance of biogenic TiO_2 NPs in dye degradation and their probable application in wastewater treatment (Thakur et al. 2024).

Likewise, Kaur et al. described, for the first time, the synthesis of biogenic TiO_2 NPs using *Elettaria cardamomum* and obtained anatase phase NPs with various nano-sizes (Pal et al. 2020; Kaur et al. 2023). XRD analysis showed the multi-shaped, nano-level elemental presence of particles in which phytochemical congregation was confirmed by FT-IR analysis, and SEM/HRTEM supported the results. These particles also showed excellent photocatalytic activity where Congo Red dye was degraded up to 97% under UV light in 150 min. Kaur et al. also pointed out the recyclability of these NPs, where they show 85% efficiency after four photocatalytic runs, making them suitable for reuse. In addition, the synthesized NPs showed antibacterial activity against bacterial strains *Staphylococcus aureus* and *Pseudomonas aeruginosa* (Fig. 12.3).

These studies establish that biogenic synthesis is a promising, eco-friendly pathway to replace chemical approaches, with improved performance in functional applications (Nhien et al. 2020). Some of these studies add value to the field of nanotechnology through plant-derived stabilizing agents, similar to providing direction towards eco-friendly methods of wastewater treatment and biomedical applications.

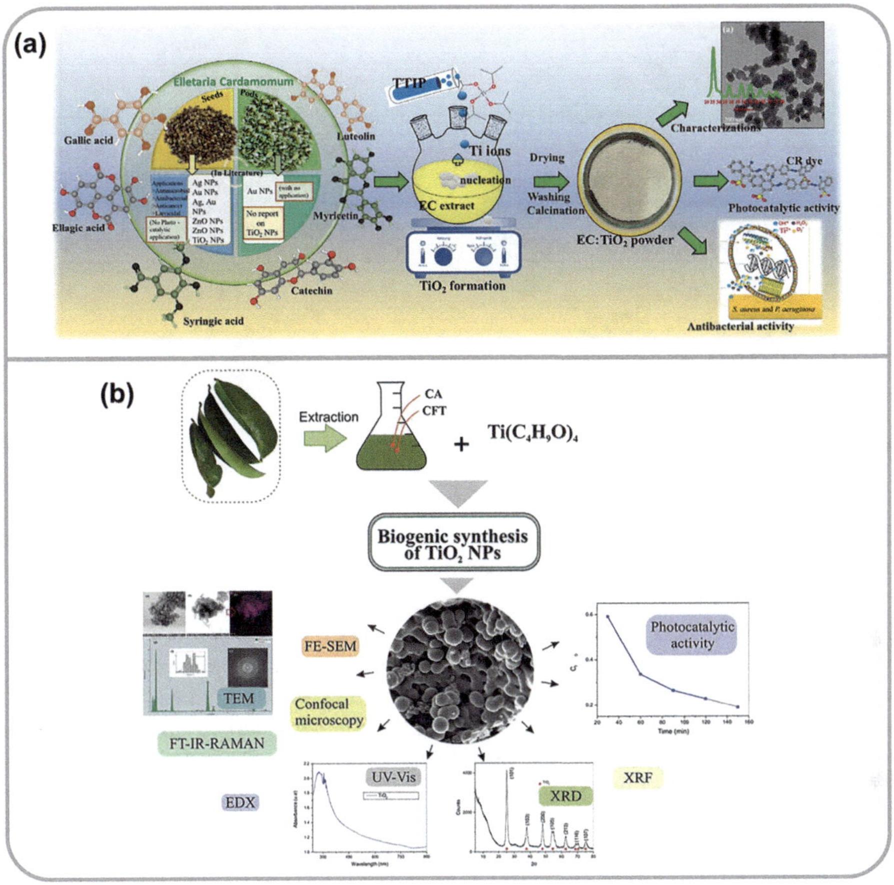

Fig. 12.3 (**a**) Biogenic synthesis of **TiO_2 NPs** using *Annona muricata L.* extract as a stabilizing agent, demonstrating significant photocatalytic activity for dye degradation. (Reproduced with permission from Asmat-Campos et al. (2023)). (**b**) *E. cardamomum*-derived **TiO_2 NPs** exhibiting multi-shaped morphology, high photocatalytic efficiency in dye degradation, and antibacterial properties, showcasing an eco-friendly synthesis approach for environmental and biological applications. (Reproduced with permission from Kaur et al. (2023))

12.2.3 Unique Catalytic Properties

In particular, BioG-NPs possess a set of catalytic characteristics that make them extremely desirable among conventionally synthesized materials. High SAV offers the largest active site exposure enhancing the catalytic activity for different syntheses (Fig. 12.4). The biogenic synthesis process means that the morphology of the nanoparticles in the resulting shape can be controlled, such as spherical or rod configurations that define the catalytic performance. Furthermore, their inherent stability in thermal and chemical environments makes them suitable for long-term use.

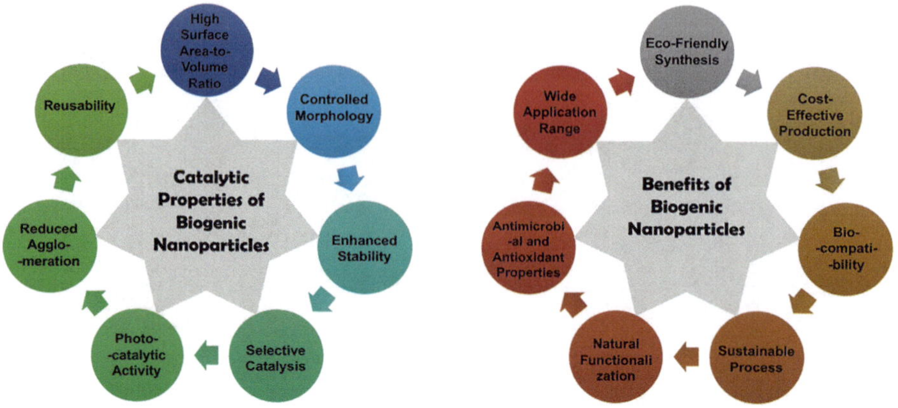

Fig. 12.4 Schematic illustration of catalytic properties and the benefits of green synthesis in catalytic applications

BioG-NPs are ideal for selective catalysis since the organic compounds originating from natural schemata enhance the specificity of the desired reaction hence enhancing its selectivity (Fig. 12.4).

In general, it could be said that one of the major merits of BioG-NPs is its non-toxicity making the product suitable for highly sensitive applications particularly in environmental solutions. The planet-friendly synthesis technique for making these compounds directly from natural biological materials such as plant extracts or microbial cultures reduces the usage of hazardous chemicals and reduces the ecological influence manyfold (Shellaiah et al. 2021). Green synthesis is normally performed under mild conditions, and thus, this method is therefore more economic and environmentally friendly. More importantly, this method inherently transfers functional biomolecules onto the NPs to prevent aggregation and, at times, to augment photocatalytic activity. For example, TiO_2 NPs synthesized from plant extracts have enhanced photocatalytic activities compared to those obtained from conventional synthesis by degrading pollutants under visible light owing to changes in the bandgap.

12.2.4 Applications in Photocatalytic Degradation

Altogether, BioG-NPs offer a great promise for attaining GNS-PCs' maximization and bringing in new specifications of energy efficiency and sustainability in the technology line. Different from common photocatalysts, including TiO_2, they mainly work under ultraviolet (UV) light while BioG-NPs can be tuned to absorb most of the solar radiation especially the visible light. This considerable change is made achievable by controlling their band-gap energies by recent synthesis techniques whereby particular dopants are used to improve their light absorption properties (Samsudin and Abd Hamid 2017).

The basis of this innovation is an important property associated with biological synthesis methods. BioG-NPs can be synthesized from naturally occurring biological sources such as fungi, bacteria, or plant extracts inherent properties of these materials make the particles bio-friendly and hazardous-free, in addition to control over the structural elasticity of the aforementioned particles at the molecular level. For instance, during the biosynthesis process, the incorporation of elements like nitrogen or sulfur in the NPs changes their electronic property by decreasing their band-gap energies by enabling them to activate under visible light (Singaram et al. 2017). This chapter focuses on the detailed processes through which these modifications are implemented and the consequences for the photocatalytic performance. It presents the dopants responsible for the change in the light absorption edge and examines how the structural and surface properties of the NPs affect their catalytic activities, as influenced by biological synthesis.

Nevertheless, the path to the large-scale use of BioG-NPs is not entirely smooth. The two most significant challenges have been identified; they include how to manufacture these devices on an industrial scale and the economics and ecological factors that come along with deploying them on a massive scale (Bonilla et al. 2018). Furthermore, the categories are stable in the long term, and reusable NPs for industrial and environmental applications provide potential for further investigation (Brar et al. 2022). To overcome these issues, this section presents some possible future research directions targeting the enhancement of synthesis and the use of BioG-NPs. The review calls for multidisciplinary approaches to materials science and bioengineering systems to improve the efficiency of these materials and photocatalytic chemistry (Gao et al. 2022). Also, it raises the possibility of scale-up of synthetic methodologies to preserve the environmentally friendly and catalytically active features of BioG-NPs, a key factor towards sustainable industrial applications.

That is why BioG-NPs are a breakthrough in the development of GNS-PCs. Thanks to the ability to manage visible light much more efficiently, they are at the vanguard of technologies that are to change the face of environmental cleanup and energy generation in the nearest future (Hegedus and Luque 2003). However, with continued sustained exploration of their useful properties and surmounting of conventional fabrication and usage hurdles, these NPs are destined for a significant role in the world's shift towards increased sustainability and energy efficiency (Pushparaj et al. 2022). Along with these advanced breakthroughs, this section also highlights the technological aspects that enable such progress and further prepares the readers for subsequent chapters that might pave the way to the complete commercialization of BioG-NPs in green PCs.

12.2.5 Visible-Light-Enabled Semiconductor Photocatalysis

One of the promising segments in GNS-PCs is visible-light-driven semiconductor PCs because they can solve such problems of traditional UV-dependent photocatalysts as the use of TiO_2. In the extension of visible light photon absorption abilities, visible-light photocatalytic materials, g-C_3N_4, ZnO, and modified TiO_2, through doping, heterojunction, and surface engineering, narrow down their bandgap and incorporate visible light photon absorption capabilities. This advancement creates new opportunities for large environmental applications such as for wastewater treatment, air purification, and clean hydrogen production (Kishore et al. 2022; Jeje et al. 2024). However, some disadvantages have not been addressed, including photogenerated charge carrier recombination and low quantum efficiencies, and thus require further research on material synthesis. In general, visible-light semiconductor PCs are ready to expand the spectrum of green technology applications in environmental and energy remedial processes to shift to sustainable solutions (Goodarzi et al. 2023).

In semiconductor PCs, when the energy of incoming photons (**hγ**) exceeds the semiconductor's bandgap energy ($\mathbf{E_g}$), it excites an electron ($\mathbf{e^-}$) from the valence band (**VB**) to the conduction band (**CB**), creating a corresponding hole ($\mathbf{h^+}$) in the **VB** (Eq. 12.1).

$$\text{Photocatalyst} + \text{hv} \rightarrow \text{e}^- + \text{h}^+ \left(\text{when}, h\gamma > \mathbf{E}_\mathbf{g}\right) \tag{12.1}$$

Once the charge carriers or excitons (e^- and h^+) are separated and reach the catalyst surface, they engage in redox reactions with pollutants adsorbed there. The holes (h^+) at the surface can oxidize OH^- or H_2O, leading to the formation of highly reactive hydroxyl radicals (·OH), a potent oxidant (Eq. 12.2).

$$\text{h}^+ + \text{H}_2\text{O} \rightarrow \text{H}^+ + \cdot\text{OH} \tag{12.2}$$

$$\cdot\text{OH} + \text{pollutant} \rightarrow \text{CO}_2 + \text{H}_2\text{O} \tag{12.3}$$

$$\text{e}^- + \text{O}_2 \rightarrow \text{O}_2\cdot^- \tag{12.4}$$

$$\text{O}_2\cdot^- + \text{H}^+ \rightarrow \cdot\text{HO}_2 \tag{12.5}$$

$$\cdot\text{HO}_2 + \cdot\text{HO}_2 \rightarrow \text{H}_2\text{O}_2 + \text{O}_2 \tag{12.6}$$

$$\text{H}_2\text{O}_2 + \text{hv} \rightarrow 2\cdot\text{OH} \tag{12.7}$$

$$\text{O}_2\cdot^- + \text{pollutant} \rightarrow \text{CO}_2 + \text{H}_2\text{O} \tag{12.8}$$

$$\cdot\text{HO}_2 + \text{pollutant} \rightarrow \text{CO}_2 + \text{H}_2\text{O} \tag{12.9}$$

On the catalyst surface, adsorbed oxygen is reduced by the electron e^- in the CB, forming superoxide radicals ($O_2{\cdot}^-$) (Eq. 12.4). These superoxide radicals may further react with H^+ ions to produce hydroperoxyl radicals ($\cdot HO_2$) (Eq. 12.5), which can subsequently form hydrogen peroxide (H_2O_2) through electrochemical reduction (Eq. 12.6). Light can then decompose H_2O_2 to generate hydroxyl radicals ($\cdot OH$) (Eq. 12.7). The resulting reactive oxygen species (**ROS**), including $\cdot OH$, $O_2{\cdot}^-$, $\cdot HO_2$, and H_2O_2, are highly effective in degrading and mineralizing persistent organic pollutants into benign end products, such as CO_2 and H_2O (Eqs. 12.3, 12.8, and 12.9) (Pelaez et al. 2012). Additionally, the photogenerated h^+ themselves act as oxidants, directly breaking down organic pollutants under specific reaction conditions and catalyst types (Dong et al. 2015). Eventually, e^-– h^+ pairs may recombine on the catalyst surface or in the medium, releasing energy as light or heat. A schematic illustration of the photocatalytic mechanism for organic pollutant degradation is presented in Fig. 12.5a.

The efficiency of a photocatalyst hinges on three crucial factors. First, the E_g: Photocatalysts with a wide bandgap are typically sensitive to UV light, while those with a narrower bandgap respond to visible light. Second, the band edge potentials, specifically the positions of the **VB** and **CB**: Reactive oxygen species (**ROS**) generation is possible only if the **CB** potential is more negative than the reduction potential of O_2 to $O_2{\cdot}^-$ (−0.33 eV vs. NHE) and/or if the VB potential is more positive than the oxidation potentials of $\cdot OH/H_2O$ (+2.68 eV vs. normal hydrogen electrode (**NHE**)) or $\cdot OH/OH^-$ (+1.99 eV vs. **NHE**) (Panneri et al. 2017). In cases where

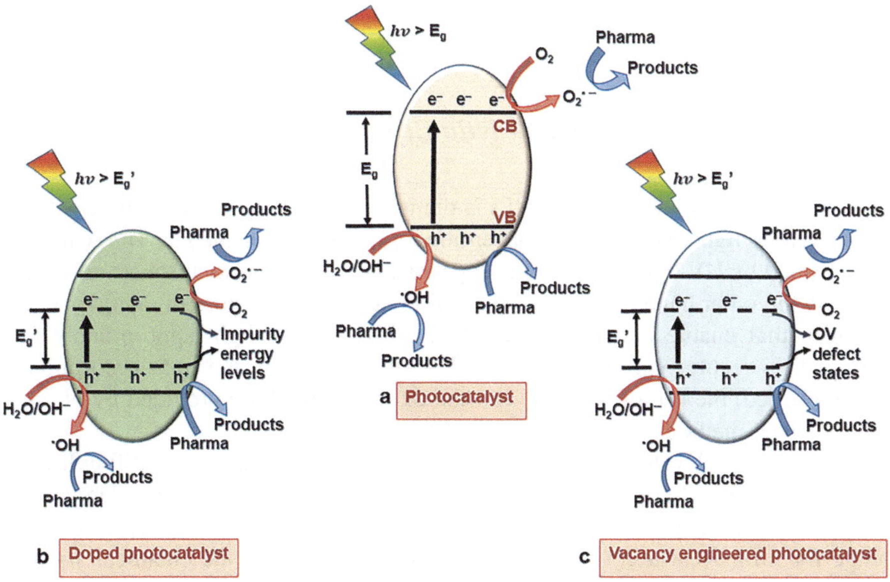

Fig. 12.5 Mechanism of visible-light-assisted pharmaceutical degradation by (**a**) semiconductor photocatalyst, (**b**) doped photocatalyst, and (**c**) vacancy-engineered photocatalyst. (Reproduced from Majumdar and Pal (2020))

these conditions aren't met, the **VB** holes (h^+) interact directly with the pollutant, with greater **ROS** generation enhancing semiconductor efficiency. Finally, the rate of electron–hole (e^-—h^+) recombination is critical: a high recombination rate decreases efficiency, as e^- rejoining **VB** limits reduction reactions for $O_2{\cdot}^-$ formation (Pelaez et al. 2012). Therefore, selecting a photocatalyst should balance these factors for optimal performance.

Semiconductor Photocatalysts Doped with Metals and Nonmetals To improve visible-light sensitivity and reduce e^-– h^+ recombination, doping wide-bandgap photocatalysts with metals (Sr, Bi, Fe) and nonmetals (C, N, S, B) creates impurity levels in the bandgap. This approach not only narrows the bandgap but also optimizes dopant levels to trap electrons, minimizing e^-– h^+ recombination (Jesudoss et al. 2016; Wang et al. 2011). However, excess dopant may accelerate recombination. Figure 12.5b illustrates the degradation mechanism of pharmaceuticals under visible light using doped photocatalysts.

Vacancy-Engineered Photocatalysts The structure and surface properties of semiconductors significantly influence their function. Defects like vacancies create energy levels in the band structure, reducing bandgap energy and expanding the semiconductor's light-absorption range. Oxygen vacancies (**OV**) are a notable example, forming intermediate energy states that enhance visible light responsiveness by narrowing the bandgap. These vacancies also serve as charge-trapping sites, promoting e^-– h^+ separation (Yu et al. 2018). Figure 12.5c depicts the mechanism of pharmaceutical degradation under visible light by vacancy-engineered photocatalysts.

12.2.6 Enhanced Photocatalytic Efficiency Under Visible Light

A significant advantage of BioG-NPs is the improvement of photocatalytic activity under visible light. Most of the conventional photocatalysts such as TiO_2 can only be activated by UV light which makes up a meager 5% of the solar spectrum. BioG-NPs, on the other hand, can be tuned to allow for specific values of energies in the bandgap that enable them to harvest and utilize visible light appropriately (Ng 2021). This section will also highlight how the change in the light absorption is accomplished, the use of dopants, and the exact morphology provided by biologically derived methods of synthesis (Fig. 12.5).

As with any development of new technologies, there are some difficulties in working with BioG-NPs. How to increase the production of these powders and, at the same time, retain the environmental and economic advantages is the major challenge for future research (Ford and Despeisse 2016). Also, the chemical stability and recovery of these NPs, upon implementation on a large scale in industries, have not been well understood. The present section will discuss the current limitations to the commercialization of BioG-NPs in GNS-PCs and suggest the future research that could help address such limitations, thereby facilitating the commercialization of these technologies. BioG-NPs are a novelty in the molecular field of GNS-PCs. For their green synthesis, ability to act as efficient catalysts, and ability to work effectively under visible light, the new nanocomposites offer great potential to revolutionize cleaning the environment and producing energy (Varma 2014). Indeed, these NPs are well-positioned to advance research into efficient and sustainable photocatalytic systems as the field progresses.

12.3 Metal-Organic Frameworks (MOFs): Pioneering Green Photocatalytic Technologies

MOFs are reinventing the essence of green photocatalysis right from introducing a brand-new class of material, rich in porosity and exhibiting high chemical variability within the Pra ze. Despite the fact that other supramolecular hosts like cyclodextrins, cucurbiturils, zeolites, clays, and hydrotalcites have garnered ample interest and have been employed in various fields such as biomedical applications, organic transformations, sensing, environment, etc., the aforementioned feature of MOFs gives them an added advantage due to the high modularity present in MOFs (Kanagaraj et al. 2023a, b; Venkatesan et al. 2023; Peng et al. 2021; Alagesan et al. 2016). These characteristics make it possible that MOFs meet certain demands in the environmental fix and green energy, in particular, the enhancement of light absorbance and catalytic activity.

Structure of MOF They are built from metal ions or clusters by the organic linkers to create a highly porous and three-dimensional network (Fig. 12.6). This structural feature enables the introduction of a large diversity of functional groups within the cavities of MOFs and makes these materials very versatile (Jiao et al. 2019). Large surface area and regulable nanopores make MOFs suitable for photocatalysis because they enhance the organic framework for varying molecular linkages (Deng et al. 2017). Due to their unique structural organization, MOFs can be easily integrated into other applications of green technologies and can play a critical role in the development of other sustainable photocatalytic systems of high performance (Swetha et al. 2022).

Figure 12.6 re-illustrates that the structure of a MOF is relatively versatile and simple, while Fig. 12.6 illustrates a few of the possible photoinduced reactions involved in photocatalytic applications of MOFs. Here, when MOFs are illuminated, you get a photoinduced charge separation process that is initiated (Thirumalaivasan et al. 2017). This process is initiated by the electronic excitation of the organic linkers that is followed by passing to the next metal nodes. So, while this has a substantial density of metal, different nodes in the structure are capable of serving as active sites for sorts of electron exchange reactions by accepting electron charges on the reactants adsorbed inside the porous structure (Venkatesan et al.

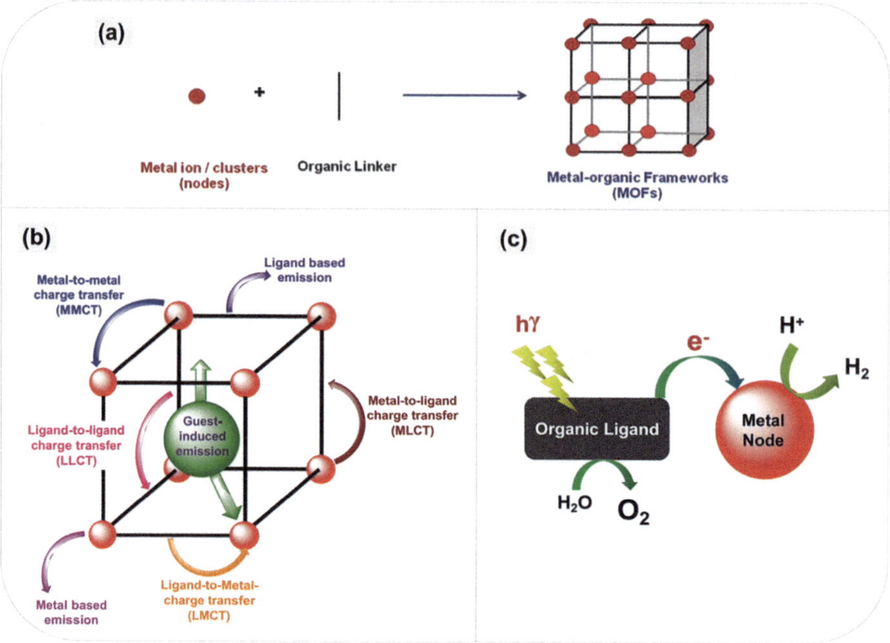

Fig. 12.6 (**a**) Structure of **MOFs** and (**b**) some possible photoinduced reaction pathways in **MOFs**. (**c**) A general mechanism of photoinduced charge separation in **MOFs** triggered by light absorption and photoinduced charge transfer to the near metal node

2019). This charge separation and transfer mechanism is very significant for the photocatalytic characteristics of MOFs since it greatly decreases the recombination rate of the electron–hole pairs and enhances the photocatalytic activity (Rajapandi et al. 2023). It indicates that MOFs are highly tunable in terms of miscibility on photocatalytic processes, including pollutant degradation, photocatalytic hydrogen production, and so on, by controlling organic linkers and metal nodes (Arumugam et al. 2023). The prospect of being designed in numerous ways combined with optimum efficiency of charge transfer characteristics makes MOFs a new concept in electro/mechanical photocatalytic technologies.

12.3.1 Design and Functionalization of MOFs

In the context of GNS-PCs, MOFs are an innovation since their given and salient structures are linked to environmental management and green power supply provisions. This part discusses some new-wave strategies in the design and functionalization of MOFs particularly with respect to photocatalytic activities and optimal exploitation of solar energy (Qian and Ma 2021). MOFs are special because the atomic structure of the network can be changed making MOFs easy to adjust in terms of pore size and transformable surface area and chemistries (Stock and Biswas 2012). This design flexibility gives opportunities to introduce organic linkers and metal nodes that can selectively absorb visible light which is so critical in provoking photocatalytic processes upon solar illumination (Wang and Zhu 2024). When such components are properly fitted and integrated, the researchers can construct MOFs with high photovoltaic efficiency and MOFs that are capable of inducing the appropriate electron transfer processes for PCs (Bai et al. 2022).

Such development is adding new components such as graphitic carbon nitride (g-C_3N_4) which is a light-capturing material that depicts material stability. Both these types of MOFs improve the photocatalytic efficiency of MOFs since they can be used for continuous outdoor environmental applications (Byrne et al. 2018). Besides the removal of general pollutants, all the PCs equipped with MOFs can have structural flexibility in their operating systems.

Furthermore, MOFs can be synthesized to respond to diverse environmental and volatile situations. For example, engineering learning was advanced, and MOFs are built in a way that they can still be stable and functional, especially for applications in high-humidity or underwater environments to enhance their constructability in real-life environmental situations. In addition to being employed as single structures exhibiting powerful photocatalytic activity, MOFs are incorporated into composites (Chen et al. 2019). These composites could achieve the desired properties of each component, for example, high light absorption and a slow electron–hole recombination process resulted in greatly improved photocatalytic activity. For instance, when MOFs combine with semiconductors like TiO_2 or conductive polymers, the formation of ROS under light exposure is much enhanced which is key in destroying contaminants (UshaVipinachandran et al. 2020).

Subsequently, the design and functionalization of MOFs will be expectedly attained with even further enhancement in the future. Emerging within this are cries and efforts to devise ways of synthesizing materials cheaper and more sustainable for the construction of MOFs (Adeel et al. 2021). In addition, great emphasis is being placed on assessing MOFs for other environmentally acceptable processes such as the photoreductive reduction of CO_2 into sustainable energy or the photochemical decomposition of water towards green hydrogen generation (Chandrasekaran et al. 2018). Such prospective improvements suggest a new generation of the GNS-PCs and MOF-based materials in technology improvement. At the same time, the specific ratios, which can be designed to fit particular uses, not only enhance the performance of the photocatalytic systems but also introduce new possible applications which is a milestone in the search for sustainable photocatalytic systems.

12.3.2 Applications in Water Splitting and CO_2 Reduction

Prominent abilities of MOFs in GNS-PCs are best captured in the area of water splitting and CO_2 reduction. These processes are not only important for the creation of a sustainable method of hydrogen production and its application, but they are also essential for the direct conversion of CO_2 into invaluable organic compounds, which may change the guidelines for clean energy and chemical manufacturing. MOFs are designed and synthesized to incorporate specific active sites which are perfectly suitable for redox reactions required in both water splitting and CO_2 reduction processes. The solar energy harvesting efficiency of the MOFs is improved by the incorporation of photocatalytically active components like graphitic carbon nitride (g-C_3N_4) (Durmus and Maijenburg 2022).

This design strategy enhances the photocatalytic activity as well as takes advantage of MOFs in the light absorption and management, so as to improve the usage efficiency of the solar energy. For instance, in the process of water splitting, the MOFs work as a catalyst to fix solar energy in splitting water into oxygen and hydrogen gas, a source of energy. This process is instrumental in the production of green hydrogen, a key element to the development of future sustainable energy systems. The use of MOFs in this capacity addresses one of the fundamental challenges in hydrogen production: the demand for a technology that can generate large amounts of hydrogen with low energy consumption that is associated with a low cost (Liu et al. 2020).

In MOFs such as UiO-66(Zr)-NH_2, the distribution of cocatalysts directly affects photocatalytic performance (Fig. 12.7) (Xiao et al. 2016). A research work was carried out which led to the discovery that the enhanced insertion of platinum nanoparticles (Pt NPs, ~3 nm) inside the cages of UiO-66(Zr)-NH_2 as compared to externally introduced Pt NPs (~25 μmol g^{-1} h^{-1}) significantly enhanced the photocatalytic efficiency for hydrogen generation (~275 μmol g^{-1} h^{-1}). Photocurrent measurements, photoluminescence, and electrochemical measurements suggested that if the Pt NPs were inside the cavities, there was better charge separation upon light irradiation. ESR spectrum analysis supported this separation as evident from the formation of

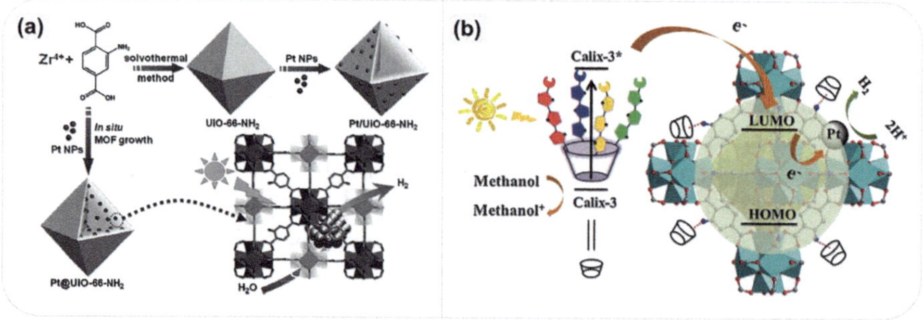

Fig. 12.7 (**a**) Illustration for the synthesis of Pt@UiO-66(Zr)-NH2 and Pt/UiO-66(Zr)-NH2, with the photocatalytic H2 production process over Pt@UiO-66(Zr)-NH_2 being highlighted. (Reproduced with permission from Xiao et al. (2016). Copyright 2016 Wiley). (**b**) Proposed mechanism of photocatalytic H_2 production over Calix-3-sensitized Pt@UiO-66-NH_2 under visible light irradiation. (Reproduced with permission from Chen et al. (2017). Copyright 2017 Elsevier)

Zr^{3+} in the nodal metal cluster (Xiao et al. 2016). Additional support for enhanced charge separation was obtained from ultrafast transient spectroscopy and time-resolved photoluminescence. Therefore, the findings of this study highlight the significance of the cocatalyst position inside MOF cavities for photocatalytic activity. Recent studies have utilized dyes like eosin B, erythrosine B, and cone-calix[4]arene (Calix-3) to sensitize **MOFs**, enhancing their photoresponse for H_2 generation under visible light and in the presence of sacrificial agents (Chen et al. 2017; Wang et al. 2017). These dyes capture visible photons, transferring excited electrons to the MOF, which then facilitates H_2 evolution, while the dye radical cation is temporarily generated. An electron donor subsequently restores the dye to its ground state (Luciani et al. 2020). Although promising photocatalytic efficiencies (e.g., 41.4 mmol g^{-1} h^{-1} using erythrosine B/UiO-66(Zr)-NH_2/rGO) have been achieved, further long-term studies are necessary to validate dye stability and practical viability.

Likewise, MOFs help lower CO_2, which is a greenhouse gas, and change it into some practical organic compounds like methane, methanol, or even various hydrocarbons (Shellaiah et al. 2020). This transformation also contributes to minimizing the effect of atmospheric CO_2 which is a greenhouse gas while at the same time contributing value by turning a gas that is deemed a pollutant into a valuable product (Anwar et al. 2020). The catalytic processes that occur are optimized to work at ambient conditions, demonstrating the feasibility of MOFs for carbon recycling and reuse within the circular economy. However, one of the most important recent developments relating to the use of MOFs in these applications is their improved stability and recyclability (Yuan et al. 2018). The stability of these frameworks is an important factor as they can go through the catalytic process more than once with efficiency and without degradation, making them suitable for scaled-up use in the long term (Howarth et al. 2016).

Regarding the matter of stability, investigators are already working to improve the conditions that support these catalysts' stability as they relate to the requirements of the industry in terms of durability and cost performance. MOFs are considered as prospective materials for PCs, and as the research community carries on the study and development of these applications, the future seems bright for these materials as leaders of sustainable energy practices (Goetjen et al. 2020). Several developments made in the design and functionalization of MOFs are predicted to improve the efficiency and applicability of MOFs for their incorporation into the current industrial processes (Silva et al. 2015). Such developments may greatly transform chemical synthesis and energy production processes toward greater sustainability and environmental responsibility.

12.3.3 *Enhancing Catalytic Performance Through Composite Strategies*

As highlighted previously, MOFs are gradually showing their potential in the field of photocatalytic applications mainly due to the structural tunability and chemical amenability (Mo et al. 2022). Figure 12.8 depicts the various composite strategies for improving the catalytic performance with major composite techniques like doping, heterojunction, and plasmonic enhancement. Nevertheless, when used together with semiconductors or carbon-related materials, the MOFs move to a new level, beyond the possibilities of individual materials, of remarkably augmented photocatalytic properties (Gulati et al. 2023). MOF composites with semiconductors also including TiO_2 or carbonaceous substances like reduced graphene oxide (rGO) form heterojunctions (Guo et al. 2017). These junctions are not just mechanical joinery but are essential improvements that contribute to aiding the separation of electrons and holes at a swifter pace. This critical change reduces the probability of electron–hole recombination, which is a major drawback in photocatalytic reactions, and increases the photocatalytic activity and efficiency of the MOFs (Sandhu et al. 2024).

Fig. 12.8 Illustration of composite strategies to enhance catalytic performance, showcasing key methods such as doping, heterojunction formation, and plasmonic enhancement

In enhancing catalytic performance through composite strategies, graphene and other carbon-enhanced materials from biomass and wastes take central stage. Such carbon sources, such as graphene, carbon nanotubes, and activated carbon, are valued for their structures as well as electrically conductive characteristics (Thirumalaivasan et al. 2024a, b, c, d; Balamurugan et al. 2017; Aryamol et al. 2024; Nair et al. 2023; Nangan et al. 2023a). They are widely applied in energy storage systems, catalysis, wastewater treatment, and the biomedical field and activities. Emerging from renewable resources, these materials improve the catalytic activity while making significant contributions to environmental sustainability due to the recycling of waste material. Due to large surface areas, electrical conductance, and controllable characteristics, they can be effectively used to enhance catalytic activity in composite systems (Thirumalaivasan et al. 2024c, d; Nair et al. 2023; Ramasubburayan et al. 2023, 2024; Chu et al. 2020; Wei et al. 2019).

The synthesis of MOF composites is directed toward the fundamental elements of photocatalytic effectiveness-charge transfer (Wu et al. 2020). However, the composite material formed by incorporating MOFs and other materials with higher electron mobility like reduced graphene oxide or MOFs with stable conduction bands like TiO_2 systems taps an enhanced carrier's path. This adjustment not only increases the rate of charge-transfer process but also prolongs the lifetime of the charge-separate state which is important for starting and maintaining photochemical reactions under light illumination (Armin et al. 2021). A typical design problem in PCs is how to capture a wider spectrum of sunlight. Since MOFs are made of metal clusters connected by organic ligands, they logically have restricted light absorption spectra. However, bridging them with other materials with similar light-absorption properties can considerably spread this range out. For example, when MOFs are incorporated with TiO_2, the absorption edge moves further into the UV region. Compared to this, when materials such as graphene are incorporated, they bring down the absorption peak into the visible and near IR regions (Han et al. 2024). Not only does this broadening maximize the effective use of solar energy, the photocatalytic activity remains viable with the full range of light essential for efficient environmental application.

The practical applications of the enhanced MOF composites present great potential in a range of environmental processes of pollutant degradation and water treatment. Because of their efficiency under actual light conditions and their versatility in various environments, these composites are ranked among the most sustainable technologies (Krauklis et al. 2021). However, simulating and amplifying these composite strategies from the lab to industrial levels presents difficult tasks. These are as follows: synthesizing these composites at low costs, making them stable and functional for long durations, and incorporating their use into the prevailing industrial procedures without drastic changes.

12.4 Emerging Trends in Photocatalytic Applications

12.4.1 Photocatalytic Water Treatment

12.4.1.1 Revolutionizing Water Purification: The Role of Green Nanostructured Photocatalysts

Green nanostructured photocatalysts in this process offer a new powerful step towards water purification and treatment with quite a difference compared to the chemical process. These catalysts harness the energy of sun radiation for the promotion and sustenance of chemical processes with self-organizing capabilities to eliminate not only organic pollutants but also pathogenic microbial loads found in water sources (Nangan et al. 2023a). Employing TiO_2, ZnO, and other advanced BioG-NPs, these photocatalysts capture solar energy in order to create electron-hole pairs (Kanagaraj et al. 2020). Table 12.1 below provides a brief on the role, benefits, and or disadvantages of the green nanostructured photocatalysts in changing the general water purification method. These reactive species are used in the degradation of pollutants at the molecular level which is enhanced by photocatalysts' efficiency under visible light—a part of the solar spectrum broader than ultraviolet.

Table 12.1 Key aspects of green nanostructured photocatalysts in revolutionizing water purification and treatment

Aspect	Description
Photocatalysts used	Titanium dioxide (TiO_2), zinc oxide (ZnO), and BioG-NPs
Mechanism	Absorption of solar energy generates electron–hole pairs, which break down pollutants and neutralize pathogens in water.
Functionality	Dual function: purifies water by removing organic contaminants and pathogens while detoxifying without additional energy inputs, relying on sunlight.
Eco-friendly production	Green synthesis using plant extracts and microorganisms minimizes ecological footprint, enhancing biocompatibility and reducing toxicity.
Advantages	Effective under visible light, low operational costs, suitable for resource-limited settings, reduced environmental impact
Challenges	Scaling up production, ensuring performance consistency in varied environmental conditions
Future research focus	Enhancing photocatalytic efficiency, stability, and durability for widespread sustainable water treatment applications
Case study impact	Pilot projects show significant reductions in chemical oxygen demand (**COD**) and microbial content in urban and industrial wastewater, proving the adaptability and scalability of green photocatalysts.
Examples of treatable contaminants	Persistent organic pollutants (**POPs**), endocrine-disrupting chemicals (**EDCs**), and other resilient environmental contaminants
Energy source	Utilizes sunlight as a natural energy source, aligning with sustainable technology goals to minimize energy and chemical footprint

The Unique Aspect of These Green Photocatalysts Lies in Their Dual Functionality They not only clear water from hazardous organic substances and pathogenic microorganisms, but also participate in detoxification without the use of extra energy, with the help of sunlight (Likodimos et al. 2010). This dual role makes them very useful especially in low-income earning areas and also helps to cut down on the costs of running water treatment plant. Furthermore, the synthesis of these photocatalysts still utilizes greener techniques like the utilization of plant extract and microorganisms for the preparation of NPs. The present green synthesis strategy not only reduces the environmental impact of the synthesis of their compounds but also improves the compatibility with biological systems and decreases the toxicity of the catalysts thus being friendly to the environment and human health (Thirumalaivasan and Wu 2020). However, there are barriers to the extensive use of green nanostructured photocatalysts, which range from large-scale fabrication issues to standard photocatalyst stability in different environments (Ahmad et al. 2016). Current research is focused on addressing these barriers, specifically in the furthering of photocatalytic performance, stability, and durability of these materials to support their utilization in advanced water infrastructure technologies.

12.4.2 Emerging Trends in Photocatalytic Applications: Energy Conversion and Storage

12.4.2.1 Energy Conversion and Storage: Harnessing Green Nanostructures for a Sustainable Future

The new prospect in the field of renewable energy technologies is to incorporate the green nanostructured photocatalysts especially for the hydrogen generation through the water splitting process. This section explores not only how these advanced materials improve the energy conversion efficiency, but also how they effectively deal with the energy storage issue—a great problem in the hydrogen economy. Green photocatalysts in nanoscale structures like MOFs and BioG-NPs are directly designed for enhancing solar energy harvesting and the light-to-energy conversion factor. Owing to their special characteristics, they can be used to initiate and advance water-splitting processes under solar light to produce hydrogen—an eco-friendly and renewable energy source. The manufacture of these materials at the nano level ensures that their capabilities of light absorption and catalytic activity are enhanced, thus increasing the efficiency of hydrogen production (Zhu and Zäch 2009).

The issue of storing hydrogen produced by photocatalytic water splitting has inherent technical limitations connected with safety and energy density. In the next table, Table 12.2, a summary has been made according to the role of green nanostructures and challenges toward the progress of sustainable energy conversion and storage applications. Scientific advances in green nanostructured materials to meet these challenges have manifested in the creation of new composite and hybrid

Table 12.2 Key aspects of harnessing green nanostructures for sustainable energy conversion and storage

Section	Description
Hydrogen production via water splitting	Green nanostructured photocatalysts (e.g., **MOFs**, **BioG-NPs**) optimize solar energy absorption for hydrogen generation.
Enhanced energy conversion efficiency	Nanostructures enable efficient solar-to-hydrogen conversion, addressing renewable energy needs.
Dual functionality in storage	Green nanostructures facilitate both hydrogen production and storage, bridging the energy generation-storage gap.
Controlled hydrogen storage	New composites and hybrids allow controlled hydrogen adsorption and desorption for safer, efficient storage.
Integrated energy system	Combining energy conversion and storage in one system enhances decentralized energy resilience.
Challenges in commercialization	Issues include scalability, cost, and integration with existing infrastructure.
Future development goals	Focus on cost-effective synthesis, material stability, and easy integration with current energy systems
Hydrogen economy support	Innovations in photocatalytic materials support hydrogen as a clean, sustainable energy carrier.
Applications beyond hydrogen production	Dual-functional materials also address environmental issues, such as CO_2 reduction and pollutant degradation.
Sustainable production methods	Environmentally friendly production using green synthesis methods, such as plant extracts and microorganisms

structures, which in addition to catalysis efficiency also enable efficient storage of hydrogen molecules.

In the future decades, green nanostructured materials are particularly expected to grow rapidly in applications of conversion and storage of energy (Balaya 2008). The innovation of even more effective and cost-efficient materials is thought to be on-going and is embarked on to expand the functionality of these entrants. Coupled with these policy elements and value chain alliances, these innovations may enable a faster advancement of green hydrogen technologies and hence the support of a sustainable and low-carbon energy matrix. Thus, nanostructured photocatalysts with the green approach are considered as a sinful frontier in the field of renewable energy. Not only do they overcome problems in the field of hydrogen economy but they also create the basis for further development in the following generation of energy sources that are both fully renewable and economically feasible (Kleijn and Van der Voet 2010).

12.4.2.2 Dual-Functional Photocatalytic Materials: A Synthesis of Efficiency and Sustainability

As for GNS-PCs, contemporary scientists are investigating the possibility of using the materials which at the same time act as a photocatalyst and hydrogen storage medium. These materials are cleverly constructed to use solar energy to phase out pollution and generate green and clean power especially in the conversion of solar energy into storable hydrogen (Winter 2009). The ability to operate in this dual capacity becomes crucial for creating a stable and ongoing process of energy generation and storage, which responds to the problems of renewable energy sources.

These latest photocatalytic materials enhance the light-trapping ability and light-to-chemical energy conversion efficiency and promote the water decomposition process to release hydrogen—a green and renewable energy source (Rafique et al. 2020; Podder et al. 2020). Through providing the means to both capture energy and store it, these materials are vital for current efforts to build a more innovative and efficient energy systems which seek to phase out the use of fossil fuels and decrease adverse environmental effects.

The arrangement of these materials is at the nano-scale, designed with intricate structural features whose dual-tasking surface spaces and electronic structures make it easier for photocatalytic activities to operate due to increased capability of capturing the light. Such innovations in this field are the incorporation of nanoscale semiconductors, metal hybrid structures, and modification of the bandgap to a suitable energy acceptable by the solar system (Volokh and Mokari 2020).

It is appreciable that MOFs and BioG-NPs have been designed and synthesized to show higher photocatalytic performance under solar radiation. These materials are developed to enhance not only the generation of hydrogen but also the storage of hydrogen through adsorption and release processes (Lim et al. 2010). This approach is very helpful in eradicating one of the leading challenges in hydrogen energy efficiency and safe storage.

However, the photocatalytic processes involved are self-cleaning methods that use solar energy; in other words, it has high propulsion towards environmental conservation. These multifunctional photocatalytic materials not only cover the application of hydrogen production but also CO_2 reduction and photocatalytic removal of pollutants, which provide plural-solution strategies for energy and environmental problems.

The field studies are currently being undertaken with the goals of improving the catalytic efficiency, stability, and affordability of these materials for practical use beyond the laboratory (Spöri et al. 2017). Through the adoption of new developments in nanotechnology and sustainability, dual-functional photocatalytic materials are poised for importance in the development of clean energy systems that signal a new epoch of sustainability and efficiency in energy supply.

12.4.2.3 Advanced Architectures for Enhanced Photocatalytic Activity

In an effort to improve the photocatalytic potential within the framework of GNS-PCs, designs for enhanced sunlight harvesting and conversion have been deployed (Shellaiah et al. 2022). This approach involves, therefore, dealing with nanostructures with high structural dependence that raises their surfaces and improves their electronic character, therefore, the photoreactivity to solar energy. Some of the proposed advancements in this area are nanoscale semiconductor, metal hybrid, and engineered bandgap (Banin et al. 2014). These materials are manmade to have high interaction with solar radiation in order to improve the rate at which the light is captured and used. Such semiconductors used are in the nanoscale which are usually TiO_2 and ZnO due to their suitable bandgap energy and resistance to photodegradation (Johar et al. 2015). However, due to such shortcomings particularly on absorption of visible light, they are usually alloyed with metals or nonmetals or combined with other materials in such a way as to alter their electronic band structure.

Metal hybrids, in which metal NPs are incorporated into semiconductor matrices, are established based on the plasmonic characteristics of metals, such as Au and Ag, improving light absorption in the visible region. This plasmonic enhancement is vital for introducing photocatalytic procedures in natural sunlight instead of UV light which is but a tiny portion of the sun's rays (Zhou et al. 2015). In addition, the design choices of registers based on the transitions of bandgaps using synthesis of composite materials or by doping methods are a crucial research frontier. These materials can be engineered and changed to respond with a specific bandgap that will trigger them under certain wavelengths, enhancing their ability to capture sunlight. For instance, when TiO_2 is doped with nitrogen or carbon, the electron structure changes so that the photocatalyst can be activated only with visible rather than UV light (Yang et al. 2008).

These structural and compositional optimizations do not only improved the photocatalytic activity of these materials but also improved their stability as well as their durability in environmental uses (Li et al. 2015). These photocatalytic architectures afford a higher degree of light absorption color and better electron–hole separation, enabling cleaner, efficient photocatalytic reactions, essential for environmental degradation and clean energy uses such as water splitting and CO_2 conversion. Indeed, these advanced materials continue to undergo constant technological development through the application of nanotechnology and other materials sciences hence indicating a great future for the development of PCs. The above technologies are still in their infancy, and the integration of these applications into full-scale real-life applications will be instrumental in combating energy sustainability problems and environmental decay.

12.4.2.4 Water Splitting: A Pathway to Green Hydrogen

Water splitting has become a key process for sustainable hydrogen production in the broad realm of GNS-PCs. When solar powered, it uses this method to dissociate water molecules into hydrogen and oxygen, generating zero-carbon emission hydrogen fuel (Kalaiselvan and Mathimani 2024). The tailored nanostructures for this process are central to the optimization of the interaction with solar radiation to maximize the hydrogen production photocatalytic efficiency and economic feasibility. At the frontiers of photocatalytic design, these nanostructures tend to be synthesized from a mixture of BioG-NPs and MOFs. Synthesis of BioG-NPs using environmentally benign processes incorporating microorganisms and plant extracts generates nanoparticles with unique catalytic properties because of their organic origin and designed morphologies. For example, these NPs can be engineered to have a modified bandgap energy and thus activated by visible light, rather than UV light, taking advantage of a greater solar energy band (Sarina et al. 2013).

MOFs also add another, much more sophisticated layer of structure. Their highly porous, customizable frameworks allow for the incorporation of a variety of functional groups and metal nodes that enhance light absorption and facilitate electron transfer processes critical for efficient PCs. Components such as graphitic carbon nitride (g-C_3N_4) have been integrated with MOFs to increase their stability and photocatalytic performance in direct sunlight (Zhao et al. 2021). More importantly, in the field of green hydrogen production, which has thus far been unable to overcome the "loss of hydrogen" phenomenon due to the scarcity of visible light, this enhanced photocatalytic activity under visible light is highly significant (Corredor et al. 2019). These advanced photocatalysts, which effectively split water molecules, produce hydrogen without the carbon dioxide emissions found in other traditional fossil fuel-based methods. In addition to demonstrating the TengraFтyit potential for using nano-engineered materials as renewable energy sources, it also elucidates the duality of these systems to serve as both agents for environmental remediation and clean energy production.

While promising advances have been made toward nanostructured photocatalytic materials for water splitting, there remain many challenges (Samanta et al. 2022). However, their economic and environmental viability for widespread commercial application and scalability of these technologies continues to be a critical area of research. These challenges can be addressed by improving the photocatalytic efficiency, reducing the cost of the materials fabrication, and ensuring long-term stability as well as reusability of catalysts (Yang et al. 2022). As a result, the future of sustainable energy production involves their strategic development of nanostructured photocatalysts for water splitting. Using the intrinsic properties of BioG-NPs and MOFs bridges the research into the development of innovative solutions that could greatly diminish our reliance on fossil fuels and push for a cleaner future (MariaJoseph et al. 2024). This work is promising not only for the production of green hydrogen but also for the applicability of photocatalytic technologies in other environmental and energy management sectors.

12.4.2.5 Overcoming the Storage Challenge

Pivotal in the hydrogen economy, hydrogen has significant (but formidable) obstacles to effective hydrogen storage (Fig. 12.9) due to its low volumetric energy density and the intricacies of safe handling (Aryamol et al. 2024). To address this crucial issue, we engineer photocatalytic materials to incorporate functionalities for both adsorption and desorption of hydrogen while optimizing both storage density and safety. The purpose of these materials is to allow reversible storage of hydrogen on the atomic or electronic scale under conditions more applicable for both mobile and stationary devices (Diao et al. 2018). Elevation of the energy density of storage

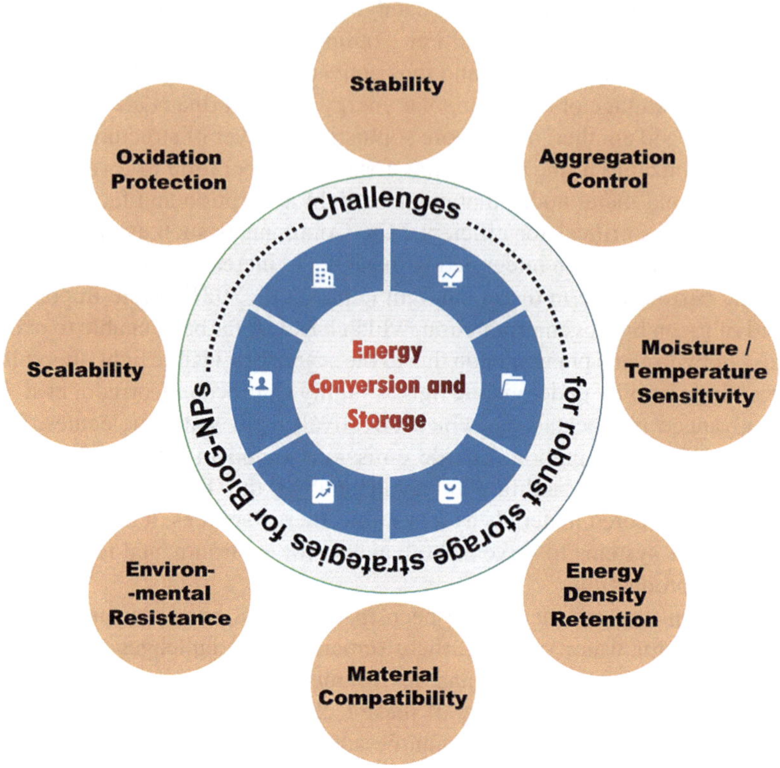

Fig. 12.9 Key challenges for robust storage strategies of **BioG-NPs** in energy conversion and storage applications

devices would naturally accommodate increased volumetric density and integrity, potentially improving the operational safety and efficiency of hydrogen storage systems, which are a critical enabler to the broad adoption of hydrogen as a clean energy carrier (Thirumalaivasan et al. 2024e).

12.4.2.6 Scaling Up: From Laboratory to Industrial Application

Many challenges exist in the transition of green nanostructured photocatalysts from laboratory achievements to industrial applications. Among these are also maintaining catalytic activity and durability over variable, often harsh conditions of real-world operation (Nangan et al. 2023b). The economic aspects of photocatalyst production, including the costs of materials and process scalability, are considerable stumbling blocks. In addition to advancing the materials, the integration of these advanced materials into existing infrastructures requires innovations in system designs and compatibility (Shellaiah et al. 2024). Realizing the potential of photocatalytic technologies in commercial settings is critical for addressing these challenges and achieving sustainable industrial practices and environmental management with contributions on commercial grounds.

12.5 Challenges and Future Directions

Challenges in the conversion of laboratory-scale prototypes to commercial-scale production of green nanostructured photocatalysts are of critical importance if broad-scale application to environmental remediation and energy generation is to be realized (Thirumalaivasan et al. 2019). To scale up, however, the photocatalysts must be adjusted in a careful manner to keep their cost effectiveness while preserving functional efficiency at higher volumes. Thus, novel ways of manufacturing must be developed to preserve the nanostructure integrity and large active surface areas of the photocatalysts, characteristics necessary for their efficiency (Jing et al. 2013).

Advanced manufacturing methods are needed to be developed so that the essential nanostructural properties are retained to make PCs effective during mass production (Lukkassen and Meidell 2003). These should provide a consistent means to reproduce nanostructures on a scale that meets industrial needs in terms of environmental and economical sustainability. Continuous flow synthesis and scalable aerosol-assisted processes could serve as viable pathways to meet these criteria in a way that maintains uniformity and functional stability of photocatalysts across batch productions (Sohrabi et al. 2020).

Furthermore, doping photocatalysts with nonmetal elements, such as nitrogen, sulfur, or phosphorus, can alter the electronic properties of the materials by reducing the bandgap and thus making the catalysts better absorb lower energy photons. It also improves their overall efficiency and makes sure that all of the solar energy is used to their maximum extent (Marschall and Wang 2014). In addition to pushing the limits of photocatalytic efficiency, these strategies open new spaces of application of these materials beyond pollution remediation, toward solar fuel production and CO_2 reduction.

12.6 Conclusion

GNS-PCs have an impressive horizon filled with infinite potential and possibilities. In this era of unprecedented scientific advancement, the design and implementation of these materials are at the cutting edge of a new technological revolution to meet some of the most pressing global energy and environmental needs. In addition to illustrating the unique functionality of these intricate materials in utilizing solar energy for simultaneous pollutant degradation and clean energy generation, this chapter postulated that they lay the groundwork for a paradigm shift in industrial and environmental sustainability. The growing advances held before represent a pivotal movement toward additional feasible, energetic, and economically suitable photocatalytic frameworks which could alleviate pollution and broaden the deployment of energy sustainability globally.

Moving and transforming, the path for GNS-PCs will be rising from limited laboratory conditions to large-scale universal practice. This transition, which is forecasted to occur over the next 10 years, will be a milestone in defining a sustainable future. But unlocking this potential will not come without significant challenges, from scaling these materials by a factor of 10,000 to making them more affordable and efficient at turning solar energy into storable hydrogen. Meeting these challenges will require an integrated, cross-disciplinary combination of chemistry, materials science, and environmental engineering expertise along with policy-development experience. If innovation is allowed to blossom in this direction, and if the pathway from research to commercialization is strategically traversed, green nanostructured photocatalysts might change environmental sustainability and energy paradigm fundamentally: a giant step towards use of clean energy sources leading to a greener planet.

References

Adeel M, Asif K, Rahman MM, Daniele S, Canzonieri V, Rizzolio F (2021) Glucose detection devices and methods based on metal–organic frameworks and related materials. Adv Funct Mater 31(52):2106023

Ahmad R, Ahmad Z, Khan AU, Mastoi NR, Aslam M, Kim J (2016) Photocatalytic systems as an advanced environmental remediation: recent developments, limitations and new avenues for applications. J Environ Chem Eng 4(4):4143–4164

Ahmad A, Noor AE, Anwar A, Majeed S, Khan S, Nisa ZU et al (2024) Support based metal incorporated layered nanomaterials for photocatalytic degradation of organic pollutants. Environ Res 260:119481

Alagesan M, Kanagaraj K, Wan S, Sun H, Su D, Zhong Z et al (2016) Enantiodifferentiating [4+4] photocyclodimerization of 2-anthracenecarboxylate mediated by a self-assembled iron tetrahedral coordination cage. J Photochem Photobiol A Chem 331:95–101

Anwar M, Fayyaz A, Sohail N, Khokhar M, Baqar M, Yasar A et al (2020) CO2 utilization: turning greenhouse gas into fuels and valuable products. J Environ Manag 260:110059

Armin A, Li W, Sandberg OJ, Xiao Z, Ding L, Nelson J et al (2021) A history and perspective of non-fullerene electron acceptors for organic solar cells. Adv Energy Mater 11(15):2003570

Arumugam S, Senthilkumar N, Thirumalaivasan N, Dharman G, Pandiaraj S, Rahaman M et al (2023) Heterogenous copper (II) schiff-base complex immobilized mesoporous silica catalyst for multicomponent biginelli reaction. J Organomet Chem 998:122804

Aryamol K, Kanagaraj K, Nangan S, Haponiuk JT, Okhawilai M, Pandiaraj S et al (2024) Recent advances of carbon pathways for sustainable environment development. Environ Res 250:118513

Asmat-Campos D, Rojas ML, Carreño-Ortega A (2023) Toward sustainable nanomaterials: an innovative ecological approach for biogenic synthesis of TiO2 nanoparticles with potential photocatalytic activity. Clean Eng Technol 17:100702

Bai Y, Liu C, Shan Y, Chen T, Zhao Y, Yu C et al (2022) Metal-organic frameworks nanocomposites with different dimensionalities for energy conversion and storage. Adv Energy Mater 12(4):2100346

Balamurugan G, Velmathi S, Thirumalaivasan N, Wu SP (2017) New phenazine based AIE probes for selective detection of aluminium (III) ions in presence of other trivalent metal ions in living cells. Analyst 142(24):4721–4726

Balaya P (2008) Size effects and nanostructured materials for energy applications. Energy Environ Sci 1(6):645–654

Banin U, Ben-Shahar Y, Vinokurov K (2014) Hybrid semiconductor–metal nanoparticles: from architecture to function. Chem Mater 26(1):97–110

Baxter J, Bian Z, Chen G, Danielson D, Dresselhaus MS, Fedorov AG et al (2009) Nanoscale design to enable the revolution in renewable energy. Energy Environ Sci 2(6):559–588

Bonilla SH, Silva HR, Terra da Silva M, Franco Gonçalves R, Sacomano JB (2018) Industry 4.0 and sustainability implications: a scenario-based analysis of the impacts and challenges. Sustainability 10(10):3740

Brar KK, Magdouli S, Othmani A, Ghanei J, Narisetty V, Sindhu R et al (2022) Green route for recycling of low-cost waste resources for the biosynthesis of nanoparticles (NPs) and nanomaterials (NMs)-a review. Environ Res 207:112202

Byrne C, Subramanian G, Pillai SC (2018) Recent advances in photocatalysis for environmental applications. J Environ Chem Eng 6(3):3531–3555

Chandrasekaran S, Bowen C, Zhang P, Li Z, Yuan Q, Ren X et al (2018) Spinel photocatalysts for environmental remediation, hydrogen generation, CO2 reduction and photoelectrochemical water splitting. J Mater Chem A 6(24):11078–11104

Chen Y-F, Tan L-L, Liu J-M, Qin S, Xie Z-Q, Huang J-F et al (2017) Calix [4] arene based dye-sensitized Pt@ UiO-66-NH2 metal-organic framework for efficient visible-light photocatalytic hydrogen production. Appl Catal B Environ 206:426–433

Chen C-G, Vijay N, Thirumalaivasan N, Velmathi S, Wu S-P (2019) Coumarin-based Hg2+ fluorescent probe: fluorescence turn-on detection for Hg2+ bioimaging in living cells and zebrafish. Spectrochim Acta A Mol Biomol Spectrosc 219:135–140

Chu C-J, Wu G-S, Ma H-I, Venkatesan P, Thirumalaivasan N, Wu S-P (2020) A fluorescent turn-on probe for detection of hypochlorus acid and its bioimaging in living cells. Spectrochim Acta A Mol Biomol Spectrosc 233:118234

Corredor J, Rivero MJ, Rangel CM, Gloaguen F, Ortiz I (2019) Comprehensive review and future perspectives on the photocatalytic hydrogen production. J Chem Technol Biotechnol 94(10):3049–3063

Deng X, Li Z, García H (2017) Visible light induced organic transformations using metal-organic-frameworks (MOFs). Chem Eur J 23(47):11189–11209

Diao H, Li T, Zhang R, Kang Y, Liu W, Cui Y et al (2018) Facile and green synthesis of fluorescent carbon dots with tunable emission for sensors and cells imaging. Spectrochim Acta A Mol Biomol Spectrosc 200:226–234

Dong S, Feng J, Fan M, Pi Y, Hu L, Han X et al (2015) Recent developments in heterogeneous photocatalytic water treatment using visible light-responsive photocatalysts: a review. RSC Adv 5(19):14610–14630

Durmus Z, Maijenburg AW (2022) A review on graphitic carbon nitride (g-C3N4)–metal organic framework (MOF) heterostructured photocatalyst materials for photo (electro) chemical hydrogen evolution. Int J Hydrog Energy 47(87):36784–36813

Ford S, Despeisse M (2016) Additive manufacturing and sustainability: an exploratory study of the advantages and challenges. J Clean Prod 137:1573–1587

Gao C, Min X, Fang M, Tao T, Zheng X, Liu Y et al (2022) Innovative materials science via machine learning. Adv Funct Mater 32(1):2108044

García-Serna J, Pérez-Barrigón L, Cocero M (2007) New trends for design towards sustainability in chemical engineering: Green engineering. Chem Eng J 133(1–3):7–30

Goetjen TA, Liu J, Wu Y, Sui J, Zhang X, Hupp JT et al (2020) Metal–organic framework (MOF) materials as polymerization catalysts: a review and recent advances. Chem Commun 56(72):10409–10418

Goodarzi N, Ashrafi-Peyman Z, Khani E, Moshfegh AZ (2023) Recent progress on semiconductor heterogeneous photocatalysts in clean energy production and environmental remediation. Catalysts 13(7):1102

Green MA, Bremner SP (2017) Energy conversion approaches and materials for high-efficiency photovoltaics. Nat Mater 16(1):23–34

Gulati S, Vijayan S, Kumar S, Harikumar B, Trivedi M, Varma RS (2023) Recent advances in the application of metal-organic frameworks (MOFs)-based nanocatalysts for direct conversion of carbon dioxide (CO2) to value-added chemicals. Coord Chem Rev 474:214853

Guo T-L, Li J-G, Sun X, Sakka Y (2017) Photocatalytic growth of Ag nanocrystals on hydrothermally synthesized multiphasic TiO2/reduced graphene oxide (rGO) nanocomposites and their SERS performance. Appl Surf Sci 423:1–12

Han C, Kundu BK, Liang Y, Sun Y (2024) Near-infrared light-driven photocatalysis with an emphasis on two-photon excitation: concepts, materials, and applications. Adv Mater 36(5):2307759

Hegedus SS, Luque A (2003) Status, trends, challenges and the bright future of solar electricity from photovoltaics. In: Handbook of photovoltaic science and engineering, pp 1–43. https://doi.org/10.1002/0470014008.ch1

Howarth AJ, Liu Y, Li P, Li Z, Wang TC, Hupp JT et al (2016) Chemical, thermal and mechanical stabilities of metal–organic frameworks. Nat Rev Mater 1(3):1–15

Ikram M, Rashid M, Haider A, Naz S, Haider J, Raza A et al (2021) A review of photocatalytic characterization, and environmental cleaning, of metal oxide nanostructured materials. Sustain Mater Technol 30:e00343

Jadhav DA, Mungray AK, Arkatkar A, Kumar SS (2021) Recent advancement in scaling-up applications of microbial fuel cells: from reality to practicability. Sustain Energy Technol Assess 45:101226

Jeje SO, Marazani T, Obiko JO, Shongwe MB (2024) Advancing the hydrogen production economy: a comprehensive review of technologies, sustainability, and future prospects. Int J Hydrog Energy 78:642–661

Jesudoss S, Vijaya JJ, Selvam NCS, Kombaiah K, Sivachidambaram M, Adinaveen T et al (2016) Effects of Ba doping on structural, morphological, optical, and photocatalytic properties of self-assembled ZnO nanospheres. Clean Techn Environ Policy 18:729–741

Jiao L, Seow JYR, Skinner WS, Wang ZU, Jiang H-L (2019) Metal–organic frameworks: structures and functional applications. Mater Today 27:43–68

Jing L, Zhou W, Tian G, Fu H (2013) Surface tuning for oxide-based nanomaterials as efficient photocatalysts. Chem Soc Rev 42(24):9509–9549

Johar MA, Afzal RA, Alazba AA, Manzoor U (2015) Photocatalysis and bandgap engineering using ZnO nanocomposites. Adv Mater Sci Eng 2015(1):934587

Kalaiselvan N, Mathimani T (2024) Solar-driven green hydrogen generation for revolutionizing the future of zero-carbon energy. Fuel 375:132538

Kanagaraj K, Liang W, Rao M, Yao J, Wu W, Cheng G et al (2020) pH-controlled chirality inversion in enantiodifferentiating photocyclodimerization of 2-antharacenecarboxylic acid mediated by γ-cyclodextrin derivatives. Org Lett 22(14):5273–5278

Kanagaraj K, Rebek J, Yu Y (2023a) Control of reactivity and selectivity in isomerization and rearrangement reactions inside confined spaces. Chem Commun 59(59):9056–9065

Kanagaraj K, Wang R, Zhao M-K, Ballester P, Rebek J Jr, Yu Y (2023b) Selective binding and isomerization of oximes in a self-assembled capsule. J Am Chem Soc 145(10):5816–5823

Kaur H, Kumar S, Kaur G, Kaur N, Badru R, Saini R (2023) An emerging expanse: novel and eco-friendly-biogenic synthesis of E. cardamomum-wrapped TiO2 nanoparticles for environmental and biological applications. Environ Res 234:116599

Kishore A, Sunil SM, Viswanathan R, Kaliaperumal A, Jaffar Ali BM (2022) Different synthetic routes and band gap engineering of photocatalysts. In: Photocatalysts and electrocatalysts in water remediation: from fundamentals to full scale applications, pp 39–80. https://doi.org/10.1002/9781119855347.ch2

Kleijn R, Van der Voet E (2010) Resource constraints in a hydrogen economy based on renewable energy sources: an exploration. Renew Sust Energ Rev 14(9):2784–2795

Krauklis AE, Karl CW, Gagani AI, Jørgensen JK (2021) Composite material recycling technology—state-of-the-art and sustainable development for the 2020s. J Compos Sci 5(1):28

Kumar S, Sharma R, Borah SJ, Gupta A, Gupta MK, Kumar R et al (2023) Catalytic heterostructured materials for CO2 mitigation and conversion into fuels: a renewable energy approach towards a sustainable environment. Sustain Energy Fuels 7(18):4354–4395

Li H, Zhou Y, Tu W, Ye J, Zou Z (2015) State-of-the-art progress in diverse heterostructured photocatalysts toward promoting photocatalytic performance. Adv Funct Mater 25(7):998–1013

Likodimos V, Dionysiou DD, Falaras P (2010) Clean water: water detoxification using innovative photocatalysts. Rev Environ Sci Biotechnol 9:87–94

Lim KL, Kazemian H, Yaakob Z, Daud WW (2010) Solid-state materials and methods for hydrogen storage: a critical review. Chem Eng Technol 33(2):213–226

Liu S, Zhang C, Sun Y, Chen Q, He L, Zhang K et al (2020) Design of metal-organic framework-based photocatalysts for hydrogen generation. Coord Chem Rev 413:213266

Luciani G, Imparato C, Vitiello G (2020) Photosensitive hybrid nanostructured materials: the big challenges for sunlight capture. Catalysts 10(1):103

Lukkassen D, Meidell A (2003) Advanced materials and structures and their fabrication processes. Narrik University College, Hin

Majumdar A, Pal A (2020) Recent advancements in visible-light-assisted photocatalytic removal of aqueous pharmaceutical pollutants. Clean Techn Environ Policy 22:11–42

Manjunathan J, Revathi M, Sowmya H, Meenambiga S, Sudha R, Prakash B et al (2023) Recent advancements in nanotechnological approaches for pollution monitoring and environmental sustainability. Clean Techn Environ Policy 26(11):3667–3683

MariaJoseph A, Nangan S, Verma D, Gnanasekaran L, Rajendran S, Natesan T et al (2024) Rational coupling of selective electrochemical oxidation and reduction reactions for in-situ value-added chemical generation. Fuel 367:131408

Marschall R, Wang L (2014) Non-metal doping of transition metal oxides for visible-light photocatalysis. Catal Today 225:111–135

Mo F, Zhou Q, Wang Q, Hou Z, Wang J (2022) The applications of MOFs related materials in photo/electrochemical decontamination: an updated review. Chem Eng J 450:138326

Mourdikoudis S, Pallares RM, Thanh NT (2018) Characterization techniques for nanoparticles: comparison and complementarity upon studying nanoparticle properties. Nanoscale 10(27):12871–12934

Nair A, Kuppusamy K, Nangan S, Natesan T, Haponiuk JT, Thomas S et al (2023) Multifunctional natural derived carbon quantum dots from Withania somnifera (L.)–antiviral activities against SARS-CoV-2 pseudoviron. Environ Res 239:117366

Nangan S, Kanagaraj K, Kaarthikeyan G, Kumar A, Ubaidullah M, Pandit B et al (2023a) Sustainable preparation of luminescent carbon dots from syringe waste and hyaluronic acid for cellular imaging and antimicrobial applications. Environ Res 237:116990

Nangan S, Natesan T, Sukmas W, Okhawilai M, Babu KJ, Tsuppayakorn-Aek P et al (2023b) Waste plastics derived nickel-palladium alloy filled carbon nanotubes for hydrogen evolution reaction. Chemosphere 341:139982

Ng KH (2021) Adoption of TiO2-photocatalysis for palm oil mill effluent (POME) treatment: strengths, weaknesses, opportunities, threats (SWOT) and its practicality against traditional treatment in Malaysia. Chemosphere 270:129378

Nhien PQ, Chou W-L, Cuc TTK, Khang TM, Wu C-H, Thirumalaivasan N et al (2020) Multi-stimuli responsive FRET processes of bifluorophoric AIEgens in an amphiphilic copolymer and its application to cyanide detection in aqueous media. ACS Appl Mater Interfaces 12(9):10959–10972

Pal A, Das R, Kumar M, Roy DN (2020) Application of nanoscience and nanotechnology in advancement of safety issues of foods. In: Biotechnological approaches in food adulterants. CRC Press, pp 282–307. https://doi.org/10.1201/9780429354557-11

Panneri S, Ganguly P, Mohan M, Nair BN, Mohamed AAP, Warrier KG et al (2017) Photoregenerable, bifunctional granules of carbon-doped g-C3N4 as adsorptive photocatalyst for the efficient removal of tetracycline antibiotic. ACS Sustain Chem Eng 5(2):1610–1618
Pelaez M, Nolan NT, Pillai SC, Seery MK, Falaras P, Kontos AG et al (2012) A review on the visible light active titanium dioxide photocatalysts for environmental applications. Appl Catal B Environ 125:331–349
Peng C, Liang W, Ji J, Fan C, Kanagaraj K, Wu W et al (2021) Pyrene-tiaraed pillar [5] arene: strong intramolecular excimer emission applicable for photo-writing. Chin Chem Lett 32(1):345–348
Podder A, Thirumalaivasan N, Chao YK, Kukutla P, Wu S-P, Bhuniya S (2020) Two-photon active fluorescent indicator for detecting NADH dynamics in live cells and tumor tissue. Sensors Actuators B Chem 324:128637
Pushparaj K, Liu W-C, Meyyazhagan A, Orlacchio A, Pappusamy M, Vadivalagan C et al (2022) Nano-from nature to nurture: a comprehensive review on facets, trends, perspectives and sustainability of nanotechnology in the food sector. Energy 240:122732
Qian Y, Ma D (2021) Covalent organic frameworks: new materials platform for photocatalytic degradation of aqueous pollutants. Materials 14(19):5600
Rafique M, Mubashar R, Irshad M, Gillani S, Tahir MB, Khalid N et al (2020) A comprehensive study on methods and materials for photocatalytic water splitting and hydrogen production as a renewable energy resource. J Inorg Organomet Polym Mater 30(10):3837–3861
Rajapandi S, Nangan S, Natesan T, Kumar A, Dharman G, Pandeeswaran M et al (2023) Ziziphus mauritiana-derived nitrogen-doped biogenic carbon dots: eco-friendly catalysts for dye degradation and antibacterial applications. Chemosphere 338:139584
Ramasubburayan R, Senthilkumar N, Kanagaraj K, Basumatary S, Kathiresan S, Manjunathan J et al (2023) Environmentally benign, bright luminescent carbon dots from IV bag waste and chitosan for antimicrobial and bioimaging applications. Environ Res 238:117182
Ramasubburayan R, Kanagaraj K, Gnanasekaran L, Thirumalaivasan N, Senthilkumar N (2024) Greener approach supported nitrogen-infused carbon dots for biocompatible cellular markers and fluorescent ink based spray-assisted fingerprint analysis. Waste Biomass Valorization 15(11):6431–6440
Salvadori MR (2019) Processing of nanoparticles by biomatrices in a green approach. In: Microbial nanobionics, vol 1, pp 1–28. https://doi.org/10.1007/978-3-030-16383-9_1. State-of-the-Art
Samanta B, Morales-García Á, Illas F, Goga N, Anta JA, Calero S et al (2022) Challenges of modeling nanostructured materials for photocatalytic water splitting. Chem Soc Rev 51(9):3794–3818
Samsudin EM, Abd Hamid SB (2017) Effect of band gap engineering in anionic-doped TiO2 photocatalyst. Appl Surf Sci 391:326–336
Samuel MS, Ravikumar M, John JA, Selvarajan E, Patel H, Chander PS et al (2022) A review on green synthesis of nanoparticles and their diverse biomedical and environmental applications. Catalysts 12(5):459
Sandhu ZA, Farwa U, Danish M, Raza MA, Talib A, Amjad H et al (2024) Sustainability and photocatalytic performance of MOFs: synthesis strategies and structural insights. J Clean Prod 470:143263
Sarina S, Waclawik ER, Zhu H (2013) Photocatalysis on supported gold and silver nanoparticles under ultraviolet and visible light irradiation. Green Chem 15(7):1814–1833
Şen Z (2004) Solar energy in progress and future research trends. Prog Energy Combust Sci 30(4):367–416
Shellaiah M, Thirumalaivasan N, Aazaad B, Awasthi K, Sun KW, Wu S-P et al (2020) Novel rhodamine probe for colorimetric and fluorescent detection of Fe3+ ions in aqueous media with cellular imaging. Spectrochim Acta A Mol Biomol Spectrosc 242:118757
Shellaiah M, Chen Y-T, Thirumalaivasan N, Aazaad B, Awasthi K, Sun KW et al (2021) Pyrene-based AIEE active nanoprobe for Zn2+ and tyrosine detection demonstrated by DFT, bioimaging, and organic thin-film transistor. ACS Appl Mater Interfaces 13(24):28610–28626
Shellaiah M, Thirumalaivasan N, Aazaad B, Awasthi K, Sun KW, Wu S-P et al (2022) An AIEE active anthracene-based nanoprobe for Zn2+ and tyrosine detection validated by bioimaging studies. Chemosensors 10(10):381

Shellaiah M, Sun KW, Thirumalaivasan N, Bhushan M, Murugan A (2024) Sensing utilities of cesium lead halide perovskites and composites: a comprehensive review. Sensors 24(8):2504
Silva P, Vilela SM, Tomé JP, Paz FAA (2015) Multifunctional metal–organic frameworks: from academia to industrial applications. Chem Soc Rev 44(19):6774–6803
Singaram B, Varadharajan K, Jeyaram J, Rajendran R, Jayavel V (2017) Preparation of cerium and sulfur codoped TiO2 nanoparticles based photocatalytic activity with enhanced visible light. J Photochem Photobiol A Chem 349:91–99
Sohrabi S, Keshavarz Moraveji M, Iranshahi D (2020) A review on the design and development of photocatalyst synthesis and application in microfluidic reactors: challenges and opportunities. Rev Chem Eng 36(6):687–722
Spöri C, Kwan JTH, Bonakdarpour A, Wilkinson DP, Strasser P (2017) The stability challenges of oxygen evolving catalysts: towards a common fundamental understanding and mitigation of catalyst degradation. Angew Chem Int Ed 56(22):5994–6021
Stock N, Biswas S (2012) Synthesis of metal-organic frameworks (MOFs): routes to various MOF topologies, morphologies, and composites. Chem Rev 112(2):933–969
Swetha S, Janani B, Khan SS (2022) A critical review on the development of metal-organic frameworks for boosting photocatalysis in the fields of energy and environment. J Clean Prod 333:130164
Thakur N, Thakur N, Kumar A, Thakur VK, Kalia S, Arya V et al (2024) A critical review on the recent trends of photocatalytic, antibacterial, antioxidant and nanohybrid applications of anatase and rutile TiO2 nanoparticles. Sci Total Environ 914:169815
Thirumalaivasan N, Wu S-P (2020) Bright luminescent carbon dots for multifunctional selective sensing and imaging applications in living cells. ACS Appl Bio Mater 3(9):6439–6446
Thirumalaivasan N, Venkatesan P, Wu S-P (2017) Highly selective turn-on probe for H 2 S with imaging applications in vitro and in vivo. New J Chem 41(22):13510–13515
Thirumalaivasan N, Venkatesan P, Lai P-S, Wu S-P (2019) In vitro and in vivo approach of hydrogen-sulfide-responsive drug release driven by azide-functionalized mesoporous silica nanoparticles. ACS Appl Bio Mater 2(9):3886–3896
Thirumalaivasan N, Kanagaraj K, Logesh K, Chandrasekaran S, Kumar S, Subramanian R et al (2024a) Exploring luminescent carbon dots derived from syrup bottle waste and curcumin for potential antimicrobial and bioimaging applications. Chemosphere 354:141592
Thirumalaivasan N, Gnanasekaran L, Kumar S, Durvasulu R, Sundaram T, Rajendran S et al (2024b) Utilization of fungal and bacterial bioremediation techniques for the treatment of toxic waste and biowaste. Front Mater 11:1416445
Thirumalaivasan N, Nangan S, Kanagaraj K, Rajendran S (2024c) Assessment of sustainability and environmental impacts of renewable energies: focusing on biogas and biohydrogen (biofuels) production. Process Saf Environ Prot 189. https://doi.org/10.1016/j.psep.2024.06.063
Thirumalaivasan N, Gopi S, Karthik K, Nangan S, Kanagaraj K, Rajendran S (2024d) Nano-PCM materials: bridging the gap in energy storage under fluctuating environmental conditions. Process Saf Environ Prot 189:1003–1021
Thirumalaivasan N, Mahapatra S, Ramanathan G, Kumar A, Raja T, Muthuramamoorthy M et al (2024e) Exploring antimicrobial and biocompatible applications of eco-friendly fluorescent carbon dots derived from fast-food packaging waste transformation. Environ Res 244:117888
UshaVipinachandran V, Rajendran S, Badagoppam Haroon KH, Ashokan I, Mondal A, Bhunia SK (2020) Detoxification of endocrine disruptors in water using visible-light-active nanostructures: a review. ACS Appl Nano Mater 3(12):11659–11687
Varma RS (2014) Journey on greener pathways: from the use of alternate energy inputs and benign reaction media to sustainable applications of nano-catalysts in synthesis and environmental remediation. Green Chem 16(4):2027–2041
Venkatesan P, Thirumalaivasan N, Yu H-P, Lai P-S, Wu S-P (2019) Redox stimuli delivery vehicle based on transferrin-capped MSNPs for targeted drug delivery in cancer therapy. ACS Appl Bio Mater 2(4):1623–1633

Venkatesan P, Govindasamy R, Periyasami G, Rahaman M, Pandiaraj S, Thiruvengadam M et al (2023) Eco-friendly, bright luminescent carbon dots and their potential applications for detecting hypochlorous acid in water and live cell imaging. J Mater Res Technol 24:6522–6532

Verma D, Okhawilai M, Senthilkumar N, Thirumalaivasan N, Incharoensakdi A, Uyama H (2024) Polyhydroxybutyrate/poly (ε-caprolactone)-based electrospun membranes loaded with amoxicillin-potassium clavulanate halloysite nanotubes for biomedical applications. Polym Int 74(1):54–65

Volokh M, Mokari T (2020) Metal/semiconductor interfaces in nanoscale objects: synthesis, emerging properties and applications of hybrid nanostructures. Nanoscale Adv 2(3):930–961

Wang L, Zhu W (2024) Organic donor-acceptor systems for photocatalysis. Adv Sci 11(10):2307227

Wang P, Yap P-S, Lim T-T (2011) C–N–S tridoped TiO2 for photocatalytic degradation of tetracycline under visible-light irradiation. Appl Catal A Gen 399(1–2):252–261

Wang Y, Yu Y, Li R, Liu H, Zhang W, Ling L et al (2017) Hydrogen production with ultrahigh efficiency under visible light by graphene well-wrapped UiO-66-NH 2 octahedrons. J Mater Chem A 5(38):20136–20140

Wei L-F, Chen C-Y, Lai C-K, Thirumalaivasan N, Wu S-P (2019) A nano-molar fluorescent turn-on probe for copper (II) detection in living cells. Methods 168:18–23

Winter C-J (2009) Hydrogen energy—abundant, efficient, clean: a debate over the energy-system-of-change. Int J Hydrog Energy 34(14):S1–S52

Wu G-S, Thirumalaivasan N, Lin T-C, Wu S-P (2020) Ultrasensitive and specific two-photon fluorescence detection of hypochlorous acid by a lysosome-targeting fluorescent probe for cell imaging. J Pharm Biomed Anal 190:113545

Xiao JD, Shang Q, Xiong Y, Zhang Q, Luo Y, Yu SH et al (2016) Boosting photocatalytic hydrogen production of a metal–organic framework decorated with platinum nanoparticles: the platinum location matters. Angew Chem Int Ed 55(32):9389–9393

Yang X, Cao C, Erickson L, Hohn K, Maghirang R, Klabunde K (2008) Synthesis of visible-light-active TiO2-based photocatalysts by carbon and nitrogen doping. J Catal 260(1):128–133

Yang W-Y, Anusuyadevi K, Lu P-H, Thirumalaivasan N, Lin W-H, Velmathi S et al (2022) A two photon fluorescent probe for highly selective detection and endogenous imaging of hydrogen sulfide. Spectrochim Acta A Mol Biomol Spectrosc 273:121043

Yu S, Zhang Y, Dong F, Li M, Zhang T, Huang H (2018) Readily achieving concentration-tunable oxygen vacancies in Bi2O2CO3: triple-functional role for efficient visible-light photocatalytic redox performance. Appl Catal B Environ 226:441–450

Yuan S, Feng L, Wang K, Pang J, Bosch M, Lollar C et al (2018) Stable metal–organic frameworks: design, synthesis, and applications. Adv Mater 30(37):1704303

Zhao G-Q, Zou J, Hu J, Long X, Jiao F-P (2021) A critical review on graphitic carbon nitride (g-C3N4)-based composites for environmental remediation. Sep Purif Technol 279:119769

Zhou N, López-Puente V, Wang Q, Polavarapu L, Pastoriza-Santos I, Xu Q-H (2015) Plasmon-enhanced light harvesting: applications in enhanced photocatalysis, photodynamic therapy and photovoltaics. RSC Adv 5(37):29076–29097

Zhu J, Zäch M (2009) Nanostructured materials for photocatalytic hydrogen production. Curr Opin Colloid Interface Sci 14(4):260–269

Correction to: Green Photocatalytic Energy Conversion

M. Muthulakshmi, T. S. Subha, and M. Muthukumaran

Correction to:
Chapter 9 in: C. M. Hussain, G. Hussain (eds.), *Green Nanostructured Photocatalysts for Clean Energy*, Clean Energy Production Technologies https://doi.org/10.1007/978-981-95-1618-6_9

The original version of the chapter "Green Photocatalytic Energy Conversion" was inadvertently published with incorrect author affiliation. The complete affiliation detail has been updated in the chapter from "M. Muthulakshmi Bharathi Women's College, Chennai, India" to "M. Muthulakshmi Ethiraj College for Women (Autonomous), Chennai-600008, India".

The updated version of this chapter can be found at
https://doi.org/10.1007/978-981-95-1618-6_9

C. M. Hussain, G. Hussain (eds.), *Green Nanostructured Photocatalysts for Clean Energy*, Clean Energy Production Technologies,
https://doi.org/10.1007/978-981-95-1618-6_13

GPSR Compliance

The European Union's (EU) General Product Safety Regulation (GPSR) is a set of rules that requires consumer products to be safe and our obligations to ensure this.

If you have any concerns about our products, you can contact us on ProductSafety@springernature.com

In case Publisher is established outside the EU, the EU authorized representative is:

Springer Nature Customer Service Center GmbH
Europaplatz 3
69115 Heidelberg, Germany

Batch number: 10755329

Printed by Printforce, the Netherlands